# NEAR-INFRARED SPECTROSCOPY IN FOOD SCIENCE AND TECHNOLOGY

# NEAR-INFRARED SPECTROSCOPY IN FOOD SCIENCE AND TECHNOLOGY

Edited By

**Yukihiro Ozaki, PhD**
Kwansei Gakuin University
Japan

**W. Fred McClure, PhD**
North Carolina State University
Raleigh, North Carolina

**Alfred A. Christy, PhD**
Adger University
Kristiansand, Norway

BICENTENNIAL
1807
WILEY
2007
BICENTENNIAL

WILEY-INTERSCIENCE
A John Wiley & Sons, Inc., Publication

For general information on our other products and services or for technical support, please contact our Customer Care Department within the United States at (800) 762-2974, outside the United States at (317) 572-3993 or fax (317) 572-4002.

Wiley also publishes its books in a variety of electronic formats. Some content that appears in print may not be available in electronic formats. For more information about Wiley products, visit our web site at www.wiley.com.

*Library of Congress Cataloging-in-Publication Data:*

Near-infrared spectroscopy in food science and technology / edited by
    Yukihiro Ozaki, W. Fred McClure, Alfred Christy.
        p.   cm.
    Includes bibliographical references and index.
    ISBN-13: 978-0-471-67201-2 (acid-free paper)
    ISBN-10: 0-471-67201-7 (acid-free paper)
    1. Food—Analysis.   2. Near infrared spectroscopy.   I. Ozaki, Y. (Yukihiro)
    II. McClure, W. F. (William F.)   III. Christy, Alfred A.

    TX547.2.I53N426 2006
    664′.07–dc22
                                                        2006040965

Printed in the United States of America

10 9 8 7 6 5 4 3 2 1

# CONTENTS

v

Near-infrared (NIR) spectroscopy has recently become increasingly important in food science and technology as a non-destructive analytical technique. An enormous number of articles and research papers are published every year that deal with applications of NIR spectroscopy in this important field. Search for a comprehensive book describing both basic principles and modern application of NIR spectroscopy food science and technology proved fruitless. This book, *Near Infrared Spectroscopy in Food Science and Technology*, fills the void. It covers principles of molecular vibrations, spectral analysis, and instrumentation for NIR spectroscopy as well as its novel applications within food science and technology. It is written to be appreciated by food and agricultural scientists and engineers as well as molecular spectroscopists. The aim of this book is to provide a basic understanding of techniques and applications that demonstrate the potential of NIR spectroscopy for researchers and users in food science and technology.

The book is suitable for students at graduate level as well as researchers and engineers in academic and industry. It may be used as a textbook for a graduate course in food science and technology or agricultural science and technology and for short courses. We hope you enjoy this book and that it will inspire you and other readers to adapt the principles and techniques discussed herein to your particular area and interests.

Yukihiro Ozaki
W. Fred McClure
Alfred Christy

# ACKNOWLEDGMENTS

The editors thank each contributor who took time from their normal duties to make this book possible. We especially thank Ms. K. Horiguchi for the preparation of manuscripts, figures, and references. We also would like to thank both families and colleagues who provided encouragement and other support – making the tasks joy rather than drudgery. We hope all our readers enjoy this book.

# ■■■■ CONTRIBUTORS

**Franklin E. Barton II,** U. S. Department of Agriculture, Agricultural Research Service, Richard B. Russell Agricultural Center, P. O. Box 5677, Athens, GA 30613, USA

**Greame D. Batten,** Farrer Centre, Charles Sturt University, LMB Bag 588 Wagga Wagga, NSW 2678, Australia

**R. Buchet,** UFR Chimie Biochimie, Universite Claude Bernard Lyon I, 43 Boulevard 11 November 1918, 69622 Villeurbanne Cedex, France

**T.M.P. Cattaneo,** Instito Sperimentale Lattiero Caseario, Via A. Lombardo, 11-26900 Lodi, Italy

**Alfred A. Christy,** Adger University, Faculty of Mathematics and Sciences, Torden-skjolds gate 65, N-4604 Kristiansand, Norway

**D. Cozzolino,** Australian Wine Institute in Glan Osmond, Adelade, Australia

**Geraed Downey,** The National Food Centre, Research & Training for the Food Industry, Dunsinea, Castleknock, Dublin 15, Ireland

**Yiping Du,** Analysis and Research Center, East China University of Science and Technology, Meilong Road 130, Shenghai 200237, China

**Janie Dubois,** Joint Institute for Food Safety and Applied Nutrition, University of Maryland and U.S. Food and Drug Administration, HFS-717, 5100 Paint Branch Parkway, College Park, MD, USA

**R. Giangiacomo,** Instituto Sperimentale Lattiero Caseario, Via A. Lombardo, 11-26900 LODI, Italy

**Kjell Ivar Hildrum,** MATFORSK, Norwegian Food Research Institute, N-1430 As, Norway

**Tomas Isaksson,** Agricultural University of Norway, Department of Food Science, P. O. Box 5036, N-1432 As, Norway

**Sumio Kawano,** Nondestructive Evaluation Laboratory, Analytical Science Division, National Food Research Institute, 2-1-2, Kannondai, Tsukuba, 305-8642, Japan

**Sandra E. Kays,** U. S. Department of Agriculture, Agricultural Research Service, Quality Assessment Research Unit, 950 College Station Rd., Athens, GA 30605, USA

**Linda H. Kidder,** Spectral Dimensions, 3416 Olandwood Court, Olney, MD 20832, USA

**Olav M. Kvalheim,** University of Bergen, Department of Chemistry, Allegaten 41, N-5007 Bergen, Norway

**G. Lachenal,** UFR Chimie Biochimie, Universite Claude Bernard Lyon I, 43 Boulevard 11 November 1918, 69622 Villeurbanne Cedex, France

**Kathryn A. Lee,** 239 Spencer Road, Basking Ridge, NJ 07920, USA

**E. Neil Lewis,** Spectral Dimensions, 3403 Olandwood Ct, Suite 102, Olney, MD 20832, USA

**W. Fred McClure,** NC State University, Biological and Agricultural Engineering Department, Campus Box 7625, Raleigh, NC 27695-7625, USA

**Shigeaki Morita,** Department of Chemistry, School of Science and Technology, Kwansei-Gakuin University, 2–1, Gakuen, Sanda, 669–1337, Japan

**Ian Murray,** Scottish Agricultural College, Craibstone, Aberdeen, AB21 9YA, UK

**Brian G. Osborne,** BRI Australia Limited, An Independent Grains Research and Development Institute, PO Box 7, North Ryde, NSW 2113, Australia

**Yukihiro Ozaki,** Department of Chemistry, School of Science and Technology, Kwansei-Gakuin University, 2-1, Gakuen, Sanda, 669-1337, Japan

**C. Sandorfy,** Departemenent de Chimie, Universite de Montreal, Montreal, Quebec, Canada H3C 377

**Sirinnapa Saranwong,** Nondestructive Evaluation Laboratory, Analytical Science Division, National Food Research Institute, 2–1–2, Kannondai, Tsukuba, 305–8642, Japan

**Vegard H. Segtnan,** MATFORSK- Norwegian Food Research Institute, Osloveien 1, N-1430 Aas, Norway

**Roumiana Tsenkova,** Faculty of Agriculture, Department of Environment Information and Bio-Production Engineering, Kobe University, 1-1 Rokkoudai, Nada-ku, Kobe, 657-8501, Japan

**Satoru Tsuchikawa,** Mechanical Engineering for Biological Materials, Biological Material Sciences, Biosphere Resources Science, Graduate School of Bioagricultural Science, Nagoya University, Furo-cho, chikusa-ku, Nagoya, 464-8602, Japan

**Phil Williams,** PDK Grain, Winnipeg, Manitoba, Canada

**Takuo Yano,** Department of Information Machine and Interfaces, Faculty of Information Sciences, Hiroshima City University, 3-4-1, Ohtsuka-Higashi, Asaminami-ku, Hiroshima, 731-3194, Japan

▰▰▰▰▰▰ **CHAPTER 1**

# Introduction

W. FRED MCCLURE

## WORLD FOOD PRODUCTION

The industrialized world consists of about 59 countries, all with a total populations of about 0.9 billion people, about one-sixth of the total world population. In contrast, about 5 billion people live in approximately 125 low- and middle-income countries. The remaining 0.4 billion live in countries in transition, which include the Baltic states, eastern Europe and the Commonwealth of Independent States (1). Today, our world produces food for 6.39 billion people (Fig. 1.1). Yet statistics show that many people go to bed hungry every night. Each year the food crisis intensifies and more and more people go hungry.

Shockingly, the push to produce more and more food is thwarted by diminishing arable land suitable for food production. Plant yields have been maximized for many crops, leaving few options for increasing food production. In the face of these seemingly insurmountable problems, scientists are beginning develop technology for maximizing *food potential*, a philosophy that calls on any means that will reduce waste.

The philosophy for maximizing food for fresh foods potential goes something like this.[1] Time of harvest for plant-based foods must be optimized in order to maximize food potential. If harvested too early, both yield and quality are reduced: Again, if crops left too long in the field, both yield and quality fall. Furthermore, between the time of harvest and the time of consumption fresh foods undergo a decaying process called senescence. Senescence can reduce food potential by 7–12%, depending on how

---

[1] "Maximizing food potential" was first introduced by W. Fred McClure at the International Conference on Planning for the Future, Newcastle University, Newcastle, UK in W. F. McClure. 1995. Biological measurements for the 21st Century. In *New Horizons, New Beginnings*, ed. Staff, 1:34–40. Newcastle University, Newcastle, UK: Newcastle University.

---

*Near-Infrared Spectroscopy in Food Science and Technology*, Edited by Yukihiro Ozaki, W. Fred McClure, and Alfred A. Christy.
Copyright © 2007 John Wiley & Sons, Inc.

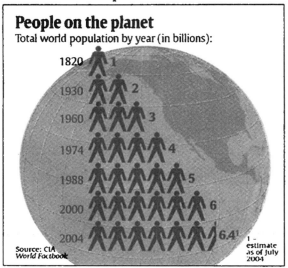

**Figure 1.1.** World population statistics for the USA, 1820–2004 published in CIA World Fact Book (USA Today Snapshots, July 2004).

quickly the food is made ready for consumption.[2] The potential exists for reducing losses if appropriate tools are made available for measuring the quality of fresh food.

Quality measurements made early in the production cycle, when fresh products are still edible, can determine whether the product goes to fresh market or to processing. The fresh market fraction with optimized quality factors is sent to fresh market; the remaining fraction is sent to facilities for further processing to make pop-tarts, jams, cooked meals and/or soups. Thus the food potential is maximized by minimizing waste.

## NEAR-INFRARED SPECTROSCOPY

Near-infrared (NIR) spectroscopy has taken its place among other proven spectroscopic tools, especially for determining chemical and physical properties of foods and food products. Covering the small region of the electromagnetic spectrum from 780 to 2500 (nm) (Sheppard, 1985 #354), producing spectra with only 860 data points spaced 2 nm apart, NIR spectroscopy has experienced phenomenal growth over its short history from 1905 (the year Coblentz produced the first official NIR publication) to the

[2] Based on a survey taken of six major grocery chains in the USA in Ibid.; reported at the conference in Newcastle, UK.

**TABLE 1.1.  Major Constituents/Properties in Foods Determined with Near-Infrared Spectroscopy**

| Number | Constituent | Comment | References |
|---|---|---|---|
| 1 | Water | Water was the first constituent to be studied with near-infrared spectroscopy. | 49 |
| 2 | Protein | Williams, Norris and Kays studied of protein, oil, water and starch in pulverized wheat and foods. | 10–15 |
| 3 | Fats (Oils) | Fats (or oils) were in foods has become a routine NIR measurement today. | 16–19 |
| 4 | Cereals: Dietary Fiber | Dietary and health issues have been studied extensively. | 20–23 |
| 5 | Scattering Properties | Any solid material demands an understanding of the scattering properties in order to obtain robust calibrations. | 24, 25 |
| 6 | Sucrose | Sucrose is a critical constituent in health issues. | 26 |
| 7 | Carbohydrates | Dietary matters call for NIR analyses of carbohydrates. | 10 |
| 8 | Energy | Energy content of food determined by NIR spectroscopy | 27 |
| 9 | Homogeneity | Homogeneity is critical to robust calibrations. | 28 |
| 10 | Condiments Sucrose, Starch, Flour | Osborne has studied the combination of these constituents in processed cereals. | 29 |
| 11 | Meats: Beef, Poultry, Fish | Beef, poultry and fish are analyzed using NIR by a number of researchers. | 30–39 |
| 12 | Fresh Foods: Fruit, Nuts | Once thought to be a very difficult determination, Kawano, Saranwong and others have demonstrated that NIR is useful for the analyses of high-moisture products. | 40–50 |
| 13 | Honey, Corn Syrup, Molasses | Even amorphous sugars in honey and other products can be measured. | 51–57 |
| 14 | Candy, Chocolate, Caramel | The same is true for candies, plus the nicotine related components in chocolates. | 54, 58–60 |

*(Continued)*

**TABLE 1.1. (*Continued*)**

| Number | Constituent | Comment | References |
|--------|-------------|---------|-----------|
| 15 | Sweets: Sucrose, Saccharin, Honey, Corn Syrup, Molasses | The spectra of sucrose and saccharin are strikingly similar. | 54, 61–65 |
| 16 | Beverages: Milk, Soft and Hard | In particular, ethanol in beer, wines and spirits is easily determined with near-infrared spectroscopy. | 66–78 |
| 17 | Bread | Osborne has been the leader in the analysis of bread using NIR spectroscopy. | 79–81 |

present. Its sensitivity to the CH, NH, and OH absorptions related to food components, its speedy response time, the simplicity of sample preparation involved, the fact that the measurement is nondestructive, and its low instrumentation cost have fixed its position along side other spectroscopies, including ultraviolet, visible, mid-infrared, Raman and others. Its expansion into the field of food production and processing is undeniable. (2–11)

NIR technology cuts across many fields (agriculture, textiles, pharmaceuticals, cosmetics, medicine, and others), with demonstrated success in all areas published in the literature. This book is not intended to address all the above areas. Rather, this book is intended to provide the reader an opportunity to understand something of the impact NIR spectroscopy has had on food science and technology. Expectedly, it does include basic principles of NIR spectroscopy (Chapter 2), characteristics of the NIR spectra (Chapter 3), instrumentation (Chapter 4), sampling techniques (Chapter 5), and chemometrics (Chapter 6). The remainder of the book discusses numerous applications of NIR technology in the food science field [agricultural and Marine (Chapter 7), food and food products (Chapter 8)] and some specialized applications (Chapter 9). The Editors and Authors all hope you will find this book to be helpful in your work.

## REFERENCES

1. *2003 World Population Data Sheet. Population Reference Bureau*. Bread for the World Institute. Accessed February 16, 2005. Electronic. Available from http://www.bread.org/hungerbasics/international.html#cite5.

2. G. S. Birth. *How light interacts with food*. In: *Quality Detection in Foods*, J. Gaffney, ed, **1**:6–11. St. Joseph, MI, ASAE, 1976.

3. A. M. C. Davies. 1984. Progress in human food analysis by near infrared. *Anal Proc* **21**: 488–491.

4. A. M. C. Davies, A. Grant. 1987. Review: near infrared analysis of food. *Int J Food Sci Technol* **22**: 191–207.

5. G. J. DeMenna, G. J. Brown. Fast Food Analysis: Don't Wait for Proximate. *Food Testing and Analysis* (**June/July**): 16–19, 1998.

6. G. C. Marten, J. S. Shenk, I. F. E. Barton. 1989. *Near Infrared Reflectance Spectroscopy (NIRS): Analysis of Forage Quality (Handbook No. 643)*. Agriculture Handbook No. 643. Springfield, VA: National Technical Information Service.

7. H. Martens, H. Russwurm(Jr). *Food Research and Data Analysis*. New York, NY: Applied Science Publishers, 1983.

8. K. H. Norris. Measurement of quality in foods and agricultural commodities by physical methods. In: *First Food Physics Symposium*, 113–124, 1956.

9. B. G. Osborne, T. Fearn. *Near Infrared Spectroscopy in Food Analysis*. Essex, UK: Longman Scientific & Technical-Wiley, 1986.

10. N. S. Sahni, T. Isakssonc, T. Næsb. In-line near infrared spectroscopy for use in product and process monitoring in the food industry. *J Near Infrared Spectrosc* **12**: 77–83, 2004.

11. P. Williams, K. Norris, eds, *Near-Infrared Technology in Agricultural and Food Industries (First Edition)*. St. Paul, MN: *Am Assoc Cereal Chemists*, 1987.

## REFERENCES ON FOOD

1. E. Anklam, F. Gadani, P. Heinze, H. Pijnenburg, G. V. D. Eede. Analytical methods for detection and determination of genetically modified organisms in agricultural crops and plant-derived food products. *Eur Food Res Technol* **214**: 3–26, 2002.

2. D. D. Archibald, S. E. Kays, D. S. Himmelsbach, F. E. BARTON-II. Raman and NIR Spectroscopic Methods for Determination of Total Dietary Fiber in Cereal Foods: A Comparative Study. *Appl Spectrosc* **52(1)**: 22–31, 1998.

3. D. D. Archibald, S. E. Kays, D. S. Himmelsbach, F. E. Barton-II. Raman and NIR Spectroscopic Methods for Determination of Total Dietary Fiber in Cereal Foods: Utilizing Model Differences. *Appl Spectrosc* **52(1)**: 32–41, 1998.

4. S. Asen, R. N. Stewart, K. H. Norris. A naturally occuring colorant for food and beverages. U. S. Department of Commerce, 1978.

5. D. Baker. The determination of fiber in processed cereal foods by near infrared reflectance spectroscopy. *Cereal Chemistry* **60**: 217–219, 1983.

6. D. Baker, K. H. Norris, B. W. Li. *Food fiber analysis: Advances in methodology*. In: *Dietary Fibers: Chemistry and Nutrition*, A. Inglett, S. I. Falkehag, eds, 68–78. New York: Academic Press, 1979.

7. R. J. Barnes, M. S. Dhanoa, S. J. Lister. Correction to the description of Standard Normal Variate (SNV) and De-Trend (DT) ransformations in Practical Spectroscopy with Applications in Food and everage Analysis–2nd Edition. *J Near Infrared Spectrosc* **1(2)**: 185–186, 1993.

8. E. O. Beasley. Light transmittance of peanut oil as an objective measurement related to quality of raw peanuts. In: *Quality Detection in Foods*, J. J. Gaffney, ed, **1**: 50–52. St. Joseph, MI: American Society of Agricultural Engineers, 1976.

9. G. S. Birth. *How light interacts with food*. ASAE Paper.

10. G. S. Birth. *Research in food instrumentation*. Instrument Society of America, 1963.

11. G. S. Birth. How light interacts with foods. In: *Quality Detection in Foods*, J. J. Gaffney, ed, **1**: 6–11. St. Joseph, MI: American Society of Agricultural Engineers, 1976.

12. G. S. Birth. The light scattering properties of food. *J Food Sci* **43**: 916–925, 1978.

13. G. S. Birth. Radiometric measurement of food quality-a review. *J Food Sci* **44**: 949–953, 957, 1979.

14. G. S. Birth, K. H. Norris. The difference meter for measuring interior quality of foods and pigments in biological tissues. Technical Bulletin No. 1341: 19-Jan, 1965.

15. G. S. Birth, K. L. Norris. The difference meter an instrument for measuring interior quality of foods. Asae, 1963.

16. O.-C. Bjarno. Meat and meat products: Multicomponent analysis of meat products. *J Assoc Off Anal Chem* **64(6)**: 1392–1396, 1981.

17. O.-C. Bjarno. Multicomponent analysis of meat products by infrared spectrophotometry: Collaborative Study. *J Assoc Off Anal Chem* **65(3)**: 696–700, 1982.

18. M. M. Brown, I. J. Ross. The use of radio frequency fields as a means of determining the concentration and volume of solutions of food components. Asae, 1967.

19. A. Brynjolfsson. The national food irradiation program concucted by the department of the army.

20. H. Buning-Pfaue. Analysis of water in food by near infrared spectroscopy. *Food Chemistry* **82**: 107–115, 2003.

21. C. S. Burks, F. E. Dowell, F. Xie. Measuring fig quality using near-infrared spectroscopy. In: Preprint, 2000.

22. D. J. Casimir, Countercurrent extraction of soluble solids from foods. *Csiro Fd Res Q* **43**: 38–43, 1983.

23. P. Chandley. The application of the DESIR technique to the analysis of beer. *J Near Infrared Spectrosc* **1(1)**: 133–139, 1993.

24. J. Y. Chen, C. Iyo, S. Kawano, F. Terada. Development of calibration with sample cell compensation for determining fat content in unhomogenized raw milk by a simple NIR transmittance method. *J Near Infrared Spectrosc* **7**: 265–273, 1999.

25. A. M. C. Davies, J. Franklin, K. M. Wright, S. M. Ring, P. S. Belton. FT – the solution to many problems. Laboratory Practice Preprint, 1985.

26. G. J. DeMenna, G. J. Brown. Fast Food Analysis: Don't Wait for Proximate. Food Testing and Analysis **(June/July)**: 16–19, 1998.

27. E. Díaz-Carrillo, A. Muñoz-Serrano, A. Alonso-Moraga, J. M. Serradilla-Manrique. Near infrared calibrations for goat's milk components: protein, total casein, as-, b- and k-caseins, fat and lactose. *J Near Infrared Spectrosc* **1(2)**: 141–146, 1993.

28. M. R. Ellekjær, K. I. Hildrum, T. Næs, T. Isaksson. Determination of the sodium chloride content of sausages by near infrared spectroscopy. *J Near Infrared Spectrosc* **1**: 65–75, 1993.

29. D. G. Evans, C. N. G. Scotter, L. Z. Day, M. N. Hall. Determination of the authenticity of orange juice by discriminant analysis of near infrared spectra: A study of pretreatment and transformation of spectral data. *J Near Infrared Spectrosc* **1(1)**: 33–44, 1993.

30. P.-G. Fyhn, E. Slinde. Measurements of monochromatic visible light changes within food products using laser and fiber optics. *Norwegian Food Research Institute* 11-Jan, 1985.

31. T. Gato. Application of near-infrared spectroscopy for predicting the moisture, total nitrogen and neutral density fiber of raw tea and tea. In: *2nd International NIRS*

*Conference*, M. Iwamoto, S. Kawano, eds, 319–328. Tsukuba, Japan: Korin Publishing Co., 1989.

32. T. Goto. Application of near infrared spectroscopy for predicting the moisture, total nitrogen and neutral detergent fiber of raw tea and tea. *Proc 2nd Intl NIRS Conf, Tsukuba, Japan* **1**: 319–328, 1989.

33. H. Abe, S. Kawano, K. Takehara, M. Iwamoto. Determination of Sucrose content in Sugarcane Juice by Near Infrared Spectroscopy, *Rep Natl Food Res Inst* **60**: 31–36. Tsukuba, Japan: Food Research Institute, 1996.

34. T. Hagiwara, H. Wang, T. Suzuki, R. Takai. Fractal Analysis of Ice Crystals in Frozen Food. *J Agric Food Chem* **50**: 3085–3089, 2002.

35. A. J. Hand, D. C. McCarthy. Interactive story: From corn to cupcakes. *Photonics Spectra* **34(3)**: 91–103, 1999.

36. A. J. Hand, D. C. McCarthy. Photonics and Food. *Photonics Spectra* **34(3)**: 89–124, 1999.

37. K. Ikegaya, M. Iwamoto, J. Uozumi, R. K. Cho. *Determination of chemical composition of Japanese green tea by near-infrared spectroscopy*. M. Iwamoto, S. Kawano, eds, *2nd International NIRS Conference. Tsukuba, Japan*: Korin Publishing Co., 1990.

38. K. Ikegaya, S. Kawano, R. K. Cho. Determination of theaflavins in black tea by near-infrared spectroscopy. In: *2nd International NIRS Conference*, M. Iwamoto, S. Kawano, eds, Tsukuba, Japan: Korin Publishing, 358, 1989.

39. T. Isaksson, Z. Wang, B. Kowalski. Optimised scaling (OS-2) regression applied to near infrared diffuse spectroscopy data from food products. *J Near Infrared Spectrosc* **1**: 85–97, 1993.

40. K. J. Kaffka, J. Farkas, Z. Seregely, L. Meszaros. Monitoring the effect of ultra-high pressure preservation technology by near infrared reflectance spectroscopy. In: *Near Infrared Spectroscopy: Proceedings of the 10th International Conference*, A. M. C Davies, R. K. Cho, eds, Preprint: 505. Kyongju, Korea: NIR Publication, UK, 2002.

41. S. Kawano, H. Abe, M. Iwamoto. Development of a calibration equation with temperature compensation for determining the Brix value in intact peaches. *J Near Infrared Spectrosc* **3**: 211–218, 1995.

42. S. E. Kays. The use of near infrared reflectance spectroscopy to predict the insoluble dietary fibre fraction of cereal products. *J Near Infrared Spectroscopy* **6**: 221–227, 1998.

43. S. E. Kays, F. E. Barton-II. Near-Infrared Analysis of Soluble and Insoluble Dietary Fiber Fractions of Cereal Food Products. *J Agric Food Chem* **50**: 3024–3029, 2002.

44. S. E. Kays, F. E. Barton-II. Rapid Prediction of Gross Energy and Utilizable Energy in Cereal Food Products Using Near-Infrared Reflectance Spectroscopy. *J Agric Food Chem* **50**: 1284–1289, 2003.

45. S. E. Kays, F. E. Barton-II, W.R. Windham. Predicting protein content by near infrared reflectance spectroscopy in diverse cereal food products. *J Near Infrared Spectrosc* **8**: 35–43, 2000.

46. G. Kisko, Z. Sertegely. Qualification of volatile oils using NIR and electronic nose. In: *10th Internation Conference on Near Infrared Spectroscopy (KOREA)*, A. M. C. Davies, ed, Preprint: 12. Kyongju, Korea: NIR Publication, UK, 2001.

47. I. M. E. Lafargue, M. H. Feinberg, J.-J. Daudin, D. N. Rutledge. Homogeneity check of agricultural and food industries samples using near infrared spectroscopy. *Anal Bioanal Cherm* **375**: 496–904, 2003.

48. I. Landa. Food constituents analysis using a monochromator with high speed scan and high energy throughput. In: *Meeting of the ASAE*:17. Winnipeg, Canada: ASAE, 1979.

49. H. Martens. Understanding food research data. *Applied Science*: May 38, 1982.

50. H. Martens, H. Russwurm(Jr). *Food Research and Data Analysis*. New York, NY: Applied Science Publishers, 1983.

51. K. A. Martin. Recent advances in near-infrared reflectance spectroscopy. In: *Applied Spectroscopy Reviews*, Jr. Edward, G. Brame, eds, **27**: 325–383. New York, NY: Marcel Dekker, Inc., 1992.

52. D. R. Massie. A high-intensity spectrophotometer interfaced with a computer for food quality measurement. In: *Quality Detection in Foods*, J. J. Gaffney, ed, **1**: 12–15. St. Joseph, MI: American Society of Agricultural Engineers, 1976.

53. D. C. McCarthy. The perfect chocolate chip cookie. *Photonics Spectra* **34(3)**: 105–111, 1999.

54. D. C. McCarthy. The perfect orange. *Photonics Spectra* **34(3)**: 113–117, 1999.

55. D. C. McCarthy. Vision shouldn't blur your beer. *Photonics Spectra* **34(3)**: 119–124, 1999.

56. W. F. McClure. Biological measurements for the 21st century: Instruments – First; Measurements - Second; Discovery - Third. In: *Agricultural and Biological Engineering: New Horizons, New Challenges*, Miron Turner, ed, 1–9. Newcastle, UK: Tynesoft Business Services, 1995.

57. W. F. McClure. Wave of the future: *Biomeasurements in the 21st Century*. Conference Talk, 1995.

58. W. F. McClure. Near-infrared Instrumentation. In: *Near-Infrared Technology in the Agricultural and Food Industries, Second Edition*, Phil Williams, Karl Norris, eds, 109–127. St. Paul, MN: American Association of Cereal Chemists, 2001.

59. S. C. Mohapatra. *World Hunger. Resource* **4 (April)**: 33, 1999.

60. N. N. Mohsenin. Application of mechanical properites of food materials in quality evaluation and control. *Asae Paper No.* 73–6510, 1973.

61. I. Murray. Nir Analysis- How Near Infrared Reflects Composition. In: *One Day Seminar on the Use of Near Infrared Reflectance (NIR) Analysis in Research, Routine and Run-Of-The-Mill Applications*, **1**: 51, School of Agriculture, Aberdeen, Scotland: Chemistry Division, School of Agriculture, 1983.

62. W. W. News. High pressure touted as improved method of food preservation, 1998.

63. S. L. Oh, R. K. Cho, B. Y. Min, D. H. Chung, S. Kawano, K. Ikegaya. Determination of nitrogen compounds in green tea infusion by near-infrared reflectance spectroscopy. In: *2nd International NIRS Conference*, M. Iwamoto, S. Kawano, eds, 376–385. Tsukuba, Japan: Korin Publishing, 1990.

64. T. P. Ojha, A. W. Farrall, A. M. Dhanak, C. M. Stine. Determination of heat transfer through powdered food products. Asae, 1966.

65. B. G. Osborne. Near infrared spectroscopic studies of starch and water in some processed cereal foods. *J Near Infrared Spectrosc* **4**: 195–200, 1996.

66. B. G. Osborne, T. Fearn. *Near Infrared Spectroscopy in Food Analysis*. Essex, UK: Longman Scientific & Technical-Wiley, 1986.

67. N. Pedretti, D. Bertrand, M. Semenou, P. Robert and R. Giangiacomo. Application of an experimental design to the detection of foreign substances in milk. *J Near Infrared Spectrosc* **1(2)**: 174–184, 1993.

68. O. Pelletier, R. Brassard. Determination of vitamin c (l-asorbic acid and dehydroasorbic acid) in food by manual and automated photometric methods. Bureau of Nutritional Sciences, Ottawa, Canada: 1–11, 1983.

69. D. D. Requena, S. A. Hale, D. P. Green, W. F. McClure, B. E. Farkas. Detection of discoloration in thermally processed blue crab meat. *Journal of the Science of Food and Agriculture* **79**: 786–791, 1999.

70. P. Robert, M. F. Devaux, A. Qannar, M. Safara. Mid and near infrared study of carbohydrates by canonical correlation analysis. *J Near Infrared Spectrosc* **1**: 99–108, 1993.

71. L. E. Rodriguez-Saona, F. S. Fry, E. M. Calvey. Use of Fourier Transform Near-Infrared Reflectance Spectroscopy for Rapid Quantification of Castor Bean Meal in a Selection of Flour-Based Products. *J Agric Food Chem* **48**: 5169–5177, 2000.

72. J. M. Roger, V. Bellon-Maurel. Improving sugar content prediction in fruits by applying genetic algorithms to near infrared spectra. *J NIRS* (Preprint for Review): 1–29, 1999.

73. S. Sasic, Y. Ozaki. Wavelength-wavelength and sample-sample two dimensional correlation analyses of short-wave near-infrared spectra of raw milk. *Appl Spectrosc* Preprint for Review, 2000.

74. V. H. Segtnan, T. Isaksson. Evaluating near infrared techniques for quantitative analysis of carbohydrates in fruit juice model systems. *JNIRS* **8**: 108–116, 2000.

75. Z. Seregely, K. J. Kaffka. Qualification of food flavors using NIR spectroscopy and chemosensor-array (electronic nose). *JNIRS* Preprint for Review: 1–7, 2001.

76. S. K. Seymour, D. D. Hamann. Design of a microcomputer-based instrument for crispness evaluation of food products. In: *Meeting of the ASAE*.

77. S. K. Seymour, D. D. Hamann. Design of a microcomputer-based instruments for crispness evaluation of food products. *Transactions of the ASAE* **27(4)**: 1245–1250, 1984.

78. O. P. Snyder. Food radiation, *the process engineering challenge*. Asae, 1966.

79. Staff. *NIR Analysis: How Near Infrared Reflects Composition*. Aberdeen, Scotland, 1983.

80. A. S. Szabo, P. Tolnay, Z. Mednyanszky. Monitoring changes in material properties of agricultural products during heating and drying by impedance spectroscopic analysis. *1st Intl Conf on Food Physics* **1**: 96–98, 1994.

81. T. Takeo, N. Okamoto. Quality control of tea by near-infrared reflectance spectroscopy. In: *2nd International NIRS Conference*, M. Iwamoto, S. Kawano, eds, 157–173. Tsukuba, Japan: Korin Publishing Co., 1989.

82. D. P. Thompson, J. C. Wolf. Available lysine losses in a stirred model food system. American Society of Agricultural Engineers, 1976.

83. K. Thyholt, T. Issaksson. Near infrared spectroscopy of dry extracts from high moisture food products on solid support: A review. Preprint, 1998.

84. U. Wahlby, C. Skjoldebrand. NIR-measurements of moisture change in foods. *J Food Engineering* **47**: 303–312, 2001.

85. B. Welz, Z. Grobenski, M. Melcher, D. Weber. *A techniques in food analysis*. Atomic Absorption Perkin Elmer Corp, 1979.

86. D. L. B. Wetzel. Analytical near infrared spectroscopy. (In: *NIR near infrared food beverage instrumentation chemometrics* (D. Wetzel, G. Charalambous, eds, Elsevier Science B. V., New York. Book: 141–194, 1998.

87. P. Williams, K. Norris. Eds, *Near-Infrared Technology in Agricultural and Food Industries (First Edition)*. St. Paul, MN: Am Assoc. Cereal Chemists, 1987.

88. P. Williams, K. Norris. Eds, *Near-Infrared Technology in Agricultural and Food Industries (Second Edition)*. St. Paul, MN: Am Assoc. Cereal Chemists, 2001.

89. R. H. Wilson. Ed, *Spectroscopic Techniques for Food Analysis (COPY of book ON SHELF)*. Edited by n/a. n/a. New York, NY: VCH Publishers, Inc., 1994.

90. J. P. Wold. Rapid quality assessment of meat and fish by using near-infrared spectroscopy, autofluoresence spectroscopy and image analysis. Doctor of Science, Agricultural University of Norway, 2000.

91. R. Yabe. Near-infrared transmittance application for determining moisture content of surimi products. In: *Proc 2nd Intl NIRS Conf, Tsukuba, Japan*:190–195, 1989.

92. H. Yang, J. Irudayaraj. Rapid determination of vitamin C by NIR, MIR and FT-Raman techniques. *J Pharmacy and Pharmocology* **54**: 1–7, 2002.

93. X. Zhao, W. Chan, M. Wong, D. Xiao, Z. Li. A fluorescnce-based iodine sensor for food analysis. *Am Laboratory* **35(11)**: 13–21, 2003.

# CHAPTER 2

# Principles of Molecular Vibrations for Near-Infrared Spectroscopy

C. SANDORFY, R. BUCHET, and G. LACHENAL

## INTRODUCTION

The fundamental vibrations of molecules lead to absorption in the infrared (200–4000 $cm^{-1}$), while their overtones and combination tones appear in the infrared and in the near infrared. The highest wavenumber associated to an infrared active fundamental vibration appears at 3998 $cm^{-1}$ for HF; therefore 4000 $cm^{-1}$ is somehow the border of the infrared and the near infrared. The near-infrared spectrum is located between the infrared and the visible, from 2500 nm to 800 nm or from 4000 $cm^{-1}$ to 12,500 $cm^{-1}$. It is sometimes called the overtone region, but, naturally, the first overtone of vibrations of lower wavenumber are in the mid-infrared. As a rule overtones and combination tones are much weaker than the fundamentals, the first ones usually by a factor of 10 to 100. Second and higher overtones and combination tones are even weaker. The near-infrared absorption of polymers, especially in the region from 4000 to 9000 $cm^{-1}$, originates from the overtones of OH, NH, CH, and SH stretching vibrations as well as from stretching-bending combinations involving these groups. The overtones and combination tones are most often affected by hydrogen bond formation, certain aspects of which can be monitored to advantage in the near infrared (1). In addition, a few electronic transitions may appear in the upper range of the near infrared, close to the visible, especially in the range of 9000 $cm^{-1}$ to 15,000 $cm^{-1}$. For example, electronic transitions of oxyhemoglobin and deoxyhemoglobin give rise to a band around 13,160 $cm^{-1}$ (760 nm). Because most of the analytical applications of near-infrared spectroscopy on polymers and biomolecules are centered in the 4000-$cm^{-1}$ to 9000-$cm^{-1}$ region, the present review focuses on overtones and combination tones of molecular vibrations.

*Near-Infrared Spectroscopy in Food Science and Technology*, Edited by Yukihiro Ozaki, W. Fred McClure, and Alfred A. Christy.

Two languages are available for the treatment of overtones: the language based on normal coordinates (2–5) and the more recent treatment based on local coordinates (6–9). Whereas the latter appears to be better adapted to higher overtones, especially when kinetics of chemical reactions are concerned, the first overtones are treated well on a normal coordinate basis. In what follows, normal coordinates will be used; then we discuss the approach based on local coordinates.

## THE DIATOMIC OSCILLATOR

The simplest possible assumption about the form of the vibrations in a diatomic molecule is that each atom moves toward or away from the other in a simple harmonic motion (2–5). Such a motion of the two atoms can be reduced to the harmonic vibration of a single mass point about an equilibrium position. In classic mechanics a harmonic oscillator can be defined as a mass point of mass $m$ that is acted upon by a force $F$ proportional to the distance $Q$ from the equilibrium position and directed toward the equilibrium position:

$$F = m\frac{\partial^2 Q}{\partial t^2} = -kQ \tag{2.1}$$

where the proportionality factor $k$ is called the force constant and depends only on the strength of the chemical bond. The solution of the differential equation is:

$$Q = Q_0 \sin(2\pi \nu t + \phi) \tag{2.2}$$

$Q_0$ is the amplitude of the vibration, $\phi$ is a phase constant dependent on the initial conditions, and $\nu$ is the vibrational frequency. The displacement of the point mass from the equilibrium position is a sine function of time. By substituting Equation 2.2 into Equation 2.1, it follows that:

$$\nu = \frac{1}{2\pi}\sqrt{\frac{k}{m}} \tag{2.3}$$

Because the force is the negative derivative of the potential energy $U$, using Equation 2.1 one obtains that:

$$U = \frac{1}{2}kQ^2 = 2\pi^2 m\nu^2 Q^2 \tag{2.4}$$

The potential energy of a harmonic oscillator is proportional to the square of the displacement from the equilibrium position. The restoring force exerted by the two atoms of a molecule on each other when they are displaced from their equilibrium position ($r_e$) is approximately proportional to the change of internuclear distance

$(r - r_e)$. For the first atom, of mass $m_1$:

$$m_1 \frac{\partial^2 r_1}{\partial t^2} = -k(r - r_e) \tag{2.5}$$

For the second atom of mass $m_2$:

$$m_2 \frac{\partial^2 r_2}{\partial t^2} = -k(r - r_e) \tag{2.6}$$

where $Q = (r - r_e)$ is the displacement from the equilibrium position, $r_1$ and $r_2$ are the distance of the two atoms from the center of gravity, $r$ is the distance of the two atoms from each other, and $r_e$ is the equilibrium distance. For a dumbbell model of a diatomic molecule:

$$r_1 = \frac{m_2}{m_1 + m_2} r \quad \text{and} \quad r_2 = \frac{m_1}{m_1 + m_2} r \tag{2.7}$$

By substituting Equation 2.7 into either Equation 2.5 or Equation 2.6 one can obtain:

$$f = \frac{m_1 m_2}{m_1 + m_2} \frac{\partial^2 Q}{\partial t^2} = -kQ$$
$$\mu = \frac{m_1 m_2}{m_1 + m_2} \tag{2.8}$$

$\mu$ is so-called reduced mass.

Equation 2.8 is identical to Equation 2.1 of the harmonic oscillator, except that $Q$ is replaced by $Q = (r - r_e)$. The vibration of the two atoms of a molecule is reduced to the vibration of a single mass point of mass $\mu$. From Equation 2.8 it follows that:

$$\nu = \frac{1}{2\pi} \sqrt{\frac{k}{\mu}} \tag{2.9}$$

Usually, the wavenumber ($\bar{\nu}$) is used in infrared and near-infrared spectroscopy. Equation 2.9 is often written as:

$$\bar{\nu} = \frac{1}{2\pi c} \sqrt{\frac{k}{\mu}} \tag{2.10}$$

The frequency of the vibrational mode depends on two parameters: the reduced mass $\mu$ and the force constant $k$. It follows from this equation that isotopic labeling is a powerful tool to assign bands in infrared and near-infrared spectra. For example, the $CH_2$ symmetric stretching of lipid is located at 2851 cm$^{-1}$, whereas the $C^2H_2$ symmetric stretching of perdeuterated lipid is at 2070 cm$^{-1}$ (10). Assuming that $k$

is not affected during the isotopic substitution and assuming that $CH_2$ and $C^2H_2$ symmetric stretching modes behave like vibrational modes of a diatomic molecule, by using Equation 2.10 and computing the reduced mass of CH and of $C^2H$, the theoretical ratio of the two wavenumbers $\bar{\nu}(CH)/\bar{\nu}(C^2H)$ can be obtained:

$$\mu(CH) = \frac{12 \times 1}{12 + 1} = 0.923$$

$$\mu(C^2H) = \frac{12 \times 2}{12 + 2} = 1.714$$

$$\bar{\nu}(CH)/\bar{\nu}(C^2H) = (\mu(C^2H)/\mu(CH))^{1/2} = 1.36$$

For comparison, the ratio of the two measured wavenumbers is $2851/2070 = 1.38$. Although the lipid is not a diatomic molecule, the CH vibrational mode behaves almost like a diatomic molecule. Some vibrational modes in polymers that are very well localized can be considered as vibrational modes of a diatomic molecule. Fortunately, most of the first overtones and combination tones in the near infrared arise from well-localized vibrational modes such as OH, NH, CH, and SH stretching modes. The force constant $k$ depends on the strength of the chemical bond, which is mainly controlled by the electronic environment surrounding atoms in molecules. Electrostatic interactions or formation of hydrogen bonds alter the force constant $k$ and therefore can shift the frequency. It is well established that infrared and near-infrared spectroscopies are very sensitive in detecting the formation of hydrogen bonds. For example, a change in hydrogen bonding distance of only 0.0002 nm shifts the wavenumber of the stretching vibration of NH of a peptide backbone by approximately 1 $cm^{-1}$. This shift is well within the experimental limit. In contrast, X-ray crystallographic results rarely have a resolution below 0.1 nm (11). Classically, only one vibrational frequency is possible according to Equation 2.9, whereas its amplitude and its energy can assume any value.

## ENERGY LEVELS

In quantum mechanics, the vibration in a diatomic molecule may also be reduced to the motion of a single point mass of mass $\mu$, whose displacement $Q$ from its equilibrium position corresponds to $(r - r_e)$. Assuming that the potential is that of a one-dimensional harmonic oscillator (4), the Hamiltonian $H$ becomes:

$$H = \frac{h^2}{8\pi^2\mu} \frac{\partial^2}{\partial Q^2} + \frac{1}{2}kQ^2 \tag{2.11}$$

Using the Schrödinger equation:

$$\left( \frac{h^2}{8\pi^2\mu} \frac{\partial^2}{\partial Q^2} + \frac{1}{2}kQ^2 \right) \psi(Q) = E\psi(Q) \tag{2.12}$$

The Hermite polynomials are the wavefunctions that satisfy Equation 2.12. For a diatomic molecule the Hermite polynomials have the form (5):

$$\psi_v = N_v e^{\frac{1}{2}Q^2} H_v(\sqrt{\alpha}Q) \tag{2.13}$$

where $\alpha = \frac{4\pi^2}{h}\mu v = \frac{2\pi}{h}\sqrt{\mu k}$, $\mu$ is the reduced mass, $h$ is the Planck constant, $k$ is the harmonic potential constant, and $N_v$ is the normalization factor. Then for successive values of the vibrational quantum number $v$ we have with $\sqrt{\alpha}Q = z$

$$H_0(z) = 1$$
$$H_1(z) = 2z$$
$$H_2(z) = 4z^2 - 2$$
$$H_3(z) = 8z^3 - 12z$$
$$H_4(z) = 16z^4 - 48z^2 + 12$$
$$H_5(z) = 32z^5 - 160z^3 + 120z$$

The solution of Equation 2.12 indicates that the vibrational energies $E(v)$ have discrete values (4).

$$E(v) = \left(v + \frac{1}{2}\right)\frac{h}{2\pi}\sqrt{\frac{k}{\mu}} \tag{2.14}$$

$$E(v) = \left(v + \frac{1}{2}\right)hv \tag{2.15}$$

where the vibrational quantum number $v$ can take only integral values, $0, 1, 2. \ldots$ These values are the only energy values allowed by theory for the harmonic oscillator. $v$ is the vibrational frequency of the diatomic molecule as defined in Equation 2.9. The lowest energy ($v = 0$) has $E(0) = (1/2)h\,v$. Thus even in the lowest state, the vibrational energy of a molecule is not zero, in contrast to the result of classic mechanics. Equation 2.14 is often written in wavenumber units as:

$$G_v = \frac{E(v)}{hc} = \left(v + \frac{1}{2}\right)\omega_e \tag{2.16}$$

## SELECTION RULES

If the diatomic molecule in its equilibrium position has a dipole moment as in heteronuclear diatomic molecules, it will in general change if the internuclear distance varies. To a first approximation it may be assumed that the change of dipole moment with internuclear distance is linear. Therefore the dipole moment changes with a frequency equal to the frequency of the mechanical vibration. On the basis of classic electrodynamics, this would lead to the emission of light of frequency $v$. Conversely, the oscillator could be set in vibration by absorption of light of frequency $v$. In quantum mechanics, emission of radiation takes place as a result of a transition of the oscillator from a higher to a lower state, and absorption is the reverse process (2). The wave number of the emitted or absorbed light is given by:

$$v^{v'v''} = \frac{E(v'')}{hc} - \frac{E(v'')}{hc} \tag{2.17}$$

where $v'$ and $v''$ are vibrational quantum numbers of the upper and the lower states, respectively. To determine which particular transitions can occur, the transition moment that corresponds to the probability of the transition from an initial to a final state must be evaluated. The transition moment for a vibrational transition is given by (2–5, 12):

$$R^{v'v''} = \int \psi_{v'}^* M \psi_{v''} \partial\tau \tag{2.18}$$

where $\psi_{v'}$ and $\psi_{v''}$ are the vibrational wave functions for the upper and lower vibrational states with their respective quantum numbers $v'$ and $v''$ and $M$ is the variable dipole moment. The well-known selection rule is $\Delta v = \pm 1$, that is, only quantum jumps by 1 are allowed. Therefore fundamental vibrations are allowed while overtones are not. This selection rule is based on the properties of the harmonic oscillator wave functions that are the normalized Hermite polynomials (Equation 2.13). Thus it only applies to the harmonic oscillator that is ruled by a quadratic potential, that is, as long as the wave functions are the exact Hermite polynomials. These conditions are never exactly met; overtones and combination bands do appear in the spectra. It is instructive to remember how this selection rule is obtained. The variable dipole moment is, if we neglect higher terms:

$$M = M_e + M_1 Q = M_e + \left(\frac{\partial M}{\partial Q}\right)_e Q \tag{2.19}$$

where $\left(\frac{\partial M}{\partial Q}\right)_e$ is taken at the equilibrium geometry. So the transition moment becomes:

$$R^{v'v''} = M_e \int \psi_{v'}^* \psi_{v''} \partial Q + \left(\frac{\partial M}{\partial Q}\right)_e \int \psi_{v'}^* Q \psi_{v''} \partial Q \tag{2.20}$$

The first term is zero because of the orthogonality of the harmonic oscillator wave functions, and the integral in the second term is, because $\partial z = \sqrt{\alpha} \partial Q$:

$$\int \psi_{v'}^* Q \psi_{v''} \partial Q = \frac{N_{v'} N_{v''}}{\alpha} \int H_{v'}(z) z H_{v''}(z) e^{-z^2} \partial z \tag{2.21}$$

Applying the recursion formula to $H_v''(z)$:

$$z H_v(z) = \frac{1}{2} H_{v+1}(z) + v H_{v-1}(z) \tag{2.22}$$

we obtain:

$$\int \psi_{v'}^* Q \psi_{v''} \partial Q = \frac{N_{v'} N_{v''}}{\alpha} \left[ \frac{1}{2} \int H_{v'}(z) H_{v''+1}(z) e^{-z^2} \partial z \right.$$

$$\left. + v'' \int H_{v'}(z) z H_{v''-1}(z) e^{-z^2} \partial z \right] \tag{2.23}$$

The first integral is different from zero only if $v' = v'' + 1$ and the second if $v' = v'' - 1$ because of the orthogonality of the Hermite polyomials. Only one of the two can apply for given values of the vibrational quantum numbers so that $\Delta v = \pm 1$, where the $+$ sign applies to absorption and the $-$ sign to emission.

In addition, $\left( \frac{\partial M}{\partial Q} \right)_e$ must also be nonzero; this cannot be the case for homoatomic diatomic molecules.

Two observations must be made:

1. The recursion formula only applies to the exact Hermite polynomials, that is, when the oscillator (and the potential) are exactly harmonic. This can never exactly apply because molecules can be dissociated; dissociating them would require infinite force for a harmonic oscillator.

2. In addition, the variation of the dipole moment must be exactly linear with the normal coordinate (in the case of a diatomic molecule this is equal to the change of the internuclear distance). Otherwise, higher powers of $Q$ appear in the second integral $R^{v'v''}$ (20) to which, again, the recursion formula does not apply.

## ANHARMONICITY

The spectral properties of near-infrared bands (frequency, intensity, breadth) depend essentially on anharmonicity. The potential function ruling the vibrational motion can be written, for small displacements from equilibrium:

$$V = V_0 + \left( \frac{\partial V}{\partial Q} \right)_e Q + \frac{1}{2} \left( \frac{\partial^2 V}{\partial Q^2} \right)_e Q^2 + \frac{1}{3!} \left( \frac{\partial^3 V}{\partial Q^3} \right)_e Q^3$$

$$+ \frac{1}{4!} \left( \frac{\partial^4 V}{\partial Q^4} \right)_e Q^4 + \ldots \tag{2.24}$$

where $V_0$ can be absorbed in the zero level of the potential energy, the second term, the force, is zero at equilibrium, $k = \frac{1}{2} \left( \frac{\partial^2 V}{\partial Q^2} \right)_e$ is the quadratic (harmonic) potential constant, and $k_3 = \frac{1}{3!}\left(\frac{\partial^3 V}{\partial Q^3}\right)_e Q^3$ and $k_4 = \frac{1}{4!}\left(\frac{\partial^4 V}{\partial Q^4}\right)_e Q^4$ are the cubic and quartic potential constants, respectively, all taken at the equilibrium distance, and $Q$ is the normal coordinate.

A second-order perturbation calculation with the sum of the cubic and quartic terms taken for the perturbing potential, neglecting higher terms, yields for the vibrational terms (in $cm^{-1}$):

$$G_v = \omega_e\left(v + \frac{1}{2}\right) + X\left(v + \frac{1}{2}\right)^2 \tag{2.25}$$

where $\omega_e$ is the unperturbed, harmonic frequency and $X$ is the anharmonicity constant (often written as $\omega_e x_e$.). $X$ is negative and numerically much smaller than $\omega_e$. For the latter the same calculation yields:

$$X = \frac{15}{4\omega_e}k_3^2 - \frac{3}{2}k_4 \tag{2.26}$$

a relationship between the anharmonicity constant and the anharmonic potential constants.

From Equation 2.25 the perturbed vibrational frequencies are for transitions from $v'' = 0$ to $v' = v$:

$$v^{0v} = G_v - G_0 = \omega_e v + X v(v + 1) \tag{2.27}$$

with

$$G_0 = \frac{1}{2}\omega_e + \frac{1}{4}X \tag{2.28}$$

Then the frequency (in wavenumber units) of the fundamental becomes

$$v^{01} = \omega_e + 2X$$

For the first overtone ($\Delta v = 2$)

$$v^{02} = 2\omega_e + 6X$$

and for the second overtone ($\Delta v = 3$)

$$v^{03} = 3\omega_e + 12X$$

so that

$$X = \frac{v^{02}}{2} - v^{01} = \frac{v^{03}}{3} - \frac{v^{02}}{2} \tag{2.29}$$

So, if the second-order approximation applies, we should obtain the same anharmonicity constant from the first overtone and the fundamental as from the second and first overtones, etc. This is a possible test for the validity of the second-order (Morse curve) approximation. The perturbed wave function will be a linear combination of the harmonic wave functions for different $v$ values giving nonzero, but usually small, transition moments to the overtones.

This anharmonicity, which is due to the nonquadratic part of the potential, is called mechanical anharmonicity. As is seen, it modifies both the frequency and the intensity of the bands.

The validity of the $\Delta v = \pm 1$ selection rule also depends on a second condition, however. Because the dipole moment $M$ varies during the vibrational motion, it can be represented by

$$M(Q) = M_e + \left(\frac{\partial M}{\partial Q}\right)_e Q + \left(\frac{\partial^2 M}{\partial Q^2}\right)_e Q^2 + \dots \tag{2.30}$$

whence

$$R^{v'v''} = M_e \int \psi_{v'}^* \psi_{v''} \partial\tau + \left(\frac{\partial M}{\partial Q}\right)_e \int \psi_{v'}^* Q \psi_{v''} \partial\tau + \frac{1}{2}\left(\frac{\partial^2 M}{\partial Q^2}\right)_e \int \psi_{v'}^* Q^2 \psi_{v''} \partial\tau \tag{2.31}$$

The first term vanishes because of the orthogonality of the harmonic oscillator wave functions. The second term is nonzero if $\Delta v = \pm 1$ and the equilibrium value of the rate of change of the dipole moment is nonzero. The third and higher terms represent a nonlinear change of the dipole moment with the normal coordinate. The integral is nonzero for $\Delta v = \pm 2$, $\Delta v = \pm 3$, ... for the respective higher terms, and the corresponding second, third, ... derivatives of $M$ must also be nonzero. The sum of these higher terms is usually called electrical anharmonicity. The electrical anharmonicity can give intensity to overtones even if the potential is perfectly harmonic. In this approximation, however, it has no influence on their frequency. In Equation 2.15 the wave functions are still those of the harmonic oscillator. So if we find an overtone whose frequency is close to the double of that of the fundamental we may guess that most of its intensity comes from electrical anharmonicity. In the general case both mechanical and electrical anharmonicities contribute to the intensities of the overtones. The sign of their contributions may or may not be the same (see below).

## THE MORSE APPROXIMATION

An analytic approximation for the effective potential $V(Q)$ of a diatomic molecule, found by Morse, has all the expected features for the molecular vibration problem. This function becomes very large for $Q \to 0$ and approaches a constant when $Q \to \infty$. It has also a minimum value at $r_e$, the equilibrium position. It leads to an exactly soluble Schrödinger equation.

$$V(Q) = D_e \left[1 - e^{-\alpha Q}\right]^2 \tag{2.32}$$

where $\alpha$ is a constant depending on the strength of the chemical bond, $D_e$ is the depth of the potential minimum relative to dissociation of the diatomic molecule, and $Q = r - r_e$.

The Hamiltonian becomes:

$$H = \frac{P_q^2}{2\mu} + D_e \left[1 - e^{\alpha Q}\right]^2 \tag{2.33}$$

Solving the Schrödinger equation, using Laguerre polynomials leads to the resulting energy levels:

$$E_v = hc\omega_e \left(v + \frac{1}{2}\right) + hc\omega_e x_e \left(v + \frac{1}{2}\right)^2$$

$$G_v = \omega_e \left(v + \frac{1}{2}\right) + \omega_e x_e \left(v + \frac{1}{2}\right)^2 \tag{2.34}$$

where   $v = 0, 1, 2, \ldots$ is the vibrational quantum number
  $\omega_e = \alpha(2D_e/\mu)^{1/2}$ (in wavenumber units)
  $x_e = (hc)(\omega_e/4D_e)$ (no units)
  $\omega_e x_e = X =$ anharmonicity constant (in wavenumber units)

The first term of the equation is obviously the ordinary harmonic oscillator approximation, and the second corresponds to the correction to the restoring potential due to anharmonicity. Because the anharmonicity constants $X$ are usually reported as negative values, the second term was kept positive as in Equation 2.34. The Morse approximation is often used, although other approximations are proposed by several authors (for more details see Ref. 13 and references therein). Ab initio calculations (14) indicated, as expected, that the harmonic treatment leads to overestimation of the experimental X-H stretching frequency. Significant improvement results from the anharmonic treatment. To take account of anharmonicity, some theoretical calculations are based on an one-dimensional Morse potential (15,16). Alternatively, some authors computed anharmonic force field by using the potential energy for a semirigid polyatomic molecule expanded in a Taylor series (17), while other authors generated potential energies from an equidistant grid (14,18). The estimation of $D_e$ for a weakly bound system is particularly problematic (19). The Morse approximation is equivalent to the second-order perturbation treatment.

## DETERMINATION OF ANHARMONICITY CONSTANTS OF POLYMERS AND OF BIOMOLECULES

The determination of anharmonicity constants $X$ (or $\omega_e x_e$) permits us not only to obtain a better approximation to the shape of potential of a diatomic molecule but also to check the band assignments, especially in the case of overlapping bands, which occur very frequently with larger molecules. The question one may ask is whether the theoretical background developed in the case of diatomic molecules is of any interest for complex cases such as that of polymers. The answer is yes, the basic equations for a diatomic molecule are still a good approximation for well-localized vibrational groups in larger molecules. In fact, a part of the absorption in the near infrared arises from overtones of well-localized vibrational groups such as N-H and O-H stretching vibrations, which behave almost like in a diatomic molecules. Table 2.1 presents fundamentals, overtones, and anharmonicity constants for some non-hydrogen-bonded species.

The NH groups of dimethylamine and $N$-methylacetamide contain well-localized NH stretching vibrations. These molecules are models for polyamines as well as for the peptide backbone in proteins. Nucleic bases and their derivatives have well-localized NH stretching vibrations, whereas methanol has a well-localized OH stretching vibration. The anharmonicity constant for the "free" OH stretching band of an alcohol X is about $-80$ to $-86$ cm$^{-1}$. It is around $-55$ to $-78$ cm$^{-1}$ for the free NH band of secondary amines, about $-59$ to $-66$ cm$^{-1}$ for the CH groups, and $-50$ cm$^{-1}$ for the SH band of thiols [32]. These values may serve to confirm the assignment of overtones. The negative sign means a decrease in wavenumber due to anharmonicity, as is usually the case (that is, when the overtone's wavenumber is less than the double of the fundamental). The determination of anharmonicity constant of a polar group that may form hydrogen bonds may provide more insight into the distortion of its energy potential caused by hydrogen bond formation (see further). The anharmonicity constants obtained from the first overtone and the fundamental or from the first overtone and the second overtone yield similar results within $\pm 5$ cm$^{-1}$ (Table 2.1), suggesting that the first and second terms of Equation 2.34 provide a good approximation for the shape of the potential energy of well-localized vibrational groups of these molecules. A Birge–Sponer plot of $\Delta E_v/v$ versus $v$ can be used to determine anharmonicity constants (30,31).

The dissociation energy ($D_e$) can be computed using Equation 2.34. For example, the dissociation energy of an axial CH group of cyclohexane (30) was obtained by using the anharmonicity constant value from Table 2.1:

$$\nu^{01} = \omega_e + 2X \qquad \text{with} \quad \nu^{01} = 2876 \text{ cm}^{-1} \text{ and } X = -65 \text{ cm}^{-1}$$

$$\omega_e = 3006 \text{ cm}^{-1}$$

$$D_e = hc(\omega_e)^2/(4X)$$

$$= 99.4 \text{ Kcal for 1 mol of CH group}$$

This value corresponds approximately to the energy of a CH chemical bond, confirming that extrapolation of the diatomic model to larger molecules is justified for the cases where the energies of vibrational modes are well centered on one specific chemical bond. The actual dissociation energy of the molecule, $D_0$ is slightly less than $D_e$: $D_e = D_0 + hc\left(\frac{1}{2}\omega_e + \frac{1}{4}X\right)$.

**TABLE 2.1. Fundamentals, Overtones, and Anharmonicity Constants of Well-Localized Vibrational Groups of Free Species**

| Molecule | Reference | Vibration | Fundamental | 1st Overtone | $X$ | 2nd Overtone | $X$ |
|---|---|---|---|---|---|---|---|
| Dimethylamine | 20 | NH stretch | 3355 | 6547 | −81 | 9589 | −78 |
| N-methylacetamide | 21 | NH stretch | 3470 | 6780 | −80 | | |
| Indole | 22 | NH stretch | 3491 | 6847 | −67 | | |
| 1-Cyclohexyluracil | 23 | NH stretch | 3395 | 6662 | −64 | | |
| Benzotriazole | 24 | NH stretch | 3447 | 6760 | −67 | 9920 | −73 |
| Ftorafur | 24 | NH stretch | 3386 | 6658 | −57 | 9785 | −67 |
| 1-Methylthymine | 25 | NH stretch | 3413 | 6662 | −55 | | |
| Phenobarbital | 25 | NH stretch | 3380 | 6649 | −56 | | |
| 2′,3′,5′-Tri-O-acetylinosine | 26 | NH stretch | 3380 | 6640 | −53 | | |
| Methanol | 27 | OH stretch | 3640 | 7120 | −80 | | |
| Phenol | 28 | OH stretch | 3612 | 7053 | −86 | | |
| Polyethylene glycol | 29 | OH stretch | 3575 | 7030 | −60 | | |
| Cyclopentane (liquid) axial | 30 | CH stretch | 2910 | | −66 | | |
| Cyclopentane (liquid) equat | 30 | CH stretch | 2937 | | −64 | | |
| Cyclohexane (liquid) axial | 30 | CH stretch | 2876 | | −65 | | |
| Cyclohexane (liquid) equat | 30 | CH stretch | 2904 | | −64 | | |
| n-Propane | 31 | CH$_3$ stretch | 2866 | | −59 | 8423 | −59 |
| n-Propane | 31 | CH$_2$ stretch | 2835 | | | | −63 |
| n-Butane | 31 | CH$_3$ stretch | 2868 | | | 8436 | −62 |
| n-Butane | 31 | CH$_2$ stretch | 2831 | | | | −68 |
| n-Pentane | 31 | CH$_3$ stretch | 2859 | | | 8396 | −60 |
| n-Pentane | 31 | CH$_2$ stretch | 2816 | | | 8275 | −64 |
| n-Hexane | 31 | CH$_3$ stretch | 2862 | | | 8392 | −61 |
| n-Hexane (liquid) | 31 | CH$_2$ stretch | 2815 | | | 8268 | −62 |
| n-Heptane (liquid) | 31 | CH$_3$ stretch | 2861 | | | 8389 | −62 |
| n-Heptane (liquid) | 31 | CH$_2$ stretch | 2814 | | | 8266 | −63 |

The first anharmonicity constant, at the first overtone ($v_{02}$), was computed by using $X = v_{01} - (v_{02}/2)$, whereas the anharmonicity constant at the second overtone ($v_{03}$) was obtained from $X = (v_{02}/2) - (v_{03}/3)$. Fundamentals, overtones and anharmonic constants are expressed in cm$^{-1}$ units. In the case of cycloalkanes (30) and hydrocarbons (31), a Birge–Sponer plot was used to determine anharmonicity constants. Axial and equat stand for axial and equatorial CH groups of cycloalkanes, respectively.

# POLYATOMIC OSCILLATORS

The theoretical background developed for the diatomic molecules explains well the essential spectral features of overtones. However, the near-infrared spectra also contain combination bands because an infrared photon can excite two (and more) distinct vibrational modes in the same molecule. For a molecule containing $N$ atoms, we now have $(3N - 6)$ normal vibrations, or $(3N - 5)$ if the molecule is linear. The potential function has the form:

$$
\begin{aligned}
V = V_0 &+ \sum_{i=1}^{3N-6} \left( \frac{\partial V}{\partial Q_i} \right)_e Q_i + \frac{1}{2} \sum_{i,j=1}^{3N-6} \left( \frac{\partial^2 V}{\partial Q_i \partial Q_j} \right)_e Q_i Q_j \\
&+ \frac{1}{3!} \sum_{i,j,k=1}^{3N-6} \left( \frac{\partial^3 V}{\partial Q_i \partial Q_j \partial Q_k} \right)_e Q_i Q_j Q_k \\
&+ \frac{1}{4!} \sum_{i,j,k,l=1}^{3N-6} \left( \frac{\partial^4 V}{\partial Q_i \partial Q_j \partial Q_k \partial Q_l} \right)_e Q_i Q_j Q_k Q_l + \dots
\end{aligned} \tag{2.35}
$$

Again, the first term can be absorbed in the zero level of the potential energy, the second term (the force) is zero at equilibrium, and the third term is the quadratic term. Subsequent terms represent mechanical anharmonicity. The cubic and quadratic potential constants are usually abbreviated as $k_{ijk}$ and $k_{ijkl}$, respectively. Cubic and quartic terms can give rise to Fermi resonance. This occurs generally between a fundamental $v_1$ and an overtone $2 v_2$ or a combination tone $v_1 + v_2$, which belong to the same symmetry species. The magnitude of the resonance, affecting the intensities of the bands, depends on the separation between the interacting vibrational levels. In case of perfect coincidence, Fermi resonance separates two interacting levels apart equally and oppositely. The weaker band in the spectrum "steals" intensity from the stronger one. Resonances caused by the quartic terms are called Darling–Dennison resonances (33).

We present an example of a triatomic model with no degeneracy, like $H_2O$. A vibrational term will have the form:

$$
\begin{aligned}
G(v_1, v_2, v_3) = \omega_1 \left( v_1 + \frac{1}{2} \right) &+ \omega_2 \left( v_2 + \frac{1}{2} \right) + \omega_3 \left( v_3 + \frac{1}{2} \right) + X_{11} \left( v_1 + \frac{1}{2} \right)^2 \\
+ X_{22} \left( v_2 + \frac{1}{2} \right)^2 &+ X_{33} \left( v_3 + \frac{1}{2} \right)^2 + X_{12} \left( v_1 + \frac{1}{2} \right) \left( v_2 + \frac{1}{2} \right) \\
+ X_{13} \left( v_1 + \frac{1}{2} \right) &\left( v_3 + \frac{1}{2} \right) + X_{23} \left( v_2 + \frac{1}{2} \right) \left( v_3 + \frac{1}{2} \right)
\end{aligned} \tag{2.36}
$$

It contains six anharmonicity constants of which three belong to one of each of the three normal vibrations and the other three are coupling constants. The wavenumber of a given vibrational transition is

$$
v = G'(v_1', v_2', v_3') - G''(v_1'', v_2'', v_3'') \tag{2.37}
$$

and for bands other than hot bands

$$v = G'(v_1', v_2', v_3') - G''(0, 0, 0) \tag{2.38}$$

It is instructive to go into some detail. From Equation 2.36 the fundamental of vibration $Q_1$ is, with $v_1 = 1, v_2 = 0, v_3 = 0$

$$v_1^{01} = \omega_1 + 2X_{11} + \frac{1}{2}X_{12} + \frac{1}{2}X_{13} \tag{2.39}$$

For a diatomic oscillator only the first two terms would appear, but now we see that the coupling constants between $Q_1$ and the other two vibrations also have a bearing on the frequency, even if it is a fundamental. The first overtone of the same vibration is, with $v_1 = 2, v_2 = 0, v_3 = 0$

$$v_1^{02} = 2\omega_1 + 6X_{11} + X_{12} + X_{13} = 2v_1^{01} + 2X_{11} \tag{2.40}$$

so that if we measure the overtone and the fundamental we can compute $X_{11}$ just as for a diatomic oscillator.

Polyatomic molecules also have combination bands. A few examples may be useful. Let $(v_1 + v_3)$ be the (observed) wavenumber of the binary combination of vibrations $Q_1$ and $Q_3$. Then:

$$(v_1 + v_3) = G'(1,0,1) - G''(0,0,0)$$

$$(v_1 + v_3) = \omega_1 + \omega_3 + 2X_{11} + 2X_{33} + 2X_{13} + \frac{1}{2}X_{12} + \frac{1}{2}X_{23} = v_1^{01} + v_3^{01} + X_{13} \tag{2.41}$$

Thus if we have the combination band and the two fundamentals we can compute the coupling constant $X_{13}$. This is a summation tone. The difference tone of $Q_1$ and $Q_3$ is obtained if $v_3$ is at level $v_3 = 1$ when the photon strikes (hot band).

$$(v_1 - v_3) = G'(1,0,0) - G''(0,0,1)$$

$$(v_1 - v_3) = \omega_1 - \omega_3 + 2X_{11} - 2X_{33} + \frac{1}{2}X_{12} - \frac{1}{2}X_{23} = v_1^{01} - v_3^{01} \tag{2.42}$$

Interestingly, the coupling constant cancels out and the wavenumber of the difference band is simply the difference of the wavenumbers of the two fundamentals. This can be useful. If for some reason we cannot measure one of the fundamentals, say $v_3^{01}$, but we have the other fundamental and a sum and a difference tone we can still compute the coupling constant:

$$X_{13} = [(v_1 + v_3) - v_1^{01}] - [v_1^{01} - (v_1 - v_3)] \tag{2.43}$$

In the harmonic oscillator, approximation combination tones are forbidden as well as overtones. Darling and Dennison (33) in their classic paper gave the anharmonicity and the potential constants for the water molecule. $v_1$ is the symmetrical stretching vibration, $v_2$ is the bending vibration, and $v_3$ is the asymmetrical stretching vibration with frequencies at 3650, 1650, and 3750 cm$^{-1}$ respectively; the anharmonicity constants are:

$$X_{11} = -43.89 \qquad X_{12} = -20.02$$
$$X_{22} = -19.5 \qquad X_{13} = -155.06$$
$$X_{33} = -46.37 \qquad X_{23} = -19.81$$

These values are by no means negligible, and the coupling constant between the two stretching vibrations is very high. The negative sign means a decrease in wavenumber due to anharmonicity, as is usually the case for overtones as well as for combination bands. This means that if, for example, a harmonic normal coordinate calculation yields the exact observed value for $v_3$, it is actually off by $2X_{33} + \frac{1}{2}X_{13} + \frac{1}{2}X_{23} = -(92.74 + 9.96 + 77.53) = -180.23$ cm$^{-1}$ (see Equation 2.39). Electrical anharmonicity is just as important for polyatomic as for diatomic molecules, and it is, of course, much more difficult to handle. If we develop the dipole moment into series we obtain:

$$M = M_e + \sum_{i=1}^{3N-6} \left( \frac{\partial M}{\partial Q_i} \right)_e Q_i + \frac{1}{2} \sum_{i,j=1}^{3N-6} \left( \frac{\partial^2 M}{\partial Q_i \partial Q_j} \right)_e Q_i Q_j$$

$$+ \frac{1}{3!} \sum_{i,j,k=1}^{3N-6} \left( \frac{\partial^3 M}{\partial Q_i \partial Q_j \partial Q_k} \right)_e Q_i Q_j Q_k + \cdots \qquad (2.44)$$

The terms higher than the linear terms constitute electrical anharmonicity. Even if we neglect all but the quadratic terms we obtain, in addition to the $(\partial^2 M/\partial Q_i^2)_e Q_i^2$, mixed terms of type $(\partial^2 M/\partial Q_i \partial Q_j)_e Q_i Q_j$. These are the terms that give intensity to the combination bands. Their contribution is probably just as important as that of mechanical anharmonicity.

Primary amines or amides have well-localized NH$_2$ symmetric ($v_1$) and antisymmetric ($v_3$) stretching vibrations, giving rise to $v_1 + v_3$ combination tones as in the case of water, which are allowed under C$_{2v}$ symmetry and even more so in molecules that have no symmetry at all. The anharmonicity constant of primary amines (Table 2.2) is generally divided about half and half between the antisymmetric and symmetric modes, as compared with the anharmonicity constants of secondary amines (Table 2.1). The anharmonic coupling constant $X_{13}$, amounted from $-91$ to $-157$ cm$^{-1}$ for substituted anilines (34), whereas it is $-155$ for water (33).

Combination tones with distinct chemical groups are observed depending on the symmetry of the vibrational mode. The near-infrared spectrum of formamide is very instructive (Table 2.3), because it is a small molecule for which normal mode calculation and assignments are relatively straightforward and it may serve as a model for near-infrared spectra of polyamides (35).

The anharmonic coupling constants are relatively large for the combination tones involving NH$_2$ symmetric ($v_1$) and antisymmetric ($v_3$) stretching, whereas the anharmonic coupling constants are smaller in the case of combination tones involving vibrations of lower wavenumber such as CH or C=O stretching vibrations. In general,

**TABLE 2.2. Anharmonicity and Coupling Constants of Free Primary Amides and Amines**

| Molecule | Reference | Vibrational Mode | Fundamental | First Overtone | Combination Tone | Anharmonicity Constants |
|---|---|---|---|---|---|---|
| Acetamide | 21 | NH$_2$ s. (1) | 3350 | 6625 | | $X_{11} = -38$ |
| | 21 | NH$_2$ as. (3) | 3450 | 6850 | | $X_{33} = -25$ |
| | 21 | (1) + (3) | | | 6710 | $X_{13} = -90$ |
| Aniline | 34 | NH$_2$ s. (1) | 3397 | 6694 | | $X_{11} = -50$ |
| | 34 | NH$_2$ as. (3) | 3484 | 6900 | | $X_{33} = -34$ |
| | 34 | (1) + (3) | | | 6730 | $X_{13} = -151$ |
| 9-Ethyladenine | 23 | NH$_2$ s. (1) | 3416 | 6743 | | $X_{11} = -45$ |
| | 23 | NH$_2$ as. (3) | 3526 | 7003 | | $X_{33} = -25$ |
| | 23 | (1) + (3) | | | 6817 | $X_{13} = -125$ |
| 8-Br-Ado | 25 | NH$_2$ s. (1) | 3413 | 6743 | | $X_{11} = -42$ |
| | 25 | NH$_2$ as. (3) | 3526 | 6997 | | $X_{33} = -28$ |
| | 25 | (1) + (3) | | | 6812 | $X_{13} = -127$ |
| 4-Aminopyrimidine | 26 | NH$_2$ s. (1) | 3422 | 6748 | | $X_{11} = -48$ |
| | 26 | NH$_2$ as. (3) | 3534 | 7013 | | $X_{33} = -28$ |
| | 26 | (1) + (3) | | | 6826 | $X_{13} = -131$ |

Anharmonicity and coupling constants were computed according to Equations (2.40) and (2.41), respectively. Fundamental, 1st overtone, and constants are expressed in cm$^{-1}$ units. s = symmetric stretching vibration; as = antisymmetric stretching vibration. 8-Br-Ado is 8-bromo-2′,3′,5′-tri-$O$-acetyladenosine.

**TABLE 2.3. Anharmonicity and Coupling Constants of Free Formamide (35)**

| Vibrational Mode | Fundamental | First Overtone | Combination Tone | Anharmonicity Constants |
|---|---|---|---|---|
| $\nu$ NH$_2$ asym., $\nu(1)$ | 3523 | 6991 | | $X_{11} = -28$ |
| $\nu$ NH$_2$ sym., $\nu(2)$ | 3406 | 6726 | | $X_{22} = -43$ |
| $\nu(1) + \nu(2)$ | | | 6810 | $X_{12} = -119$ |
| $\nu$ CH, $\nu(3)$ | 2874 | | | |
| $\nu$ C=O, $\nu(4)$ | 1710 | | | |
| $\nu(3) + \nu(4)$ | | | 4572 | $X_{34} = -12$ |
| $\delta$ NH$_2$, $\nu(5)$ | 1583 | | | |
| $\nu(1) + \nu(5)$ | | | 5085 | $X_{15} = -21$ |
| $\nu(2) + \nu(5)$ | | | 4977 | $X_{25} = -12$ |
| $\delta$ CH, $\nu(6)$ | 1391 | | | |
| $\nu(1) + \nu(6)$ | | | 4914 | $X_{16} = 0$ |
| $\nu(2) + \nu(6)$ | | | 4778 | $X_{26} = -19$ |

Anharmonicity and coupling constants were computed according to Equations 2.40 and 2.41, respectively. Fundamental, 1st overtone, and constants are expressed in cm$^{-1}$ units. s = symmetric stretching vibration; as = antisymmetric stretching vibration. The indices of anharmonicity constant corresponded to the indices of vibrational modes involved in overtones or combination tones, as used by Tanaka and Machida (35).

hydrogen vibrations of high wavenumber have the highest anharmonicity constants (27,32,35–38). In the ether+hydrogen fluoride system it probably reaches $-145$ or $-200$ cm$^{-1}$ for the FH stretching vibration. Carbonyl stretching bands are not very anharmonic ($X_{11} = -5$ or $-10$ cm$^{-1}$). The coupling constant is probably about $-70$ cm$^{-1}$ for the ether+HF system, whereas it is much smaller in the case of formamide. It is often believed that low-frequency vibrations of weak bonds are necessarily very anharmonic, but this is not so in reality. The effects of hydrogen bonds on anharmonicity are discussed below. Table 2.4 gives the approximate absorption regions for the most important groups and vibrations. The assignments of overtones and combination tones reported by several authors (39–44) are consistent with the overall positions of overtone and combination tones in Table 2.4.

## LOCAL MODE TREATMENT

Normal modes of vibration in conjunction with normal coordinates describe well the first vibrational levels, for example, $v = 1$ or $v = 2$ (fundamental or first overtone). The effects of anharmonicity on the shape of the potential energy and on wave functions are relatively small for the low vibrational levels. This model becomes less satisfactory for higher vibrational states, where the effects of anharmonicity become more important. The normal mode treatment fails when the energy approaches the bond dissociation limit. In the case of molecules such as $CH_4$, $CH_2Cl_2$, $C_2H_4$, $C_6H_6$ ... containing identical CH bonds, the normal mode model gives a different picture from that obtained from the measured spectra. For the benzene molecule, the

**TABLE 2.4. Overall Positions of Overtones and Combination Tones of Groups in Polymers**

| Wavenumber(cm)$^{-1}$ | Assignment |
|---|---|
| 4400–4200 | $\nu + \delta$ ($CH_3$ and $CH_2$) |
| ~4500 | $\nu$ ($CH_3$, $CH_2$) $+\nu$ ($C{=}O$) |
| 4900–4600 | $\nu + \delta$(NH) |
| 6000–5600 | $2\nu$ (1st overtones of $CH_3$ and $CH_2$) |
|  | $\nu + 2\delta$($CH_3$, $CH_2$) |
| ~6200 | $\nu$ ($CH_3$, $CH_2$) $+2\nu$ ($C{=}O$) |
| 6600–6300 | $2\nu$ (hydrogen-bonded NH, 1st overtone) |
| 6700–6500 | $2\nu$ (free NH) |
| 6900–6700 | $2\nu$ (associated $NH_2$) |
| 7000–6700 | $2\nu$ (free $NH_2$) |
| 7000–6200 | $2\nu$ (hydrogen-bonded OH, 1st overtone) |
|  | $\nu$ (OH) $+\nu$ (CH) |
| 7200–7000 | $2\nu$ (free OH) |
|  | $2\nu + \delta$ ($CH_3$, $CH_2$) |
| 8600–8200 | $3\nu$ ($CH_3$, $CH_2$, 2nd overtones) |
|  | $2\nu + 2\delta$ |
| 10,000–9000 | $3\nu$ (2nd overtone of hydrogen-bonded OH) |
| ~10,300 | $3\nu$ (free OH) |

$\nu$ = stretching; $\delta$ = bending (deformation).

H-stretching vibrations yield 6 states for $V = 1$, 21 states for $V = 2$, 56 states for $V = 3$, 126 states for $V = 4$, and 252 states for $V = 5$, giving rise to a more complex spectrum than the observed spectra. As underlined by Mills (45), the increasing simplicity in overtone spectra is quite surprising when we consider the total density of vibrational states. Not only do the higher overtone bands appear simpler and less overlapped, but the observed wavenumbers of the C-H stretching overtones of polyatomic molecules follow a simple rule analogous to that observed in the case of diatomic molecules. At high levels of excitation of stretching vibrations in some molecules, the vibrational energy seems to be more concentrated or localized on a single chemical bond, and this leads to the concept of local mode treatment. This concept has been discussed by several authors (43–55). A long time ago, Mecke et al. (56,57) proposed the concept of localized vibrational modes to interpret the spectra. Siebrand and Williams suggested that the vibration of highly excited molecules can be described in terms of local modes (58). Wallace used a local mode model based on the Morse oscillator (59) to explain some characteristics of overtones. Finally, the local mode model was introduced by Hayward and Henry (49), who analyzed the bandshape of high overtones of dichloromethane. The principal peaks of the near-infrared spectrum, mostly dominated by combination and overtones of CH vibrational modes, can be interpreted in terms of vibrations localized in a single CH chemical bond. The local mode description applies very well to X-H stretching vibrations whose fundamentals are located around and above 3000 cm$^{-1}$, far away from fundamentals originating from other types of vibrational modes. Furthermore, it is usually assumed that couplings between two fundamentals are very weak. These theoretical approaches were examined by Siebrand and Williams, who analyzed the C-H vibrations of benzene (58). Based on the normal mode concept, the excitation of these vibrational modes to the highest vibrational quantum number would ultimately lead to the simultaneous rupture of all six C-H bonds. Indeed, the vibrational energy is equally distributed over the six bonds, providing there is no energy dissipation due to anharmonic coupling between different normal modes. This would require six times the C-H bond dissociation energy ($\sim 6 \times 412$ kJ$\cdot$ mol$^{-1}$). Thus the dissociation of benzene into $C_6H_5$ and H would occur with a finite probability as soon as the vibrational energy exceeds the dissociation energy $D_0$ of one C-H chemical bond. However, no normal mode of benzene leads to any CH stretching vibration localized in one single bond, although the dissociation of chemical bonds of benzene is essentially a local process. In the case of other polyatomic molecules such as $H_2O$, $NH_3$, and $CO_2$, again local excitation could be treated as a quasi-steady-state phenomenon and the state corresponding to it can be described as a set of oscillators with *negligible harmonic coupling* and all vibrational *energy concentrated* in only one of them (58). Chemists are familiar with the concept of localized vibrational modes, especially for the assignment of group frequencies, whereas spectroscopists use the normal mode concept because most spectral analysis of spectra involves fundamentals and the first overtones.

The local mode description with only one bond such as X-H being in motion at a time leads us to consider the polyatomic molecule as a single diatomic molecule M-H where H is hydrogen and M denotes the full rest of the molecule. The main advantage of this description is the dramatic reduction of the number of coupling constants. The anharmonic oscillator potential adopted in the local mode model is the Morse potential, which takes account of the dissociation energy $D_e$ (53).

The normal and local parameters required to describe the X-H stretching vibrational modes in three types of molecules can be summarized (53) as indicated in Table 2.5:

The reduction of the number of constants using the local mode approach increases with increasing molecule size.

Henry (50,51) and Child and Halonen (47) proposed a two-parameter local model giving rise to the following equation, which is similar to the equation for a diatomic molecule (Equation 2.24):

$$G(v_a, v_b) = \sum_{v=a,b} \left[ (v_V + 1/2)\bar{\omega} - (v_V + 1/2)^2 \bar{\omega}x \right] \qquad (2.45)$$

**TABLE 2.5. Reduction in the Number of Parameters from Normal Modes to Local Modes (see Ref. 53)**

| Molecule | Symmetry | Normal Mode | Local Mode |
|----------|----------|-------------|------------|
| $H_2O$ | $C_{2v}$ | 6 | 3 |
| $CH_3X$ | $C_{3v}$ | 8 | 3 |
| $H_2C=CH_2$ | $D_{2h}$ | 21 | 5 |

For example, in the case of $H_2O$, if we neglect the bending mode there are only two stretching vibrations, the symmetric stretching and the antisymmetric stretching. At a high level of excitation the vibrational energy is concentrated in only one bond and the local mode description corresponds well to the observed results. However, at lower vibrational energies, the motions of atoms along chemical bonds are not independent from each other but are coupled. A coupling parameter $\lambda$ (in the case of water ) $\lambda = \frac{1}{2}(v_3 - v_1) = 49.7 \text{ cm}^{-1}$) was defined to take account of the coupling between different vibrational modes (52). When the anharmonicity constant $X_M$ is greater than the coupling constant $\lambda$, the local mode concept is appropriate, whereas the normal mode description is more accurate for larger coupling constants. The actual situation is a compromise between these two extremes, as in the case of $H_2O$ (53,54). Stretching vibration energy levels for $H_2O$ up to 4 quanta of excitation are presented in the normal mode basis and in the local mode treatment together with the intermediate situations in Figure 2.1.

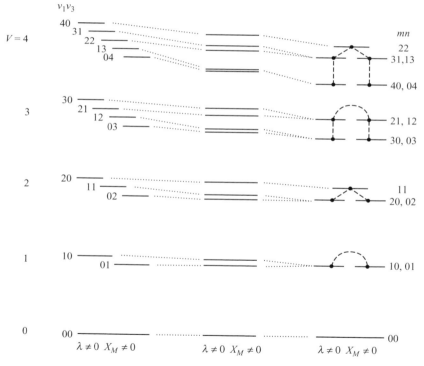

**Figure 2.1.** Schematic stretching vibrational energy levels for the water molecule. The levels on the left side represent the normal mode limit and are indicated with quantum numbers $v_1$, $v_3$. The levels on the right side represent the local mode limit, and they are labeled with the quantum numbers $m,n$. The true energy levels are shown in the center of diagram. From reference 53. Reproduced by permission from Elsevier Science B.V.

The energy levels on the left-hand side, reflecting normal mode levels, were computed assuming that anharmonicity $X_M$ is zero and that the coupling constant $\lambda$ is nonzero. They are labeled $(v_1 v_3)$ with quantum numbers $v_1$ and $v_3$. The energy levels on the right-hand side, corresponding to local mode levels, were obtained with l is equal to zero and $X_M$ nonzero. They are labeled $mn$ with the quantum numbers $m$ and $n$. With $\lambda = 0$, there are independent vibrations along the two chemical bonds. The total vibrational quantum number is $V$, given according to:

$$V = v_1 + v_3 = n + m \qquad (2.46)$$

Figure 2.1 shows that, for $V = 2$, the level of the combination band $n_1 + n_2$ and the two overtone levels (02) and (20) are close but separated as within the normal mode concept, whereas in the local mode limit there are two degenerate levels (02) and (20). Between these two extremes, the real energy levels are shown in the center of the diagram. They may be obtained by introducing a nonzero anharmonicity constant $X_M$ and a nonzero coupling constant $\lambda$. The relationship of normal modes is illustrated with the detailed analysis of anharmonicity and coupling constants of the water molecule:

$$x_{11} = -43, 89$$

$$x_{13} = -155, 06$$

$$x_{33} = -46, 37$$

$$K_{1133} = -155 \text{ cm}^{-1}$$

$$X_{11} \approx X_{33} \approx 1/4 X_{13} \approx 1/4 K_{1133} \approx 1/2 X_m \qquad (2.47)$$

where $x_{11}$ et $x_{33}$ are the anharmonicity constant for the symmetric $(v_1)$ and antisymmetric $(v_3)$ stretching, $x_{13}$ is the coupling constant between the two normal modes, and $K_{1133}$ is the Darling–Dennison (quartic) coupling constant between two harmonic levels (33). The three anharmonicity parameters of water determined by using local mode analysis are (53,54) $x_{13} = x_{11} = -42$ and $x_{13} = K_{1133} = -168 \text{ cm}^{-1}$.

The effect of the anharmonicity constant $X_M$ on the potential energy is a function of the square of $v$, in contrast to the effect of the coupling constant l on the potential energy, which is only $v$ dependent. This explains that the local mode is a poor description at the fundamental level for coupled oscillators. At a higher vibrational level the local mode description becomes more realistic because the anharmonicity $X_M$ dominates. Generally, the local mode description is an oversimplification. Other effects such as Darling–Dennison coupling (33) or Fermi resonance (60–61) must be taken into account. The Hamiltonian can be written more generally:

$$H = H_{L(q)} + H_{N(Q)} + H_{NL(q,Q)} \qquad (2.48)$$

where $H_L$ is the Hamiltonian for the local mode approach and is written in terms of the local displacement $q$, $H_N$ is the Hamiltonian for the normal mode approach and is written in terms of the normal displacement $Q$, and $H_{NL}$ corresponds to the mixing between local and normal modes. This is a compromise between two limiting cases, the normal mode and the local mode model, depending on the relative magnitudes of the Morse anharmonicity constant and coupling constants. In practice, it is found that even 2 or 3 quanta of stretching vibrational energy may be sufficient to bring the true vibration wavenumber closer to the local mode than the normal mode limit (54).

The question of how the vibrational energy spreads over the molecule from one vibrational mode to another (dealing with the breaking of a chemical bond) is of great interest for chemistry, photochemistry, and dynamics. This is the object of the intramolecular molecular vibrational redistribution (IVR) theory (62).

## ANHARMONICITY AND HYDROGEN BONDING

Hydrogen bonds are ubiquitous in nature, in both the living and the mineral worlds. They are weaker but almost as important as covalent bonds. In living organisms we invariably encounter them when specificity is needed. Hydrogen bond enthalpies range between 0.5 and 50 kcal/M. This is accompanied by very different types of potentials and very different spectral features.

In the weaker hydrogen bonds the proton is much closer to one of the "bridgeheads" than to the other, so they appear to have a one-well potential. In the case of medium strong and strong hydrogen bonds double-well or broad single-well potentials may exist. According to the classification proposed by Novak (63), 1) weak hydrogen bonds have their $\nu(OH)$ band at wavenumbers higher than 3200 cm$^{-1}$, the O...O distance is longer than 2.70 Å, and the $\Delta H$ value is less than 5 kcal/M like in water or alcohols; 2) medium strong hydrogen bonds have their $\nu(OH)$ band between 3200 and 2800 cm$^{-1}$, the O...O distance is 2.70–2.60 Å, and $\Delta H$ is 6–8 kcal/M, cyclic dimers of carboxylic acids being a typical example; 3) strong hydrogen bonds have broad $\nu(OH)$ bands anywhere between 2800 and 700 cm$^{-1}$, the O...O distance is 2.60–2.40 Å, and the $\Delta H$ value is often much greater than 8 kcal/M; examples are the acid salts, $M^+H(RCOO)_2^-$ (See Refs. 63,64).

Self-associated alcohols such as cyclohexanol are a good example of relatively weak hydrogen bonds (65). At 0.01 M in carbon tetrachloride solution, only the free (non-hydrogen bonded) $\nu(OH)$ band is seen at 3620 cm$^{-1}$. At 0.03M the O-H...: O-H dimer band appears at 3485 cm$^{-1}$. At higher concentrations the intensity of the so-called polymer band, 3360–3320 cm$^{-1}$, increases steadily, and at 1.0 M the dimer band disappears under the wing of the strong polymer band. The latter is due to those OH groups that are proton acceptors as well as proton donors, O-H...:O-H...:O-H...:O-H.... The band is broad, and this is characteristic of all but the weakest hydrogen bond systems. Actually, it is the analysis of this breadth phenomenon that provides the most insight into the nature of hydrogen bonding. In the particular case of self-associated alcohols one of the causes of the breadth is the existence of a variety of associated species in the solution, trimers, tetramers, pentamers, . . . , both linear and cyclic. The effect of this can be partly counterbalanced by deuteration-isolation experiments (65). However, the bands are broad even when only one associated species is present. In other cases a number of satellite bands accompany the main $\nu(XH)$ band, another important characteristic of hydrogen bond systems. The origin of these bands has been the object of much discussion (see below).

It is natural to invoke anharmonic coupling in attempts to explain the breadth of $v(XH)$ and the satellite bands. The coupling could be between $v(XH)$ ($X = N, O \ldots$) and a low-frequency bridge vibration ($X \ldots Y$), essentially an intermolecular stretching or bending motion of the two molecules forming the hydrogen bond. For weak and, perhaps, medium strong hydrogen bonds second-order perturbation theory is applicable; therefore the results of anharmonic coupling would be the appearance of combination bands—both summation and difference tones—that could appear as a fine structure on the main $v(XH)$ band. Classic examples of this are gaseous mixtures of ether-HCl and ether-HF complexes (and others), which were examined by Millen and his coworkers in the vapor phase (66–70). These bands form a series, $v(XH) \pm n v_\sigma(XY)$, $v_\sigma$ being the low-frequency bridge stretching vibration and $n = 1, 2 \ldots$, and the spacing between the bands is approximately equal to the $v(XY)$ frequency (usually between 100 and 200 cm$^{-1}$). Bevan et al. (71) found them at the level of the first overtone as well, $v^{02}(XH) \pm n v_\sigma(XY)$, in the near infrared, thereby confirming that these bands are not due to combinations of other vibrations or to other associated species. There are other combinations that might appear in this fine structure, among others those involving the low-lying bridge bending vibration (around 50–100 cm$^{-1}$). Another type of important combination band is the following. Let us take $v_1$ for the high-frequency XH vibration and $v_3(= v_\sigma)$ for the low-frequency bridge vibration. Then:

$$\{(v_1 + v_3) - v_3\} = G'(1, 0, 1) - G''(0, 0, 1)$$

$$= \omega_1 + 2X_{11} + \frac{1}{2}X_{12} + \frac{3}{2}X_{13} \tag{2.49}$$

$$= v_1^{01} + X_{13}$$

The combination is just $X_{13}$ apart from the $v_1^{01}$ fundamental. Furthermore, because the coupling constant might be either positive or negative, we might find this hot band at either the high- or the low-frequency side of $v_1^{01}$. Other examples need little comment:

$$\{(v_1 + 2v_3) - 2v_3\} = G'(1, 0, 2) - G''(0, 0, 2)$$

$$= \omega_1 + 2X_{11} + \frac{1}{2}X_{12} + \frac{5}{2}X_{13} \tag{2.50}$$

$$= v_1^{01} + 2X_{13}$$

$$\{(v_1 + 2v_3) - v_3\} = G'(1, 0, 2) - G''(0, 0, 1)$$

$$= \omega_1 + \omega_3 + 2X_{11} + 4X_{33} + \frac{1}{2}X_{12} + 3X_{13} + \frac{1}{2}X_{23} \tag{2.51}$$

$$= v_1^{01} + v_3^{01} + 2X_{33} + 2X_{13}$$

These bands are hot bands, but in view of the low frequency of the $v_3$ they have high Boltzmann constants. They can make important contributions to the breadth of the main band. For strong hydrogen bonds, second-order perturbation theory no longer applies. Then the high- and low-frequency modes are not even approximately independent from one other and more advanced perturbation treatments have to be worked out. Such treatments were given by Maréchal and Witkowski (72), Sokolov and Savelev (73), and others.

The $\nu(XH) \pm n\nu_\sigma(XY)$ combinations are not the only ones that must be considered, however. Summation tones, binary, ternary, etc., of medium-frequency fundamentals of the molecules forming the hydrogen bonds fall very often near enough to the main $\nu(XH)$ band and may appear on the envelope of it. Their intensity might be boosted by Fermi resonance or higher anharmonic resonance. Many such "satellite" bands appear in the spectra of carboxylic acids (74) and amine salts (75,76) and in other cases.

Every theory of hydrogen bonding involves anharmonicity, either explicitly or not. The first of these was given by Bratos and Hadzi in their seminal paper of 1957 (77). Pimentel and McClellan reviewed the experimental data and concluded that the hydrogen bonds slightly decrease or do not affect anharmonicities (78). For example, the overtones of dimeric dimethyl amine lead to normal anharmonicities (20). However, this is not the case for the alcohol polymers, where hydrogen bonds generally increase the coupling constants from about $-80$ (free alcohol) to $-95$ or $-120$ cm$^{-1}$ or more (27). The anharmonicity constant of $\nu(NH)$ of intramolecular hydrogen bonded *ortho* N-methyl-O-nitroaniline is by 33 cm$^{-1}$ greater than that of free *para*-substituted N-methyl-O-nitroaniline (34). Intermolecular hydrogen bonds increased slightly the anharmonicity constant of symmetric $NH_2$ groups ($\nu_1$) of 8-bromoadenosine from $-43$ to $-52$ cm$^{-1}$ and also increased anharmonicity constant of antisymmetric $NH_2$ group ($\nu_3$) from $-27$ to $-46$ cm$^{-1}$ (25). Table 2.6 indicates some anharmonicity and coupling constants for hydrogen-bonded X-H groups.

The examples discussed above can explain the complexity of the water band. The complexity originates from the ability of water molecules to form hydrogen bonds in various manners. The understanding and analysis of near-infrared spectra can provide information on the nature of the water environment. The wavenumbers of the stretching-bending combination of the water adsorbed in some polymers and composites are summarized in Table 2.7.

Several papers have dealt with the effects of the water content on the NIR spectra of polymers. Useful information can be obtained about the nature of adsorbed water and the process of diffusion of water in the polymers. The band positions and the number of water bands observed in the NIR region are modified by the strength of hydrogen bonding interactions between the hydroxyl groups and the different polymers (Table 2.7). In addition, the effects of absorbed water molecules on some bands of the functional groups of the polymers can be important. In general, the combination of stretching and bending OH vibrations, $\nu_{asOH} + \delta_{OH}$, (about 5200 cm$^{-1}$) and the first OH overtone (about 7000 cm$^{-1}$) are used, but the 5200-cm$^{-1}$ band is often less overlapped than the 7000-cm$^{-1}$ band .

For hydrated PET or PET-glass composites, the water band absorbs near 5240 cm$^{-1}$, indicating weakly bonded water molecules (79,80), in good agreement with mid IR measurements on PET films. The peak position of this combination band is not affected by the amount of water. In contrast, the band of water is shifted to near 5150 cm$^{-1}$ for a glass fiber-nylon 6 composite, indicating strong hydrogen bonding, and

**TABLE 2.6. Anharmonicity and Coupling Constants of Hydrogen-bonded X-H Groups**

| Molecule | Reference | Vibrational Mode | Fundamental | First Overtone | Combination Tone | Anharmonicity Constants |
|---|---|---|---|---|---|---|
| Methanol | 27 | OH stretch | 3340 | 6450 | | $X_{11} = -115$ |
| Ethanol | 27 | OH stretch | 3250 | 6258 | | $X_{11} = -121$ |
| Phenol | 28 | OH stretch | 3422 | 6650 | | $X_{11} = -97$ |
| Indole | 22 | NH stretch | 3489 | 6808 | | $X_{11} = -85$ |
| Dimethylamine | 26 | NH stretch | 3288 | 6432 | | $X_{11} = -72$ |
| 8-Br-Ado | 25 | $NH_2$ s. (1) | 3413 | 6743 | | $X_{11} = -42$ |
| | 25 | $NH_2$ as. (3) | 3490 | 6889 | | $X_{33} = -46$ |

Anharmonicity and coupling constants were computed according to Equations 2.40 and 2.41, respectively. Fundamental, 1st overtone, and constants are expressed in $cm^{-1}$ units. s = symmetric stretching vibration; as = antisymmetric stretching vibration; 8-Br-Ado is 8-bromo-2′,3′,5′-tri-$O$-acetyladenosine.

**TABLE 2.7. Combination Tone of OH Antisymmetric Stretching and OH Deformation ($\nu_{asOH} + \delta_{OH}$) of Water Molecules Trapped in Different Types of Polymers**

|  | Wavenumber, cm$^{-1}$ |
| --- | --- |
| Water vapor | 5320 |
| Epoxy resins | 5240 |
| Epoxy-glass fiber | 5220 |
| Epoxy-glass fiber, printed circuit substrate | 5240 |
| PET | 5245–5240 |
| PET long-fiber glass coating 1 | |
| PET short-fiber glass coating 2 or 3 | 5240–5150 (bands overlapped) |
| Nylon 6 | 5150 |
| PA 6-glass fiber composite | |
| Gelatin | 5180 |
| Cellulose | 5190 |
| Polyurethane | 5200 |
| PES-EPOXY blends | 5220–5230 |
| Poly(vinyl alcohol) | 5140 |

this is in good agreement with the NIR spectra of a nylon matrix (81,82). In addition, the amide I + $\nu$(NH) and amide II + $\nu$(NH) combination bands near 5000 cm$^{-1}$ are also affected by the water content. After immersion in water, epoxy resin cured with aromatic diamine induced modifications of the NH combination band as well as the CH first overtone band (83). Basset et al. have observed the combination band of water band at 5190 cm$^{-1}$ in cellulose (84). Glatt et al. assigned the NIR band at 5140 cm$^{-1}$ to bound water in poly(vinyl alcohol) (85). Miller et al. attributed the 5200-cm$^{-1}$ band to water in polyurethanes (86). In some composites, the water band shapes differ somewhat from those of the polymeric matrix without fillers, suggesting that new sites are accessible to water molecules because of to matrix modifications or fiber interactions (82). Changes in shape and position of NH combination bands of diamine, used as hardener for epoxy resins, provided more insight into the molecular interactions (87,88).

Detailed near-infrared spectra of PET exposed to different relative humidities indicated three different subbands of the first overtone of water at 7080 cm$^{-1}$, 7010 cm$^{-1}$, and 6810 cm$^{-1}$. The comparison with the water spectrum of bulk water suggested that most of the water is only weakly bonded with PET (89). The analysis of difference spectra of dry nylon and nylon exposed to different humidities, indicated that there were distinct populations of hydrogen-bonded water in it (90). Recently, Musto et al. (91) investigated the nature of molecular interactions of water in epoxy resins by means of near-infrared spectroscopy as proposed by Fukuda et al. (89,90). They found three subbands at 7076 cm$^{-1}$, 6820 cm$^{-1}$, and 6535 cm$^{-1}$, evidencing two kinds of water adsorbed in the polymer (mobile water localized in micro vide and water molecules firmly bonded to the network). However, hydroxyl groups of epoxy may complicate the analysis of water content in polymers because they absorb also in the same overtone region as water.

Using principal component analysis, Dreassi et al. (92) investigated interactions between skin and water or lipids. Displacements of the peaks were observed in the region attributable to the first overtone bands of water, whereas no shift was observed for the combination band.

Dziki et al. (93) used near-infrared spectroscopy to characterize the mobility of water within the sarafloxacin crystal lattice; differences in the location or orientation of the water molecules within the crystal were detected. The presence or absence of water in the crystal lattice can affect physical properties and processing ability. Analysis of near-infrared spectra of polymer samples allows us to distinguish between acceptable and unacceptable batches for formulation purposes.

## RELATIVE FREE/ASSOCIATION BAND INTENSITIES

A surprising fact concerns the relative intensities of free and association bands of $\nu(OH)$, $\nu(NH)$, and possibly other vibrations taken fundamental and overtone levels. The relative intensity of the free band is much greater for the overtones than for the fundamentals, a fact observed by Luck and Ditter (94) and by Burneau and Corset (95). Figure 2.2 gives two examples (96).

In the IR spectrum of a 2.0 M solution of tertiary butanol a very weak free band is present at about 3620 cm$^{-1}$, while the association band centered at about 3360 cm$^{-1}$ is enormous. At the level of the first overtone the free band is prominent at about 7060 cm$^{-1}$, while the association band is lost in the background. The Raman spectrum is intermediate from this point of view: There the free and association fundamentals are of comparable intensity. (Although Raman spectra are not treated in this review, it should be remembered that their intensity does not depend on the rate of change of the dipole moment, but on the rate of change of the polarizability.) Similar observations can be made for self-associated polyamide (97) (Fig. 2.3).

The phenomenon seems to be quite general. The ratios of intensities of fundamentals of free and associated species are lower than the ratios of intensities of overtones of free and associated species. We observed that for stretching-bending combination bands (about 5000 cm$^{-1}$ for OH) the free and association bands have comparable intensities (98).

Di Paolo et al. (99) made an attempt to explain this phenomenon. For this they used the results by Herman and Schuler (100), who computed the integrated intensities of both the fundamental ($A^{1,0}$) and first overtone ($A^{2,0}$) taking into account both mechanical and electrical anharmonicities.

$$A^{1,0} = \frac{8\pi^3 N}{3hc}\omega^{1,0}\left\{ M_1\left(\frac{1}{\sqrt{2}} - \frac{3}{2\sqrt{2}}g + \frac{11}{4\sqrt{2}}b^2\right) \right.$$
$$\left. + M_2\left(-\frac{5b}{\sqrt{2}} + \frac{88}{\sqrt{2}}bg - \frac{715}{12\sqrt{2}}b^3\right)\right\}$$

$$A^{2,0} = \frac{8\pi^3 N}{3hc}\omega^{2,0}\left\{ \sqrt{M_1}\left(\frac{b}{2} - \frac{111}{8}bg - \frac{71}{48}b^3\right) \right.$$
$$\left. + \frac{M_2}{\sqrt{2}}\left(1 - \frac{15}{4}g + \frac{3}{8}b^2\right)\right\}^2 \tag{2.52}$$

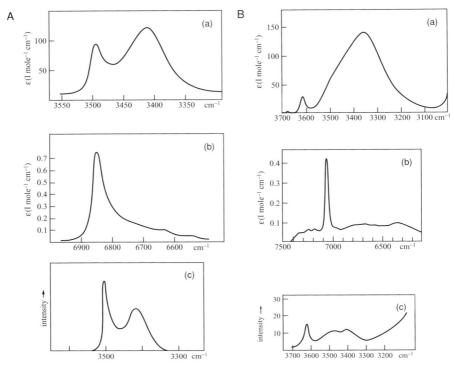

**Figure 2.2.** A. the infrared fundamentals (a), infrared overtones (b), and Raman spectra (c) of solutions of pyrrole in $CCl_4$ at room temperature (concentrations 3.0 M, 2.50 M, and 2.44 M, respectively). From Reference 96, p. 155. Reproduced by permission from the North-Holland Publishing Company. B. the infrared fundamentals (a), infrared overtones (b), and Raman spectra (c) of a 2.0 M solution of tertiary butanol in $CCl_4$ at room temperature. From Reference 96, p. 155. Reproduced by permission from Elsevier Science B. V.

where $\omega$ is the wave number in $cm^{-1}$: $b \equiv k_3/\omega_e$, $g \equiv k_4/\omega_e$, and $N$ is the number of molecules per $cm^3$ in the lower vibrational level.

Considering $M_1$ and $M_2$, if the first and second derivatives of the dipole moment have the same sign [for the fundamental, because $b(k_3)$ is usually negative], the largest (first) terms in each brackets have the same sign. $k_3$ and $k_4$ were varied within wide limits, and it was found that the contribution of anharmonicity to $A^{1,0}$ does not exceed 20%.

The situation is different for the overtone. There, naturally, the whole intensity comes from anharmonic contributions. The first terms in each bracket are again the largest. If $M_1$ and $M_2$ have the same sign, then because of the negative sign of $b(k_3)$, the two terms will have opposite effect on the intensity of the overtone. This applies for hydrogen-bonded systems, which are strongly polar and therefore have a large $M_1$ value.

These facts have practical importance. When in a concentrated solution, or in a liquid, we are looking for proof for the presence of some free molecules this may not

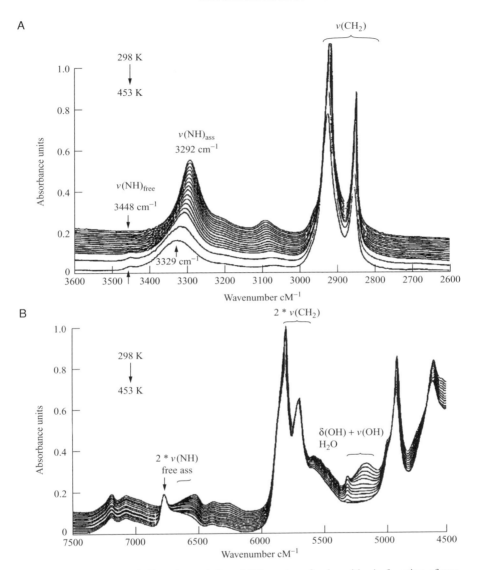

**Figure 2.3.** Mid-infrared (A) and near-infrared (B) spectra of polyamides in function of temperature. Reproduced by permission from Ellis Horwood Limited.

be seen at the fundamental level because of the weakness of the free band. There is a good chance to see it at the overtone level. This effect has been used to advantage in investigations on the structure of liquid water (101–103) and on the interactions between water and polymers (89–91).

Numerous cases of Fermi and higher anharmonic resonances occur in the near infrared, leading to characteristic bands, useful for the chemical analysis of polymers. Other applications of the near infrared include the determination of equilibrium constants. The determinations of hydrogen-bond equilibrium constants are more exact in the overtone region because band overlapping is much smaller in the overtone region as compared to the mid-infrared (104). Very often, the band selected to determine hydrogen bond complexes is the first overtone of free X-H stretching vibration. This band has the advantage of monitoring only one chemical species, namely the free molecule, in contrast to associated bands that may contain several types of hydrogen-bonded complexes, giving rise to overlapping bands. Identical equilibrium constants were obtained for nucleic base pairs from either mid-infrared or near-infrared spectra (23,25). For example, the equilibrium constant between the 8-bromoadenosine derivative and 1-methylthymine is $22 \pm 1.5 \, M^{-1}$ as measured with a free $NH_2$ antisymmetric stretching vibration ($3526 \, cm^{-1}$), whereas it amounted to $24 \pm 1.3 \, M^{-1}$ as determined with the first overtone of free $NH_2$ antisymmetric stretching vibration ($6997 \, cm^{-1}$) (25). Within experimental error, the equilibrium constants were identical when measured either with the combination band of the free $NH_2$ group of the adenosine derivative (located at $6812 \, cm^{-1}$ ) or with the first overtone of the free $NH_2$ symmetric stretching vibration ($6743 \, cm^{-1}$) (25). In this case, near–infrared spectra of mixtures containing adenosine derivatives had three bands available for the determination of equilibrium constant, whereas in the mid-infrared range, only one band can be used because the free $NH_2$ symmetrical stretching band is overlapped by other bands. As pointed out above, the ratio of intensities of free species: bonded species is higher in the overtone region compared to the fundamental region. This was also observed in the case of adenosine derivative. The absorption coefficient is $126 \, M^{-1} \, cm^{-1}$ for the free $NH_2$ antisymmetric stretching vibration ($3526 \, cm^{-1}$), and $114 \, M^{-1} \, cm^{-1}$ for the bonded $NH_2$ antisymmetric stretching vibration ($3490 \, cm^{-1}$), giving rise to an intensity ratio close to unity with slight band overlapping. However, in the overtone region, the absorption coefficient is $0.4 \, M^{-1} \, cm^{-1}$ for the first overtone of the free $NH_2$ antisymmetric stretching vibration ($6997 \, cm^{-1}$), while it is $0.08 \, M^{-1} \, cm^{-1}$ for the bonded $NH_2$ antisymmetric stretching vibration ($6889 \, cm^{-1}$), leading to an intensity ratio of 5:1, with negligible band overlapping.

## CONCLUSIONS

The most prominent bands in the near infrared are overtones and combination tones of O-H, N-H, C-H, and S-H stretching vibrations and stretching-bending combinations. Numerous cases of Fermi and higher anharmonic resonances occur. Knowledge of the frequencies of such overtones and combination tones is necessary to determine anharmonicity and anharmonic coupling constants that make it possible to gain knowledge concerning the potential functions ruling these vibrations. Furthermore, such data are a most important input for investigations on the nature of hydrogen bonding. The near infrared offers a good opportunity for the search for free (non-hydrogen bonded) species in concentrated solutions or liquids, because of the relative strength of the free overtones and combination tones.

# REFERENCES

1. C. Sandorfy. The near-infrared. A reminder. *Bull Polish Acad Sci Chem* **43**: 7–24, 1995.

2. G. Herzberg. *Molecular Spectra and Molecular Structure*, vol. 1, *Spectra of Diatomic Molecules*. Van Nostrand, New York, 1950.

3. G. Herzberg. *Molecular Spectra and Molecular Structure*, vol. 2, *Infrared and Raman Spectra of Polyatomic Molecules*. Van Nostrand, New York, 1945.

4. E. B. Wilson, J. C. Decius, P. C. Cross. *Molecular Vibrations*. McGraw-Hill, New York, 1955.

5. P. Barchewitz. *Spectroscopie infrarouge*, vol. 1. Gauthier-Villars, Paris, 1961.

6. B. R. Henry. Use of local modes in the description of highly vibrationally excited molecules. *Acc Chem Res* **10**: 207–213, 1997.

7. B. R. Henry. The local mode model and overtone spectra: a probe of molecular structure and conformation. *Acc Chem Res* **20**: 429–435, 1987.

8. H. J. Kjaergaard, D. M. Turnbull, B. R. Henry. Intensities of CH- and CD-stretching overtones in 1,3-butadiene and 1,3-butadiene-$d_6$. *J Chem Phys* **99**: 9438–9452, 1993.

9. B. I. Niefer, H. G. Kjaergaard, B. R. Henry. Intensity of CH- and NH-stretching transitions in the overtone spectra of cyclopropylamine. *J Chem Phys* **99**: 5682–5700, 1993.

10. R. Mendelsohn, H. H. Mantsch. *Fourier transform infrared studies of lipid-protein interaction*. in: *Progress in Protein-lipid interactions*. Vol. 2, A. Watts, J. J. H. H. M. De Pont, eds. Elsevier, Amsterdam, 103–146, 1986.

11. M. S. Braiman, K. J. Rotschild. Fourier transform infrared techniques for probing membrane protein structure. *Ann Rev Biophys Biophys Chem* **17**: 541–570, 1988.

12. C. Sandorfy. *Overtones and combination tones: application to the study of molecular associations*. In: *Infrared and Raman spectroscopy of biological molecules*. T. M. Theophanides, D. Reidel, eds. Dordrecht, The Netherlands, 305–318, 1979.

13. M. Tinkham. *Group Theory and quantum mechanics*. McGraw-Hill, New York, 1964.

14. J. E. Del Bene, M. J. T. Jordan. A comparative study of anharmonicity and matrix effects on the complexes $XH:NH_3$, X = F,Cl and Br. *J Chem Phys* **108**: 3205–3212, 1998.

15. H. C. W. Tso, D. J. W. Geldart, P. Chylek. Anharmonicity and cross section for absorption of water dimer. *J Chem Phys* **108**: 5319–5328, 1998.

16. P. Blaise, O. Henri-Rousseau. Spectral density of medium strength H-bonds. Direct damping and intrinsic anharmonicity of the slow mode. Beyond adiabatic approximation. *Chem Phys* **256**: 85–106, 2000.

17. A. Manani, E. Cané, P. Palmieri, A. Trombetti, N. C. Handy. Experimental and theoretical anharmonicity for benzene using density functional theory. *J Chem Phys* **112**: 248–259, 2000.

18. D. Luckhaus, M. J. Coffey, M. D. Fritz, F. F. Crim. Experimental and theoretical vibrational overtone spectra of $\nu CH = 3, 4, 5$ and 6 in formaldehyde ($H_2CO$). *J Chem Phys* **104**: 3472–3478, 1996.

19. D. T. Anderson, S. Davis, D. J. Nesbitt. Hydrogen bond spectroscopy in the near infrared: out-of-plane torsion and antigeared bend combination bands in $(HF)_2$. *J Chem Phys* **105**: 4488–4503, 1996.

20. M.-C. Bernard-Houplain, C. Sandorfy. Low temperature infrared study of association in dissolved dimethylamine. *J Chem Phys* **56**: 3412–3417, 1972.

21. S. E. Krikorian, M. Mahpour. The identification and origin of HH overtone and combination bands in the near-infrared spectra of simple primary and secondary amides. *Spectrochim Acta* **29A**: 1233–1246, 1973.

22. V. Stefov, Lj. Pejov, B. Soptrajanov. The influence of NH...$\pi$ hydrogen bonding on the anharmonicity of the n(N-H) mode and orientational dynamics of nearly continuously solvated indole. *J Mol Struct* **555**: 363–373, 2000.

23. G. M. Nagel, S. Hanlon. Higher order of adenine and uracil by hydrogen bonding. *I. Self-association of 9-ethyladenine and 1-cyclohexyluracil*, **11**: 816–823, 1972.

24. R. Buchet, A. Dion. Hydrogen bond equilibrium constants for the complexes of benzotriazole and some nucleoside derivatives. A near infrared study. *J Biomol Struct Dynamics* **4**: 231–241, 1986.

25. R. Buchet, C. Sandorfy. The effects of barbiturates on the hydrogen bonds of nucleotide base pairs. *J Phys Chem* **88**: 3282–3287, 1984.

26. R. Buchet, L. Beauvais, C. Sandorfy. Hydrogen bond equilibrium constants of some unusual nucleotide base pairs. *J Biomol Struct Dynamics* **2**: 221–232, 1984.

27. C. Bourdéron, C. Sandorfy. Association and the assignment of the OH overtones in hydrogen bonded alcohols. *J Chem Phys* **59**: 2527–2536, 1973.

28. S. Melikova, D. Shchepkin, A. Koll. Effect of electrical and mechanical anharmonicity on vibrational spectra of H-bonded complexes phenol...B (B=acetonitrile, pyridine) systems. *J Mol Struct* **448**: 239–246, 1998.

29. D. L. Snavely, J. Dubsky. Near-infrared spectra of polyethylene, polyethylene glycol, and polyvinylethyl ether. *J Polymer Sci A Polymer Chem* **34**: 2575–2579, 1996.

30. J. S. Wong, R. A. MacPhall, C. B. Moore , H. L. Strauss. Local mode spectra of inequivalent C-H oscillator in cycloalkanes and cycloalkenes. *J Phys Chem* **86**: 1478–1484, 1982.

31. W. R. A. Greenlay, B. R. Henry. The discrete excitation of nonequivalent CH oscillators— a local mode analysis of the high energy overtone spectra of alkanes. *J Chem Phys* **69**: 82–91, 1978.

32. R. Bicca de Alencastro, C. Sandorfy. A low temperature infrared study of self-association in thiols. *Can J Chem* **50**: 3594–3600, 1972.

33. B.T. Darling, D.M. Dennison. The water vapor molecule. *Phys Rev* **57**: 128–139, 1940.

34. J. H. Lady, Whetsel, Kermit. B. New assignments for the first overtone N-H and N-D stretching bands of anilines and the effect of intramolecular hydrogen bonding on the anharmonicities of N-H vibrations. *Spectrochim Acta* **21**: 1669–1679, 1965.

35. Y. Tanaka, K. Machida. Near-infrared spectra of formamide and its anharmonic potential. *J Mol Struct* **63**: 306–316, 1976.

36. M. Asselin, C. Sandorfy. A low temperature study of self-associated alcohols in the near infrared. *J Mol Struct* **8**: 145–158, 1971.

37. M.-C. Bernard-Houplain, C. Sandorfy. Low temperature infrared study of association in dissolved dimethylamine. *J Chem Phys* **56**: 3412–3417, 1972.

38. C. Sandorfy. Anharmonicity and hydrogen bonding. In: *The Hydrogen Bond*. vol. 2, P. Schuster, G. Zundel, C. Sandorfy, eds. North-Holland, Amsterdam, 613–654, 1976.

39. K.-Z. Liu, M. Jackson, M. G. Sowa, H. Ju, I. M. C. Dixon, H. H. Mantsch. Modification of the extracellular matrix following myocardial infarction monitored by FTIR spectroscopy. *Biochim Biophys Acta* **1315**: 73–77, 1996.

40. C. P. Schultz, H. Fabian, H. Mantsch. Two-dimensional mid-IR and near-IR correlation spectra of ribonuclease A: Using overtones and combination modes to monitor changes in secondary structure. *Biospectroscopy* **4**: S19–S29, 1998.

41. Y. Liu, Y. Ozaki, I. Noda. Two dimensional Fourier-transform near-infrared correlation spectroscopy study of dissociation of hydrogen-bonded *N*-methylacetamide in the pure liquid state. *J Phys Chem* **100**: 7326–7332, 1996.

42. Y. Wu, B. Czarnik-Matusewicz, K. Murayama, Y. Ozaki. Two-dimensional near-infrared spectroscopy study of human serum albumin in aqueous solutions: using overtones and combination modes to monitor temperature-dependent changes in the secondary structure. *J Phys Chem* **B 104**: 5840–5847, 2000.

43. J. J. Burmeister, H. Chung, M. A. Arnold. Phantoms for noninvasive blood glucose sensing with near infrared transmission spectroscopy. *Photochem Photobiol* **67(1)**: 50–55, 1998.

44. G. Lachenal. Structural investigations and monitoring of polymerisation by NIR spectroscopy. *J Near Infrared Spectroscopy* **6**: 299–306, 1998.

45. I. M. Mills. *Understanding spectra of highly excited vibrational states*. In: *Making Light Work: Advances in Near Infrared Spectroscopy*. I. Murray and I Cowe, VCH, Weinheim, eds. 1992.

46. I. M. Mills I.Min. *Molecular Spectroscopy: Modern Research*, vol 1, K. N. Rao, C. W. Matthews, eds. Academic Press, Boston, 115–140, 1972.

47. M. S. Child, L. Halonen. Overtone frequencies and intensities in the local mode picture. *Adv Chem Phys* **57**: 1–58, 1984.

48. M. S. Child. Local mode overtone spectra. *Acc Chem Res* **18**: 45–50, 1985.

49. R. H. Hayward, B. R. Henry. A general local-mode theory for high energy polyatomic overtone spectra and application to dichloromethane. *J Mol Spectrosc* **57**: 221–235, 1975.

50. R. H. Henry. Use of local modes in the description of highly vibrationally excited molecules. *Acc Chem Res* **10**: 207–213, 1977.

51. R. H. Henry. *The local mode model*. In: *Vibrational Spectra and Structure*, vol. 10. J. R. Durig, eds. Elsevier, Amsterdam, 269–319, 1981.

52. M. Quack. Spectra of dynamics of coupled vibrations in polyatomics molecules. *Annu Rev Phys Chem* **41**: 839–874, 1990.

53. J. L. Duncan. The determination of vibrational anharmonicity in the molecules from spectroscopic observations. *Spectrochim Acta* **47A**: 1–27, 1991.

54. I. M. Mills, A. G. Robiette. On the relationship of normal modes to local modes in the molecular vibrations. *Mol Phys* **56/4**: 743–765, 1985.

55. G. Lachenal, H. W. Siesler. La spectroscopie dans le proche infrarouge. *Spectra Anal* **176**: 28–32, 1994.

56. W. Baumann, R. Mecke R. Das Rotationsschwingungsspektrum des Wasserdampes. *Zeitschrift Phys* **81**: 445–464, 1933.

57. K. Freudenberg, R. Mecke. Das Rotationsschwingungsspektrum des Wasserdampes III. *Zeitschrift Phys* **81**: 465–481, 1933.

58. W. Siebrand, D. F. Williams. Radiationless transitions in polyatomic molecules. *J Chem Phys* **49/4**: 1860, 1968.

59. R. Wallace. A theory of nuclear motion in polyatomic molecules based upon the Morse oscillator. *Chem Phys* **11**: 189–199, 1975.

60. L. Halonen, E. Kauppi. Fermi resonances in overtone spectra of bromoform. *J Chem Phys* **92/6**: 3277–3282, 1990.

61. T. Lukka, E. Kauppi, L. Halonen. Fermi resonances and local modes in pyramidal HX3 molecules. *J Chem Phys* **102**: 5200–5206, 1995.

62. I. M. Mills. *Understanding spectra of highly excited vibrational states*. In: *Making Light Work: Advances in Near Infrared Spectroscopy*. I. Murray, I. Cowe, eds. VCH, Weinheim.

63. A. Novak. *Hydrogen bonding in solids. Correlation of spectroscopic and crystallographic data*. In: *Structure and Bonding*. Springer, Berlin **18**: 177–212, 1974.

64. G. Zundel. *Easily polarizable hydrogen bond - Their interactions with the environment— IR continuum and anomalous large proton conductivity*. In: *The Hydrogen Bond*, vol. 2, P. Schuster, G. Zundel, C. Sandorfy, eds. North-Holland, Amsterdam, 683–766, 1976.

65. R. N. Jones, C. Sandorfy. *The application of Infrared and Raman spectrometry on the elucidation of molecular structure*. In: *Weissberger's Technique of Organic Chemistry*, vol. 9, W. West, eds. Interscience, New York, 247–580, 1956.

66. A. C. Legon, D. J. Millen. Gas-phase spectroscopy and the properties of hydrogen-bonded dimers: HCN . . . HF as the spectroscopic prototype. *Chem Rev* **86**: 635–657, 1986.

67. J. Arnold, D. J. Millen. Hydrogen bonding in gaseous mixtures. Part II. Infrared spectra of ether-hydrogen fluoride systems. *J Chem Soc* 503–509, 1965.

68. J. E. Bertie, D. J. Millen. Hydrogen bonding in gaseous mixtures. Part I. Infrared spectra of ether-hydrogen chloride systems. *J Chem Soc* 497–503, 1965.

69. A. C. Legon, D. J. Millen. Gas-phase spectroscopy and the properties of hydrogen-bonded dimers: HCN . . . HF as the spectroscopic prototype. *Chem Rev* **86**: 635–657, 1986.

70. A. C. Legon, D. J. Millen. Directional character, strength, and nature of the hydrogen bond in gas-phase dimers. *Acc Chem Res* **20**: 39–46, 1987.

71. J. W. Bevan, B. Martineau, C. Sandorfy. Gas phase observation of the first overtone of the H-F stretching fundamental in hydrogen bonded complexes. *Can J Chem* **57**: 1341–1349, 1979.

72. Y. Maréchal, A. Witkowski. Infrared spectra of H-bonded systems. *J Chem Phys* **48**: 3967–3705, 1968.

73. N. D. Sokolov, V. A. Savelev. Dynamics of the hydrogen bond: two-dimensional model and isotope effects. *Chem Phys* **22**: 383–399, 1977.

74. M. Haurie, A. Novak. Spectres de vibration des molécules $CH_3COOH$, $CH_3COOD$, $CD_3COOH$ et $CD_3COOD$. Spectres infrarouges et Raman des dimères. *J Chim Phys* **62**: 146–157, 1965.

75. R. D. Waldron. The infrared spectra of three solid phases of methyl ammonium chloride. *J Chem Phys* **21**: 734–741, 1953.

76. A. Cabana, C. Sandorfy. The infrared spectra of solid methylammonium halides. Spectrochim Acta **18**: 843–861, 1962.

77. S. Bratos, D. Hadzi. Infrared spectra of molecules with hydrogen bonds. *J Chem Phys* **27**: 991–997, 1957.

78. G. C. Pimentel, A. L. McClellan. *The Hydrogen Bond*. W. H. Freeman, San Francisco, California, 111–114, 1960.

79. G. Lachenal. Dispersive and FT-NIR spectroscopy of polymers. *Vibr Spectrosc* **9(1)**: 93, 1995.

80. G. Lachenal. Characterization of poly(ethylene terephtalate) using near and far FTIR spectroscopy. *Int J Polym Anal Charact* **3**: 145, 1997.

81. G. Lachenal. Structural investigations and monitoring of polymerization by NIR spectroscopy. *J Near Infrared Spectrosc* **6**: 1998.

82. G. Lachenal, Y.Ozaki. Advantage of near infrared spectroscopy for the analysis of polymers and composites. *Macromol Symp* **141**: 283, 1999.

83. J. Derouault. In: *Advances in Practical Spectroscopy, Proceedings of ESIS.* G. Lachenal, H.W. Siesler, eds. UCB, Lyon, France, **41**: 1994.

84. K. H. Basset, C. Y. Liang, R. H. Marchessault. The near infrared spectrum of cellulose. *J Polym Sci* **A1**: 1687, 1963.

85. L. Glatt, D. S. Weber, C. Seaman, J. W. Ellis. The perturbed OH and the CH modes in polyvinyl alcohol. *J Chem Phys* **18**: 413, 1950.

86. C. E. Miller, B. E. Eichinger. Analysis of RIM molded polyurethanes by near infrared diffuse reflectance spectroscopy. *J Appl Polym Sci* **42**: 2169–2190, 1991.

87. G. Lachenal, I. Stevenson. *Interactions and hydrogen bonding studied by near FT-IR spectroscopy NIR news.* **6/2**: 10, 1995.

88. B. Chabert, G. Lachenal C. Vinh-Tung. Epoxy resins and epoxy blends studied by near infrared spectroscopy. *Macromol Symp* **94**: 145–158, 1995.

89. M. Fukuda, M. Miyagawa, H. Kawai, N. Yagi, O. Kimura, T. Otha. FT IR study on the moisture sorption isotherm of nylon 6. *J Polym* **19**: 785, 1987.

90. M. Fukuda, H. Kaway, N. Yagi, O. Kimura, T. Otha. FT IR Study on the nature of water sorbed in poly(ethylene terephthalate). *J Polym* **31**: 295, 1990.

91. P. Musto, G. Ragosta, L.Mascia. Vibrational spectroscopy evidence for the dual nature of water sorbed into epoxy resins. *Chem Mater* **12**: 1331–1341, 2000.

92. E. Dreassi, G. Ceramelli, L. Fabbri, F. Vocioni, P. Bartalini, P. Corti. *Analyst* **122/8**: 767, 1997.

93. W. Dziki, J. F. Bauer, J. J. Szpylman, J. E. Quick, B. C. Nichols. NIR spectroscopy to monitor the mobility of water within sarafloxacin crystal lattice. *J Pharma Biomed Anal* **22**: 829–848, 2000.

94. W. A. P. Luck, W. Ditter. Zur bestimmung der wasserstoffbrückenbindung im oberschwingungs-gebiet (Hydrogen bond determination in the overtone region). *J Mol Struct* **1**: 261–282, 1967–1968.

95. A. Burneau, J. Corset. Vibrational spectra and anharmonicity of $H_2O$, $D_2O$, and HOD in dilute solutions. *Chem Phys Lett* **9**: 99–102, 1971.

96. M. C. Bernard-Houplain, C. Sandorfy. On the similarity of the relative intensities of Raman fundamentals and infrared overtones of free and hydrogen bonded X-H stretching vibrations. *Chem Phys Lett* **27**: 154–156, 1974.

97. U. Eschenauer, D. Henck, M. Hühne, P. Wu, I. Zebger, H. W. Siesler. *Near infrared spectroscopy in chemical research, quality assurance and process control.* In: *Near Infrared Spectroscopy, Bridging the Gap between Data Analysis and NIR Applications* K.I Hildrum, T. Isaksson, T. Naes, A. Tandberg, eds. Ellis Horwood, Chichester, 11–18, 1992.

98. G. Trudeau, K. C. Cole, R. Massuda, C. Sandorfy. Anesthesia and hydrogen bonding. A semi-quantitative infrared study at room temperature. *Can J Chem* **56**: 1681–1686, 1978.

99. T. Di Paolo, C. Bourdéron, C. Sandorfy. Model calculations on the influence of mechanical and electrical anharmonicity on infrared intensities: relation to hydrogen bonding. *Can J Chem* **50**: 3161–3166, 1972.

100. R. C. Herman, K. E. Schuler. Vibrational intensities in diatomic infrared transitions. The vibrational matrix elements for CO. *J Chem Phys* **22**: 481–490, 1954.

101. W. A. P. Luck, W. Ditter. Approximate methods for determining the structure of $H_2O$ and HOD using near-infrared spectroscopy. *J Phys Chem* **74**: 3687–3695, 1970.

102. W. C. McCabe, S. Subramanian, H. F. Fisher. A near-infrared spectroscopic investigation of the effect of temperature on the structure of water. *J Phys Chem* **74**: 4360–4369, 1970.

103. J. J. Péron, C. Bourdéron, C. Sandorfy. On the existence of free OH groups in liquid water. *Can J Chem* **49**: 3901–3903, 1971.

104. L. England-Kretzer, M. Fritze, W. A. Luck. The intensity change of IR OH bands by H-bonds. *J Mol Struct* **175**: 277–282, 1988.

# CHAPTER 3

# Spectral Analysis

YUKIHIRO OZAKI, SHIGEAKI MORITA, and YIPING DU

## INTRODUCTION

NIR spectra contain a great deal of physical and chemical information about molecules (1–6). However, this information cannot always be extracted straightforwardly from the spectra. There may be two major reasons that make the extraction of information difficult. One is that an NIR spectrum consists of a number of bands arising from overtones and combination modes that overlap heavily with each other, and thus so-called multicollinearity is very strong in the NIR region. This is an intrinsic reason for the difficulty in the analysis of NIR spectra, and another is a more practical reason; NIR spectroscopy deals quite often with "real-world" samples, which yield rather poor signal-to-noise (SN) ratio, baseline fluctuations, and severe overlapping of bands due to various components. Thus spectral analysis in the NIR region usually must overcome these two difficulties (4–6).

As in the case of IR and Raman spectroscopy, band assignments give rise to the base for spectral analysis in the NIR region (4–6). Conventional band assignment methods such as those based on group frequencies, spectra-structure relationship, and comparison of the NIR spectrum of a compound with those of related compounds are, of course, very useful for the analysis of NIR spectra. Moreover, general spectral analysis methods like second derivative and different spectra are employed in NIR spectroscopy. However, to unravel complicated NIR spectra these band assignment methods and spectral analysis methods are not always enough. Chemometrics has most often been used to extract rich information from NIR spectra (7–11). A main part of chemometrics is multivariate data analysis, which is essential for qualitative and quantitative assays based on NIR spectroscopy. Multivariate data analysis is very useful for spectral analysis in the NIR region. For example, it allows one to overcome the problem of multicollinearity. Besides well-known multivariate data analysis methods

*Near-Infrared Spectroscopy in Food Science and Technology*, Edited by Yukihiro Ozaki, W. Fred McClure, and Alfred A. Christy.

such as principal component analysis/regression (PCA/PCR) and partial least squares regression (PLSR), self-modeling curve resolution (SMCR), which is used to predict pure component spectra and pure component concentration profiles from a set of NIR spectra, is also becoming important. It is always recommended that chemometrics is used with the band assignment methods.

In addition to the conventional spectral analysis methods and chemometrics, two-dimensional (2D) correlation spectroscopy has recently been introduced to NIR spectroscopy (4, 12–16). In this method spectral peaks are spread over a second dimension to simplify the visualization of complex spectra consisting of many overlapped bands and to explore correlation between the bands. There are two kinds of 2D correlation spectroscopy used in NIR spectroscopy. One is statistical 2D correlation proposed originally by Barton et al (16). This method employs cross-correlation based on the least-squares linear regression analysis to assess spectral changes in two regions, such as the NIR and mid-IR regions, that arise from variations in sample composition (16). In another 2D correlation spectroscopy proposed by Noda (12, 13), 2D spectra are constructed from a set of spectral data collected from a system under an external physical perturbation, which induces selective alterations in spectral features.

In this way, nowadays we can use conventional spectral analysis methods, chemometrics, and 2D correlation spectroscopy to analyze NIR spectra, depending upon the purposes. However, in any case it is rather rare in NIR spectroscopy that original spectra are subjected to spectra analysis without any pretreatment or data transformation (5–11). Usually proper pretreatments are imposed on the experimental data to reduce noise, to correct baseline variations, to enhance apparent spectral resolution and/or to normalize the data. The pretreatment methods can be divided into four categories. One method is noise reduction. Smoothing is a representative method for noise reduction. Another method is baseline correction. The second derivative and multiplicative scatter correction (MSC) are most frequently employed for baseline correction. The third method is centering and normalization, and the last is resolution enhancement. Difference spectra, mean centering, and second derivative are used in NIR spectroscopy as resolution enhancement methods. The purpose of this chapter is to outline spectral analysis in the NIR region. This chapter consists of four sections; (1) characteristics and examples of NIR spectra, (2) pretreatment methods in NIR spectroscopy, (3) band assignments in NIR spectroscopy, and (4) 2D correlation spectroscopy. Chemometrics is described in Chapter 6.

## CHARACTERISTICS AND SOME EXAMPLES OF NIR SPECTRA

In this section four examples of NIR spectra are shown to discuss the characteristics of NIR spectra. Figure 3.1, A and B, shows two examples of NIR transmission spectra; NIR spectra of (A) aqueous solutions of human serum albumin (HSA) with the concentrations of 1.0, 2.0, 3.0, 4.0, and 5.0 wt% (17), and (B) cow milk (18). Figure 3.2, A and B, depicts two examples of NIR diffuse-reflectance (DR) spectra; DR spectra of (A) pellet samples of 16 kinds of linear low-density polyethylene (LLDPE) and one kind of high-density polyethylene (HDPE) (19) and (B) serum albumin and

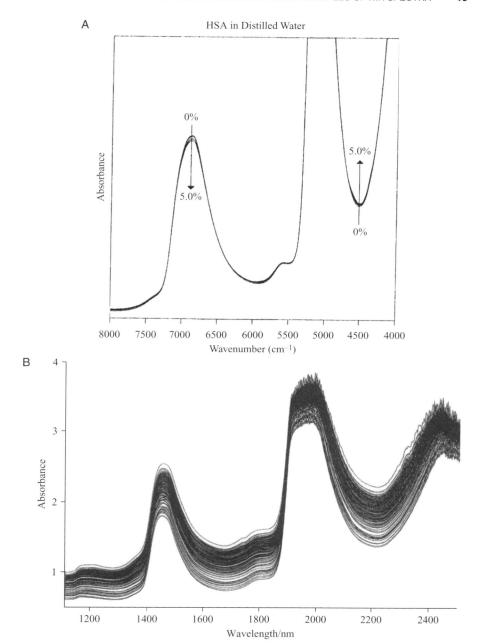

**Figure 3.1.** A. NIR spectra of aqueous solutions of human serum albumin (HSA) with concentrations of 1.0, 2.0, 3.0, 4.0, and 5.0 wt%. [Reproduced from Ref. 17 with permission. Copyright (2000) Society for Applied Spectroscopy.] B. NIR spectra of milk [Reproduced from Ref. 18 with permission. Copyright (1999) Society for Applied Spectroscopy].

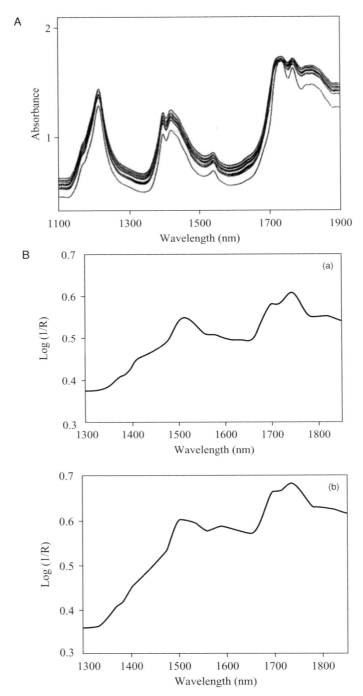

**Figure 3.2.** A. NIR DR spectra of 16 kinds of LLDPE and one kind of HDPE in pellets. [Reproduced from Ref. 19 with permission. Copyright (1998) John Wiley & Sons.] B. NIR DR spectra of (a) serum albumin and (b) $\gamma$-globulin in solid states. [Reproduced from Ref. 20 with permission. Copyright (1998) Elsevier].

$\gamma$-globulin in solid states (20). The NIR spectra of HSA in aqueous solutions are very close to an NIR spectrum of water (4). A broad feature near 7000 cm$^{-1}$ is largely due to the combination of OH symmetric and antisymmetric stretching modes of water. It is almost impossible to draw useful information about the protein in aqueous solutions directly from the NIR spectra. It is noted in the NIR spectra of milk that the baseline changes from one spectrum to another. The baseline changes are caused by light-scattering particles in the form of fat globules and protein micelles contained in milk (18). These particles introduce erroneous information into NIR spectral measurements. Fat globules induce much severe baseline variations than protein micelles. We discuss below how we can overcome these baseline fluctuations.

In Figure 3.2A bands are rather broad and the baseline varies from one sample to another. The spectra may be devided into three spectral regions, the 1100 to 1250-nm region, the 1350 to 1600-nm region, and the 1650 to 1900-nm region (19). The first region contains bands due to the second overtones of CH stretching modes of the CH$_3$ and CH$_2$ groups and those assigned to their combinations. The second region solely includes bands arising from the combinations of CH vibrations. The third region is concerned with bands due to the first overtones of CH stretching modes of CH$_3$ and CH$_2$ groups and those ascribed to their combinations. The spectra in Figure 3.2B have the typical appearance of a protein NIR spectrum (20). A broad feature near 1500 nm is ascribed to the first overtone of the NH stretching mode of the protein, and a doublet in the 1690 to 1750-nm region is due to the first overtone of CH stretching modes of CH, CH$_2$, and CH$_3$ groups.

The characteristics of NIR bands can be summarized as follows (1–6).

1. NIR bands are much weaker than IR bands, and the former are usually broad.
2. NIR bands, which have their origins in overtones and combinations of fundamentals, strongly overlap with each other, yielding severe multicollinearity.
3. The assignments of NIR bands are often difficult because of the overlapping of bands, complicated combinations of vibrational modes, and possible Fermi resonance.
4. The NIR region is dominated by bands ascribed to functional groups that contain a hydrogen atom (e.g., OH, CH, NH). This occurs partly because the anharmonic constant of an XH bond is large and partly because the fundamentals of XH stretching vibrations are of high frequency.
5. As in the case of a mid-IR spectrum, a hydrogen bonding causes a band shift for particular bands. The shift is much larger for a NIR band than for the corresponding mid-IR band.

From the above characteristics one may consider that NIR spectroscopy has properties that appear to be disadvantages rather than advantages because NIR bands are weak and broad and overlap heavily. However, NIR spectra are still very rich in inherent information, and NIR spectroscopy has a number of advantages. For example, it is suitable for nondestructive and in situ analysis, and it allows analysis using an optical fiber. Thus spectral analysis in the NIR region is very important and has been a focus of great attention. One must note that the precise and proper spectral analysis of NIR spectra permits one to extract useful information from NIR spectra whereas wrong or improper spectral analysis may lead to the extraction of wrong information.

## PRETREATMENT METHODS IN NIR SPECTROSCOPY

In modern NIR spectroscopy there have been intensive discussions on the subject of data pretreatment because NIR spectra quite often suffer from the problems of unwanted spectral variations and baseline shifts (1–6). These may have the following sources.

1. Light scattering from solid samples or cloudy liquids
2. Poor reproducibility of NIR spectra caused by, for example, pathlength variations
3. Variations in temperature, density, and particle size of samples
4. Various kinds of noises such as those from a detector, an amplifier, or an AD converter.

The interferences described above may easily violate the assumptions on which chemometrics equations are based. For example, the simple linear relationship as stated by Beer's law does not hold completely, and the additivity of individual spectral responses is not guaranteed. Therefore, data pretreatment is often crucially important. Whenever one tries to improve SN ratio or to correct baseline fluctuations, one should investigate the cause of poor SN ratio and of baseline changes. Otherwise, one may not be able to find proper pretreatment methods. One good example of this sort of studies is the studies by Geladi et al. (21). They modeled the reflectance spectra of milk by optical effects and chemical light absorption effects. The optical effects cause changes in the direction of the light, and the chemical light absorption effects are related to light absorption. In some cases, the former brings about more prominent changes to spectra than the latter. The response of the spectral data to the physical effects is significant baseline changes. Based on this study Geladi et al. (21) proposed multiplicative scattering collection (MSC) as a preprocessing tool to correct the light scattering problems in the NIR spectra of milk.

In this section, the four kinds of pretreatment methods, noise reduction methods, baseline correction methods, resolution enhancement methods, and centering and normalization methods are introduced (2–11).

### Noise Reduction Methods

There are several kinds of noise arising from various interfering physical and/or chemical processes. One is high-frequency noise associated with the instrument's detector and electronic circuits. There are other forms of noise as well, for example, low-frequency noise and localized noise. Low-frequency noise is caused, for example, by instrument drift during the scanning measurements. Usually, reduction of the low-frequency noise is more difficult because it often resembles the real information in the data.

The most generally used method for improvement of SN ratio in spectra is accumulation-average processing that requires increase of the accumulation number and calculation of an average. This reduces the effects of high-frequency noise but technically is not a "pretreatment" but a normal, integrated part of collecting spectra. If the noise reduction by the accumulation average is still not enough, one can use smoothing to remove high-frequency noise. The most commonly used smoothing methods are the moving-average method and the Savitzky–Golay method.

**Smoothing** In the moving-average method, which is the simplest type of smoothing, the reading $A_i'$ ($A$ is, for example, absorbance) at each variable $i = 1, 2, \cdots k$ is replaced by a weighted average of itself and its nearest neighbors. From $i - n$ to $i + n$:

$$A_i = \sum_{k=-n}^{n} w_k A_{i+k} \qquad (3.1)$$

$w_k$, defining the smoothing, is called the convolution weights.

The Savitzky–Golay method came from the idea that in the vicinity of a measurement point a spectrum can be fitted by low-degree polynomials. In practice, $w_k$ is determined by fitting the spectrum with low-degree polynomials by using least squares regression. Savitzky and Golay calculated $w_k$ for the different orders of polynomials and $N(N = 2n + 1)$. One can find these calculated convolution weights in a numeral table. For example, when $N$ is equal to 5, smoothed values can be obtained by substituting $w_k = -3/35, 12/35, 17/35, 12/35, -3/35$ ($k = -2, -1, 0, 1, 2$) into Equation 3.1. It must be kept in mind that if one tries to increase the effect of smoothing by increasing the number of points of $w_k$, a band shape would be distorted. This distortion may lead to the decrease of spectral resolution and band intensity.

There are other methods for noise reduction such as wavelets, eigenvector reconstruction, and artificial neural networks (ANN) (22, 23).

**Wavelets** The wavelet transform is superior to smoothing in that it removes both high- and low-frequency noise as well as localized noise due to phenomena like scattering. Smoothing, as described above, can only remove high-frequency data. Low-frequency noise, though relatively uncommon in modern NIR spectrometers, may come from a variety of sources like a poor interferometer in an FT-NIR instrument or poor detector electronics. Wavelets operate by taking the spectrum and transforming it into the wavelet domain and returning it to the spectral domain (22, 23). This is similar in practice to applying the Fourier transform to data to reduce noise, except that wavelets use a much more sophisticated function to model the data than the sine/cosine of Fourier analysis. Wavelets are able to denoise data by only modeling data above a certain user-selected threshold. When the data are reconstructed, all the data beneath the noise threshold level are lost. As a result, one obtains a cleaner, more reliable spectrum for analysis.

## Baseline Correction Methods

As described, baselines change for various reasons in NIR spectra (1–4). Baseline shifts are also induced by the influence of optical fiber cables.

An observed NIR spectrum, $A(\lambda)$, can be represented as follows

$$A(\lambda) = \alpha A_0(\lambda) + \beta + e(\lambda) \tag{3.2}$$

Here, $A_0(\lambda)$, $\alpha$, $\beta$, and $e(\lambda)$ are a real spectrum, multiplicative scatter factor (amplification factor), additive scatter factor (offset deviation), and noise, respectively. There are several methods to eliminate or reduce the effects of $\alpha$ and $\beta$. Here, we describe three of them.

***Derivative Methods*** Derivative methods have long been used in NIR spectroscopy as pretreatment methods for resolution enhancement as well as baseline correction (24, 25). A derivative spectrum is an expression of derivative values, $d^n A/d\lambda^n (n = 1, 2, \cdots)$, of a spectrum $A(\lambda)$ as a function of $\lambda$. The second derivative, $d^2 A/d\lambda^2$, is most often used. The superimposed peaks in an original spectrum turn out as clearly separated downward peaks in a second derivative spectrum. Another important property of the second derivative method is the removal of the additive and multiplicative baseline variations in an original spectrum. On the other hand, a drawback of the derivative methods is that the SN ratio deteriorates every time a spectrum is differentiated.

Because, in general, spectrum data take discrete values, the calculation of derivatives of various order is carried out by algebraic differences between data taken at closely spaced wavelengths. Transformation to first and second derivatives are then

$$dA_i = A_{i+k} - A_{i-k} \tag{3.3a}$$

$$d^2 A_i = d(A_{i+k} - A_{i-k})$$

$$= A_{i+2k} - 2A_i + A_{i-2k} \tag{3.3b}$$

Figure 3.3 displays the second derivative obtained with the Savitzky–Golay method of the spectra shown in Figure 3.2A (19). It is noted that the calculation of the second derivative makes a number of bands clearly detectable. Moreover, it can be seen from the second derivative spectra that the second derivative is powerful in removing additive and multiplicative baseline variations of the spectra.

***Multiplicative Scatter Correction (MSC)*** MSC is a powerful method for correcting vertical variations of the baseline (additive baseline variation) and inclination of the baseline (multiplicative baseline variation) (21). The basis of MSC lies in the fact that light scattering has a wavelength dependence different from that of chemically based light absorbance (21). Thus we can use data from many wavelengths to distinguish between light absorption and light scattering.

MSC corrects spectra according to a simple linear univariate fit to a standard spectrum; $\alpha$ and $\beta$ are estimated by least squares regression using the standard spectrum. As the standard spectrum, a spectrum of a particular sample or an average spectrum is used.

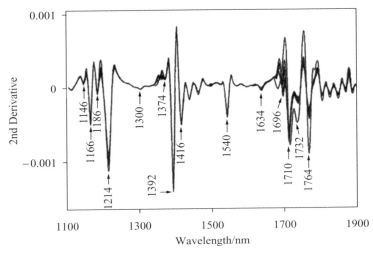

**Figure 3.3.** Second derivatives of the NIR spectra shown in Fig. 3.2A. [Reproduced from Ref. 19 with permission. Copyright (1998) John Wiley & Sons].

Figure 3.4 shows the NIR spectra shown in Figure 3.2A after the MSC treatment (19). The spectra demonstrate the potential of MSC in correcting offset and amplification in NIR spectra. In general, MSC improves essentially the linearity in NIR spectroscopy. Although it is generally an excellent technique, care must be exercised because the use of MSC may generate unwanted artifacts.

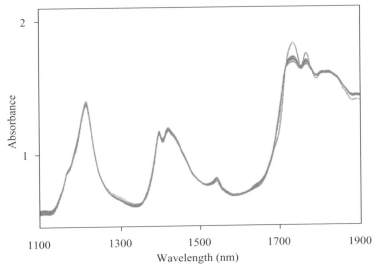

**Figure 3.4.** NIR spectra shown in Figure 3.2A after the MSC treatment. [Reproduced from Ref. 19 with permission. Copyright (1998) John Wiley & Sons].

***Orthogonal Signal Correction (OSC)***  The OSC procedure was introduced by Wold and coworkers (26) in the setting of calibration to remove the interfering variations present in the spectra. The idea of OSC is to remove a small number of factors that account for as much as possible of the total variation in the spectral matrix **X** and are orthogonal to the dependent variable to be modeled. They also present a modified nonlinear iterative partial least squares (NIPALS) algorithm to construct these factors. Very recently, Fearn (27) suggested an alternative algorithm for OSC based on the ideas of the work of Wold et al (26). The algorithm of OSC can be described briefly as follows. The first step to compute the OSC component is the calculation of the first principal component score **t** of **X**. This vector is then orthogonalized to **y**, giving the actual correction vector **t**\*. Subsequently, PLS weights **w** are computed such that **X w** becomes as close as possible to **t**\*. The correction vector **t**\* is then processed in the NIPALS algorithm through **X** to give an updated score **t**, which is then again orthogonalized to **y**. This computational procedure is repeated until convergence, yielding the correction vector **t**\* and a load vector **p**. The correction vector **t**\* in some sense has the property that it is orthogonal to **y** and accounts for as much as possible of the variations of **X**. The interfering variations are then removed by subtracting **t**\* **p**$^T$ from **X**. After the first OSC step, a second or more OSC steps may be performed. The residual matrix finally obtained after OSC is then the filtered spectral matrix without interfering variations.

## Resolution Enhancement Methods

Resolution enhancement methods play very important roles in unraveling overlapping bands and elucidating the existence of obscured bands. In NIR spectroscopy derivative methods, difference spectra, mean centering, and Fourier self-deconvolution are used as resolution enhancement methods (4, 5). Interestingly, PCA loading plots are often useful for resolution enhancement (19). 2D correlation spectroscopy is also capable of resolution enhancement (14, 15), but because it is not a pretreatment method it is discussed elsewhere. Mean centering is discussed in the section on centering and normalization.

***Difference Spectra***  The difference spectrum between a spectrum of sample *a* and that of sample *b* can be calculated by subtracting the spectrum of sample *a* from that of sample *b* (24, 25). It is very effective in detecting slight differences and changes between two spectra. The calculation of difference spectra is useful to analyze perturbation-dependent NIR spectra such as temperature-dependent, pH-dependent, or concentration-dependent spectra. To calculate reliable difference spectra one must measure spectra with very high wavelength accuracy.

Figure 3.5A depicts NIR spectra in the 7400 to 6700-cm$^{-1}$ region of octanoic acid in the pure liquid state measured over a temperature range of 15 to 92°C (28). Figure 3.5B displays difference spectra in the 7100 to 5900-cm$^{-1}$ region over a temperature range of 20 to 92°C (28). In the original spectra several bands overlap with each other. It is noted that the intensity of a band at 6920 cm$^{-1}$ increases as a function of temperature, whereas those of other bands in the 7100 to 6800-cm$^{-1}$

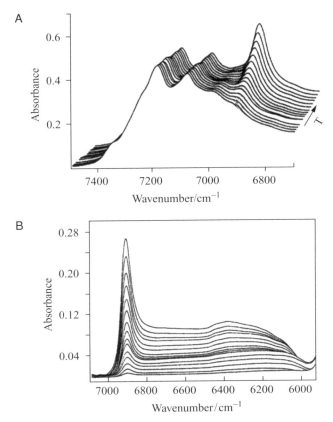

**Figure 3.5.** A. FT-NIR spectra of octanoic acid in the pure liquid state over the temperature range of 15–90°C (after density correction). B. Difference spectra of octanoic acid over the temperature range of 20–92°C. The spectrum at 15.1°C was taken as a reference spectrum. [Reproduced from Ref. 28 with permission. Copyright (1993) Society for Applied Spectroscopy].

region are almost temperature independent. The band at 6920 cm$^{-1}$ is assigned to the first overtone of the OH stretching mode of the monometric species of octanoic acid, and the rest is due to combinations of CH vibrations (28). To investigate the temperature-dependent intensity vibrations of the OH stretching band more clearly, the difference spectra in Figure 3.5B were calculated by subtracting the spectrum measured at 15°C from all the other spectra (28). Note that the OH band is clearly separated from the CH bands in the difference spectra. This is a good example to demonstrate the usefulness of difference spectra.

## Centering and Normalization Methods

Centering and normalization are often useful in chemometrics analysis of NIR data. Mean centering is simply an adjustment to a data set to reposition the centroid of the data to the origin of the coordinate system (7–11). Normalization is an adjustment to a data set that equalizes the magnitude of each sample (7–11).

**Centering Methods**  Mean centering is an operation in which from every element of the $j$th spectrum (row) the column mean is subtracted:

$$X_{jcent} = X_j - \left( \frac{1}{n} \sum_{j=1}^{n} X_{ij} \right) \tag{3.4}$$

$X_j$ and $X_{ij}$ are an element of the $j$th spectrum and of a data matrix $X$, respectively. After this step, all means are zero and variances are spread around zero. Each mean centering spectrum can be regarded as a difference spectrum between the individual spectrum and an averaged spectrum. Mean centering is often powerful in resolution enhancement. Recently, mean centering has been used as a pretreatment procedure for 2D correlation spectroscopy (14, 15). Figure 3.6A shows FT-NIR spectra in the 6000 to 5500-cm$^{-1}$ region of nylon 12 measured over a temperature range from 30 to 150°C, and Figure 3.6B displays their mean centered spectra (29 ). The mean-centered spectra reveal that the intensity of a band at 5770 cm$^{-1}$ due to the first overtone of the CH$_2$ stretching mode changes largely with temperature.

**Normalization**  There are two popular normalization procedures in common practice. Most normalization methods use vectors normalized to a constant euclidean, norm, that is,

$$x_{j,\text{norm}} = x_j / ||\mathbf{x}|| \tag{3.5}$$

where $||\mathbf{x}||$ is the euclidean norm of the spectral vector $\mathbf{x}$. This normalization transforms the spectral points on a unit hypersphere, and all data are approximately in the same scaling. A good property of this normalization is that the similarity between two spectral vectors may be measured by the scalar product of these two vectors. However, the normalization causes a geometric configuration of the data points, either the clustering structure or the spreading directions, substantially different from the original one, which may result in a misleading understanding of the data in exploratory data analysis. Moreover, the variation in spreading directions has a significant effect on PCA-related analysis. Therefore, one must be careful in using normalization in situations where exploratory data analysis and PCA-related procedures such as PCA, PLS, and so on are concerned.

Another normalization procedure is so-called mean normalization, where all points of the $j$th spectrum are divided by its mean value

$$X_{jnorm} = X_j / \left( \frac{1}{m} \sum_{i=1}^{m} X_{ij} \right) \tag{3.6}$$

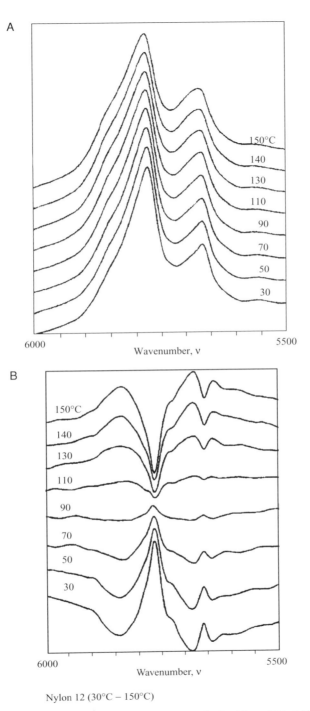

A

150°C
140
130
110
90
70
50
30

6000                    5500

Wavenumber, ν

B

150°C
140
130
110
90
70
50
30

6000                    5500

Wavenumber, ν

Nylon 12 (30°C – 150°C)

**Figure 3.6.** A. Temperature-dependent FT-NIR spectra obtained from 30 to 150°C in the 6000 to 5500-cm$^{-1}$ region of nylon 12. B. Mean centering of the spectra shown in A. [Reproduced from Ref. 29 with permission. Copyright (1997) American Chemical Society].

where $m$ is the total number of spectral points. After mean normalization all the spectra have the same area. Essentially, mean normalization is equivalent to normalize the spectral vectors to a constant 1-norm, that is, the sum of spectral values (always positive) equals to a constant. This means that the geometry of mean normalization is to transform the spectral points to be contained in a convex set, and the dimensionality of the spectral space is thus decreased by 1. This transformation is very useful in SMCR.

## CONVENTIONAL SPECTRAL ANALYSIS METHODS IN NIR SPECTROSCOPY

Conventional spectral analysis methods used in NIR spectroscopy are summarized as follows.

1. Spectral analysis based on group frequencies
2. Spectra-structure correlations
3. Spectral analysis based on perturbation
4. Derivative spectra
5. Difference spectra
6. Fourier self-deconvolution
7. Curve fitting
8. PCA loading plots
9. Spectral interpretation by polarization measurements
10. Isotope exchange experiments
11. Theoretical calculations of frequencies of bands due to overtones and combination modes

Spectral analysis based on group frequencies and the spectra-structure correlation method is well established in IR and Raman spectroscopy. These methods are also becoming popular in NIR spectroscopy, and recently Weyer and Lo (30) wrote a valuable review article on the relationship between NIR bands and chemical structure. They compared NIR spectra of selected aliphatic hydrocarbons, those of selected aromatics, those of selected alcohols, and so on and discussed detailed spectra-structure correlations in the NIR region. Tables for group frequencies in the NIR region are also available in some NIR books (3, 4).

In this chapter we will discuss group frequencies and spectra-structure correlations for selected alcohols. Figure 3.7 shows NIR spectra of methanol, ethanol, 1-propanol, 2-propanol, 1-butanol, and $t$-butanol in the pure liquid state. Comparisons of the NIR spectra of the six alcohols enable one to make band assignments in the 12,000 to 5000-cm$^{-1}$ region. A broad feature in the 6900 to 6200-cm$^{-1}$ region is assigned to the first overtone of OH stretching modes of monomer and various hydrogen-bonded

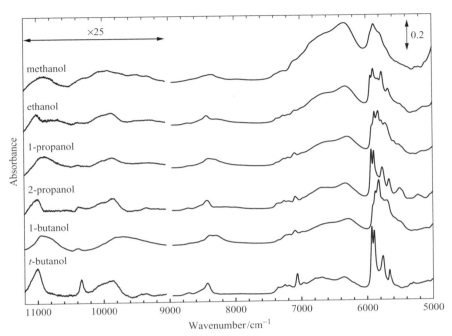

**Figure 3.7.** NIR spectra of methanol, ethanol, 1-propanol, 2-propanol, 1-butanol, and $t$-butanol in the pure liquid state.

species of alcohols. The band is so broad because there are many hydrogen-bonded species in the alcohols. The corresponding second overtones are seen in the 10,200 to 9200-cm$^{-1}$ region. A group of bands in the 6000 to 5500-cm$^{-1}$ region are attributed to the first overtones of $CH_3$ and $CH_2$ stretching modes and their combination modes. Bands in the 8800 to 8000-cm$^{-1}$ region arise from the second overtones of $CH_3$ and $CH_2$ stretching modes and their combination modes, and those in the 7400 to 7000-cm$^{-1}$ region are assigned to CH combination modes. In this way, by comparing NIR spectra of a series of similar molecules one can discuss band assignments.

Spectral analysis by use of perturbation is also a traditional method in vibrational spectroscopy. Temperature, pH, and concentration are typical examples of useful perturbations for spectral analysis. The perturbation method is very useful for spectral analysis because some bands are very sensitive to a perturbation while some others are less sensitive or not sensitive to it. For example, bands due to OH and NH groups are often sensitive to a temperature variation because they are often concerned with a hydrogen bonding, but those arising from CH groups are usually not sensitive to it. Figure 3.5A yields a good example of perturbation-dependent spectral changes (28). The OH band at 6920 cm$^{-1}$ shows the marked temperature-dependent changes whereas the CH bands do not. In this way, the perturbation, temperature in the present case, enables one to discriminate the band due to the OH group from those ascribed to the CH groups.

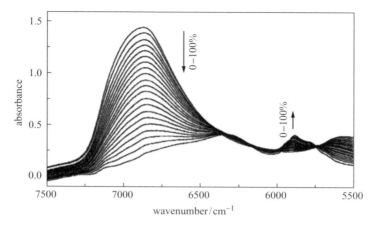

**Figure 3.8.** NIR spectra in the 7500 to 5700-cm⁻¹ region of water-methanol mixtures with a methanol concentration of 0–100 wt% at increments of 5 wt% at 25°C. [Reproduced from Ref. 31 with permission. Copyright (2002) NIR Publications].

The usefulness of the second derivative has already been demonstrated, but in Figures 3.8–3.10 another good example showing the usefulness of the second derivative is given together with the potential of a curve fitting method.

Figure 3.8 displays NIR spectra in the 7500 to 5500-cm⁻¹ region of water-methanol mixtures with a methanol concentration of 0–100 wt% at increments of 5 wt% at 25°C (31). A broad band centered near 6900 cm⁻¹ is composed of a number of overlapped bands due to the combinations of OH antisymmetric and symmetric stretching modes of various water species and the first overtones of the OH stretching modes of free and hydrogen-bonded species of methanol. Weak features in the 6000 to 5700-cm⁻¹ region are due to the first overtones and combinations of the $CH_3$ stretching modes of methanol (31). Figure 3.9A shows the enlargement of the 6000 to 5700-cm⁻¹ region of the spectra in Figure 3.8 (31). It is noted that this region shows significant changes with the concentration. To investigate the spectral variations in this region in detail both the second derivative and the curve fitting method were employed (31).

Figure 3.9B shows the second derivative in the 6000 to 5700-cm⁻¹ region of the spectra shown in Figure 3.9A (31). The second derivative spectra reveal that two bands near 5940 and 5900 cm⁻¹ due to the first overtones of the $CH_3$ asymmetric stretching modes show a downward shift by about 30 cm⁻¹. These shifts are not clear in the original spectra. Figure 3.10 depicts the result of the curve fitting for the NIR spectrum of methanol (31). The number of bands was estimated from the second derivative spectra. A Gaussian distribution function was assumed for each isolated band. A residual calculated by subtracting the trial function from the original spectrum is sufficiently small, meaning that the trial function consisting of the five Gaussian bands could reproduce the raw spectrum in the 6000 to 5700-cm⁻¹ region. Katsumoto et al. (31) proposed the following band assignments for the NIR spectra of

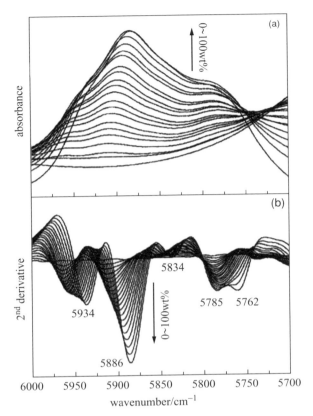

**Figure 3.9.** A. NIR spectra in the 6000 to 5700-cm$^{-1}$ region represented in Fig. 3.8. B. The second derivative in the 6000 to 5700-cm$^{-1}$ region of the spectra shown in Fig. 3.8. [Reproduced from Ref. 31 with permission. Copyright (2002) NIR Publications].

methanol: 5929 and 5884 cm$^{-1}$, first overtones of $CH_3$ asymmetric stretching modes; 5845 and 5799 cm$^{-1}$, combinations of $CH_3$ asymmetric and symmetric stretching modes; 5756 cm$^{-1}$, first overtone of $CH_3$ symmetric stretching mode.

Isotope exchange experiments and polarization measurements are powerful experimental methods for band assignments in the NIR region as well as in the mid-IR region. Recently, Wu and Siesler (32) reported a good example of NH/ND-deuteration experiments and polarization measurements to interpret NIR spectra of polyamide II. Spectral analysis of this sort of molecules is of importance because not only polyamides, but also proteins, peptides, and poly amino acids, have –NH-CO- groups in common that show several key bands in the NIR region. A very simple example for the potential of isotope exchange experiments in NIR spectroscopy is demonstrated in Figure 3.11, where NIR spectra of $CH_3CH_2OH$, $CH_3CH_2OD$, and $CD_3CH_2OH$ are shown. By comparing the three spectra, one can easily make band assignments

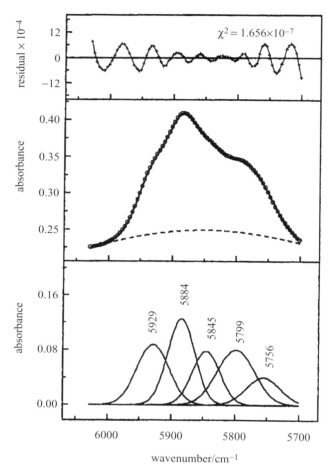

**Figure 3.10.** Result of curve fitting for the NIR spectrum of methanol in the 6000 to 5700-cm$^{-1}$ region. [Reproduced from Ref. 31 with permission. Copyright (2002) NIR Publications].

for OH, CH$_3$, and CH$_2$ groups. For example, two bands near 8700 and 8400 cm$^{-1}$, which show a shift in the spectrum of CD$_3$CH$_2$OH, are attributed to the CH$_3$ group, and a band at 8250 cm$^{-1}$, which does not show a shift, is ascribed to the CH$_2$ group.

## TWO-DIMENSIONAL CORRELATION SPECTROSCOPY

As mentioned above, for NIR spectroscopy there are two kinds of 2D correlation spectroscopy; one by Noda (12–15) and the other by Barton et al. (16). In this section only 2D correlation spectroscopy by Noda is explained.

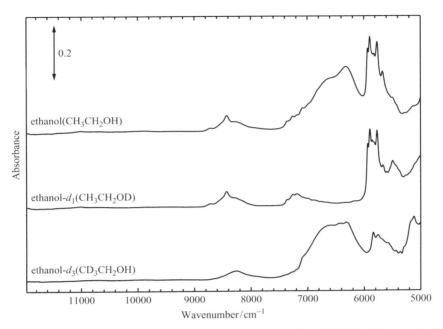

**Figure 3.11.** NIR spectra of $CH_3CH_2OH$, $CH_3CH_2OD$, and $CD_3CH_2OH$ in the pure liquid state.

2D correlation spectroscopy was first realized in NMR spectroscopy about twenty years ago. The 2D correlation spectroscopy in vibrational spectroscopy was proposed by Noda (12, 15) in 1986 as 2D mid-IR correlation spectroscopy. In this 2D mid-IR, a system is excited by an external perturbation that induces dynamic fluctuations of mid-IR signals, and a simple cross-correlation analysis is applied to sinusoidally varying dynamic mid-IR signals to generate a set of 2D mid-IR correlation spectra (12, 15). Because dynamic 2D mid-IR spectra are powerful in emphasizing spectral features not readily observable in conventional one-dimensional spectra, 2D mid-IR correlation spectroscopy has been successful in the investigations of systems stimulated by a small-amplitude mechanical or electrical perturbation (14, 15). However, in this original approach the time-dependent behavior (i.e., waveform) of dynamic spectral intensity variations must be a simple sinusoid to effectively employ the original data analysis scheme (12). Therefore, in 1993 Noda (13) presented a more generally applicable, yet reasonably simple, mathematical formalism to construct 2D correlation spectra from any transient or time-resolved spectra having an arbitrary waveform. He named this new 2D correlation spectroscopy generalized 2D correlation spectroscopy (13). The newly proposed formalism can be applicable to various types of spectroscopy, including NIR and Raman spectroscopy (14, 15).

   In the generalized 2D correlation method, a variety of external perturbations can be applied to stimulate a system of interest (13), for example, electrical, thermal, magnetic, chemical, acoustic, or mechanical excitations. When an external perturbation is applied to a system, various chemical components of the system are selectively excited. The excitation and subsequent relaxation process caused by a perturbation can be monitored with many different types of electromagnetic probes. To construct generalized 2D correlation spectra from perturbation-induced dynamic fluctuations of spectroscopic signals, a set of dynamic spectra must be calculated first (see, for example, Figure 3.6B). The dynamic spectra are then transformed into 2D spectra by a correlation method.

The advantages of generalized 2D correlation spectroscopy lie in the following points (13, 15).

1. Enhancement of spectral resolution by spreading peaks over the second dimension. In this respect, asynchronous spectra are particularly useful.
2. Simplification of complex spectra consisting of many overlapped peaks
3. Band assignments through correlation analysis of bands
4. Investigations of various inter- and intramolecular interactions through selective correlation of peaks
5. Probing the specific order of the spectral intensity changes
6. So-called heterospectral correlation, i.e., the investigation of correlation among bands in two different types of spectroscopy, for example, the correlation between NIR bands and mid-IR bands

### Principles of Generalized Two-Dimensional Correlation Spectroscopy

Detailed mathematical treatment for generalized 2D correlation spectroscopy is provided in references 13 and 15. In this section, it is described briefly. Let us consider a perturbation-induced change of spectral intensity $y(v, t)$ observed during an interval of some external variable $t$ between $T_{min}$ and $T_{max}$. Although the external variable $t$ is often treated as the chronological time, it can also be any other reasonable measure of physical quantity, such as temperature, pressure, pH, polarization angle, etc. The variable $v$ can be any appropriate physical variable, for example, NIR wavenumber (wavelength), Raman shift, or even X-ray diffraction angle (13, 15). To construct generalized 2D correlation spectra, it is necessary to calculate the dynamic spectrum first. It is defined as

$$\tilde{y}(v, t) = \begin{cases} y(v, t) - \bar{y}(v) & \text{for } T_{min} \leq t \leq T_{max} \\ 0 & otherwise \end{cases} \tag{3.7}$$

where $\bar{y}(v)$ is the reference spectrum. Although the selection of reference spectrum is somewhat arbitrary, $\bar{y}(v)$ is usually set to be the stationary or averaged spectrum defined as

$$\bar{y}(v) = \frac{1}{T_{max} - T_{min}} \int_{T_{min}}^{T_{max}} y(v, t)dt \tag{3.8}$$

As the next step, one must Fourier transform the dynamic spectra measured in the time-domain into the frequency domain (18). The forward Fourier transform $\tilde{Y}_1(\omega)$ of the dynamic spectral intensity fluctuations $\tilde{y}(\nu_1, t)$ observed at $\nu_1$ is given by

$$\tilde{Y}_1(\omega) = \int_{-\infty}^{\infty} \tilde{y}(\nu_1, t)e^{-i\omega t}dt = \tilde{Y}_1^{\text{Im}}(\omega) + i\tilde{Y}_1^{\text{Im}}(\omega) \qquad (3.9)$$

where $\tilde{Y}_1^{\text{Re}}(\omega)$ and $\tilde{Y}_1^{\text{Im}}(\omega)$ are the real and imaginary components of the complex Fourier transform of $\tilde{y}(\nu_1, t)$. The Fourier frequency $\omega$ represents the individual frequency component of the time-dependent variation of $\tilde{y}(\nu_1, t)$. Similarly, $\tilde{Y}_2^*(\omega)$, the conjugate of the Fourier transform of dynamic spectral intensity $\tilde{y}(\nu_2, t)$ at spectral variable $\nu_2$, is given by

$$\tilde{Y}_2^*(\omega) = \int_{-\infty}^{\infty} \tilde{y}(\nu_2, t)e^{+i\omega t}dt = \tilde{Y}_2^{\text{Re}}(\omega) + i\tilde{Y}_2^{\text{Im}}(\omega) \qquad (3.10)$$

Now, one can define the complex 2D correlation intensity between $\tilde{y}(\nu_1, t)$ and $\tilde{y}(\nu_2, t)$. 2D correlation is nothing but a quantitative comparison of spectral intensity variations observed at two different spectral variables over some finite observation internal between $T_{\text{min}}$ and $T_{\text{max}}(T = T_{\text{max}} - T_{\text{min}})$. Equation 3.11 yields the complex 2D correlation intensity.

$$\Phi(\nu_1, \nu_2) + i\Psi(\nu_1, \nu_2) = \frac{1}{\pi(T_{\text{max}} - T_{\text{min}})} \int_0^{\infty} \tilde{Y}_1(\omega)\tilde{Y}_2^*(\omega)d\omega \qquad (3.11)$$

The real and imaginary components of the complex 2D correlation intensities, $\Phi(\nu_1, \nu_2)$ and $\Psi(\nu_1, \nu_2)$, are referred to, respectively, as the generalized synchronous and asynchronous correlation spectra of the dynamic spectral intensity variations. The synchronous spectrum represents the simultaneous or coincidental changes spectral intensities at $\nu_1$ and $\nu_2$, whereas the asynchronous spectrum represents sequential or unsychronized variations.

## Synchronous and Asynchronous Two-Dimensional Correlation Spectra

Figure 3.12A and B, depicts schematic contour maps of synchronous and asynchronous 2D correlation spectra, respectively (the terms synchronous and asynchronous are always used even when the spectral variation is measured as a function of not time but another physical variable) (13). A one-dimensional refernce spectrum is provided at the top and left side of the contour map to show the basic feature of spectra of the system during experiment. A synchronous spectrum is symmetric with respect to a diagonal line corresponding to spectral coordinates $\nu_1 = \nu_2$. Peaks

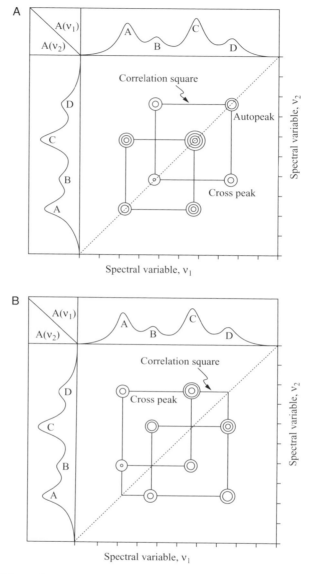

**Figure 3.12.** Schematic contour maps of synchronous (A) and asynchronous (B) spectra. A one-dimensional reference is also provided at the top and side of the 2D map. [Reproduced from Ref. 13 with permission. Copyright (1993) Society of Applied Spectroscopy].

observed in a synchronous spectrum are classified into those at diagonal positions and those at off-diagonal positions. The intensity of peaks located at diagonal positions corresponds to the autocorrelation function of spectral intensity variations observed during a period $T$. Those peaks are therefore referred to as autopeaks. The autopeaks represent the overall extent of dynamic variations of spectral intensity. Thus regions of a dynamic spectrum that changes intensity to a greater extent show stronger autopeak, whereas those remaining constant give little or no autopeaks.

Synchronous cross peaks appearing at off-diagonal positions represent the simultaneous changes of spectral signals at two different wavenumbers. In other words, a synchronous change suggests the possible existence of a coupled or related origin of the spectral intensity variation. Positive cross peaks indicate that the spectral intensities at corresponding wavenumbers are either increasing or decreasing together as functions of physical variable during the observation period, whereas negative cross peaks (usually marked by shading) show that one of the intensities is increasing while the other is decreasing.

An asynchronous 2D correlation spectrum, which consists exclusively of off-diagonal cross peaks and is antisymmetric with respect to the diagonal line, provides information complementary to the synchronous spectrum. It represents sequential, or unsynchronized, changes of spectral intensities measured at $v_1$ and $v_2$. An asynchronous cross peak develops only if the intensities of two spectral peaks vary out of phase (i. e., delayed or accelerated) with each other. This feature of asynchronous spectrum is powerful in differentiating overlapped NIR bands arising from different spectral origins or moieties. The sign of an asynchronous cross peak becomes positive if the intensity change at $v_1$ occurs *predominantly* before $v_2$ in the sequential order of $t$. It becomes negative, on the other band, if the change occurs after $v_2$. This rule, however, is reversed if $\Phi(v_1, v_2) < 0$.

## Two-Dimensional NIR Correlation Spectroscopy

The use of generalized 2D correlation spectroscopy was extended into the NIR region in 1995 by Noda et al. (33). They investigated temperature-dependent NIR spectral variations of oleyl alcohol in the pure liquid state by generalized 2D correlation analysis. 2D correlation spectroscopy is powerful in the analysis of NIR spectra because it has the advantages of resolution enhancement for highly overlapped bands and selective interband correlations to assist band assignments. Generalized 2D NIR correlation spectroscopy provides the intriguing possibility of correlating various overtone and fundamental bands to establish unambiguous assignments in the NIR region (15). 2D correlation spectroscopy solves to some extent the problem of multicollinearity that NIR spectra often encounter. It is also possible to investigate correlations between bands in an NIR spectrum and those in other spectra such as mid-IR and Raman spectra (heterospectral correlation).

Generalized 2D NIR correlation spectroscopy has been applied to study, for example, temperature-dependent spectral variations of various compounds such as $N$-methylacetamide (NMA) (34) and nylon 12 (29), concentration-dependent spectral changes in milk (18) and protein solutions at various temperatures (35, 36), composition-dependent spectral changes in polymer blends (37), and depth-dependent spectral variations of a polymer film (38). Examples of heterospectral correlation are 2D NIR-mid IR heterospectral correlation analysis of nylon 11 (39) and 2D NIR-Raman correlation analysis of polymer blends (40).

## REFERENCES

1. P. Williams, K. Norris, eds. *Near Infrared Technology in the Agriculture and Food Industries*, 2nd Ed. American Association of Cereal Chemists, St. Paul, 1990.

2. D. A. Burns, E. W. Ciurczak, eds. *Handbook of Near-Infrared Analysis*. Marcel Dekker, New York, 1992.

3. B. G. Osborne, T. Fearn, P. H. Hindle. *Practical Near Infrared Spectroscopy with Applications in Food and Beverage Analysis*. Longman Scientific and Technical, Harlow, 1993.

4. H. Siesler, Y. Ozaki, S. Kawata, M. Heise. *Near infrared spectroscopy—Principles, Instruments, and Applications*. Wiley-VCH, Weinheim, 2002.

5. Y. Ozaki, S. Šašic, J.H. Jiang. How can we unravel complicated near infrared spectra?—Recent progress in spectral analysis methods for resolution enhancement and band assignments in the near infrared region. *J Near Infrared Spectrosc* **9**: 63, 2001.

6. Y. Ozaki, Y. Katsumoto, J. H. Jiang, Y. Liang. Spectral analysis in the NIR region, *in Useful and Advanced Information in the Field of near Infrared Spectroscopy*, Ed. S. Tsuchikawa Research Signpost, Trivandrum, 2003.

7. H. Martens, T. Næs. *Multivariate Calibration*. John Wiley & Sons, Chichester, 1991.

8. H. Mark. *Principle and Practice of Spectroscopic Calibration*, John Wiley & Sons, New York, 1991.

9. R. Kramer. *Chemometric Techniques for Quantitative Analysis*, Marcel Dekker, New York, 1998.

10. B. G. M. Vandenginste, D. L. Massart, L. M. C. Buydens, S. Jong, P. J. Lew, J. Smeyers-Verbeck. *Handbook of Chemometrics and Qualimetrics*. Elsevier, Amsterdam, 1998.

11. T. Næs, T. Isaksson, T. Fearn, T. Davies. *A User-Friendly Guide to Multivariate Calibration and Classification*. NIR Publications, Chichester, 2002.

12. I. Noda. Two-dimensional infrared spectroscopy, *Bull Am Phys Soc* **31**: 520, 1986.

13. I. Noda. A Generalized two-dimensional correlation method applicable to infrared, Raman, and other types of spectroscopy. *Appl Spectrosc* **47**: 1329, 1993.

14. Y. Ozaki, I. Noda. *Two-Dimensional Correlation Spectroscopy*, AIP Conference Series, American Institute of Physics, New York, **503**: 2000.

15. I. Noda, Y. Ozaki. *Two-Dimensional Correlation Spectroscopy*. John Wiley & Sons, Chichester, 2004.

16. F. E. Barton II, D. S. Himmelsbach, J. H. Duckworth, M. J. Smith. Two-dimensional vibration spectroscopy: correlation of mid- and near-infrared regions. *Appl Spectrosc* **46**: 420, 1992.

17. K. Murayama, B. Czarnik-Matusewicz, Y. Wu, R. Tsenkova, Y. Ozaki. Comparison between conventional spectral analysis methods, chemometrics, and two-dimensional correlation spectroscopy in the analysis of near-infrared spectra of protein. *Appl Spectrosc* **54**: 978, 2000.

18. B. Czarnik-Matusewicz, K. Murayama, R. Tsenkova, Y. Ozaki. Analysis of near-infrared spectra of complicated biological fluids by two-dimensional correlation spectroscopy: protein and fat concentration-dependent spectral changes of milk. *Appl Spectrosc* **53**: 1582, 1999.

19. M. Shimoyama, T. Ninomiya, K. Sano, Y. Ozaki, H. Higashiyama, M. Watari, M. Tomo. Near infrared spectroscopy and chemometrics analysis of linear low-density polyethylene. *J NIR Spectrosc* **6**: 317, 1998.

20. K. Murayama, K. Yamada, R. Tsenkova, Y. Wang, Y. Ozaki. Near-infrared spectra of serum albumin and $\gamma$-globulin and determination of their concentrations in phosphate buffer solutions by partial least squares regression. *Vib Spectrosc* **18**: 33, 1998.

21. P. Geladi, D. MacDougall, H. Martens. Linearization and scatter-correction for near-infrared reflectance spectra of meat. *Appl Spectrosc* **39**: 491, 1985.

22. F. T. Chau, T. M. Shih, J. B. Gao, C. K. Chan. Application of the fast wavelet transform method to compress ultraviolet-visible spectra. *Appl Spectrosc* **50**: 339, 1996.

23. C. L. Stork, D. J. Veltkamp, B. R. Kowalski. Detecting and identifying spectral anomalies using wavelet processing. *Appl Spectrosc* **52**: 1348, 1998.

24. P. R. Griffiths, J. A. Pieerce, G. Hongjin. In *Computer-Enhanced Analytical Spectroscopy*, H. L. C. Meuzelaar, T. L. Isenhour, eds. Plenum Press, New York, 29, 1987.

25. P. R. Griffiths. In *Laboratory Methods in Vibrational Spectroscopy* 3rd ed. H. A. Willis, J. H. Van der Maas, R. G. J. Miller, eds. John Wiley & Sons, Chichester, 121, 1987.

26. S. Wold, H. Antii, F. Lindgren, J. Ohman. *Chemom Intell Lab Syst* **44**: 175, 1998.

27. T. Fearn. On orthogonal signal correction. *Chemom Intell Lab Syst* **50**: 47, 2000.

28. M. A. Czarnecki, Y. Liu, Y. Ozaki, M. Suzuki, M. Iwahashi. Potential of fourier transform near-infrared spectroscopy in studies of the dissociation of Fatty acids in the liquid phase. *Appl Spectrosc* **47**: 2162, 1993.

29. Y. Ozaki, Y. Liu, I. Noda. Two-dimensional near-infrared correlation spectroscopy study of premelting behavior of nylon 12. *Macromolecules* **30**: 2391, 1997.

30. L. G. Weyer, S. C. Lo. Spectra-structure correlation in the near-infrared. In: *Handbook of Vibrational Spectroscopy,* J. M. Chalmers, P. R. Griffiths, eds. John Wiley & Sons, Chichester, **3**: 1817, 2002.

31. Y. Katsumoto, D. Adachi, H. Sato, Y. Ozaki. Usefulness of a curve fitting method in the analysis of overlapping overtones and combinations of CH stretching modes, *J. NIR Spectrosc.* **10**: 85, 2002.

32. P. Wu, H. W. Siesler. The assignment of overtone and combination bands in the near infrared spectrum of polyamide 11. *J Near Infrared Spectrosc* **7**: 65, 1999.

33. I. Noda, Y. Liu, Y. Ozaki, M. A. Czarnecki. Two-dimensional Fourier transform near-infrared correlation spectroscopy studies of temperature-dependent spectral variations of oleyl alcohol. *J Phys Chem* **95**: 3068, 1999.

34. Y. Liu, Y. Ozaki, I. Noda. Two-dimensional Fourier-transform near-infrared correlation spectroscopy study of dissociation of hydrogen-bonded *N*-methylacetamide in the pure liquid state. *J Phys Chem* **100**: 7326, 1996.

35. Y. Wang, K. Murayama, Y. Myojo, R. Tsenkova, N. Hayashi, Y. Ozaki. Two-dimensional Fourier transform near-infrared spectroscopy study of heat denaturation of ovalbumin in aqueous solutions. *J Phys Chem B* **102**: 6655, 1998.

36. Y. Wu, B. Czarnik-Matusewicz, K. Murayama, Y. Ozaki. Two-dimensional near-infrared spectroscopy study of human serum albumin in aqueous solutions: using overtones and combination modes to monitor temperature-dependent changes in the secondary structure. *J Phys Chem B* **104**: 5840, 2000.

37. Y. Ren, T. Murakami, T. Nishioka, K. Nakashima, I. Noda, Y. Ozaki. Two-dimensional near-infrared correlation spectroscopy studies of compatible polymer blends: composition-dependent spectral variations of blends of atactic polystyrene and poly- (2, 6-dimethyl-1, 4-phenylene ether). *J Phys Chem B* **104**: 679, 2000.

38. I. Noda, G. M. Story, A. E. Dowrey, R. C. Reeder, C. Marcott. Applications of two-dimensional correlation spectroscopy in depth profiling photoacoustic spectroscopy, near-infrared dynamic rheooptics, and spectroscopic imaging microscopy. *Makromol Chem Macromol Symp* **119**: 1, 1997.

39. M. A. Czarnecki, P. Wu, H. W. Siesler. 2D FT-NIR and FT-IR correlation analysis of temperature induced changes of nylon 12. *Chem Phys Lett* **283**: 326, 1998.

40. A. Matsushita, Y. Ren, K. Matsukawa, H. Inoue, Y. Minami, I. Noda, Y. Ozaki. Two-dimensional Fourier-transform Raman and near-infrared correlation spectroscopy studies of poly (methyl methacrylate) blends. 1. Immiscible blends of poly (methyl methacrylate) and atactic polystyrene. *Vib Spectrosc* **24**: 171, 2000.

# CHAPTER 4

# INSTRUMENTATION

# Instruments

W. F. MCCLURE and SATORU TSUCHIKAWA

## INTRODUCTION

Near-infrared (NIR) spectrometry is an instrumental method for acquiring spectra of foods and other materials. The acquired spectra are used for determining both qualitative and quantitative characteristics. NIR spectrometry is popular because of its four main advantages: speed (a spectrum can be acquired in as little as a tenth of a second), little or no sample preparation (if any preparation is required, it is usually quite simple), multiple analyses from a single scan (it is not necessary to scan the sample for each constituent), and a nondestructive measurement process (allowing for the return of the analyzed subsample to the original lot) (28).

Currently, NIR spectrometry has three shortcomings. First, it is a technology that must be trained. That is, the instrumentation must be calibrated by scanning a set of sample with known qualitative/quantitative parameters. Second, the known levels often involve expensive and complicated reference methods closely tied to wet chemistry that demand the input of highly skilled personnel. Third, modern-day calibration methods rely on rather sophisticated chemometric techniques, thus calling into play the assistance of personnel who are highly trained in chemometrics (or statistics in chemistry).

In recent years, costs of wet chemistry have risen at a staggering rate. The need for trained personnel to do the wet chemistry further exacerbates the problems. At the same time, an increased need for more chemical information overloads many laboratories to the point where turnaround of analyses is no longer timely. Little wonder that researchers have worked feverishly to develop short-cut wet chemistry methods, typified by autoanalyzers and autosamplers. Even then, the problems of multiplicative wet chemistry analyses leave modern laboratories in a state of frustration.

*Near-Infrared Spectroscopy in Food Science and Technology*, Edited by Yukihiro Ozaki, W. Fred McClure, and Alfred A. Christy.
Copyright © 2007 John Wiley & Sons, Inc.

NIR technology has come to the forefront for food analysis with tremendous potential for minimizing the impact of the above problems. Despite the fact that NIR spectrometry is a *trained tool* requiring calibration, one dominant fact remains. It is an instrument-based technology that offers tremendous savings, both for production and for processing within the food industry.

The discovery of the NIR spectrum was reported in 1800 by William Herschel (17, 19), yet the NIR spectrum did not appear as a potential analytical tool until the middle of the twentieth century (7, 52, 65). The contributions of the pioneers cannot be ignored. People like Coblentz (5), Ellis (9), Wulf and Liddel (68), and Collins (6) were among the first to study NIR spectroscopy. After 1950 and well into the 1960s, NIR spectrometry became even more popular with researchers like Kaye (25), Willis (66), Goddu (13), Wetzel (59), Wheeler (58), and Norris (44) who made contributions to both NIR spectroscopy and spectrometry. It was Norris who first demonstrated that NIR spectrometry could be calibrated with multiple linear regression.

This chapter discusses spectrometry applied to the food industry. There is a brief treatment of generalized spectrometry, with definitions and ground rules for the rest of the chapter. Primary attention is focused on wavelength-isolating technology—filter, prism, grating, acousto-optical tunable filters (AOTF), emitting diode array (EDA), diode-array detection, lasers, FT-NIR, and Hadamard. A commercial example of each technology is presented, with a little attention to food applications. No chapter on instrumentation would be complete without consideration being given to trends and futuristic concepts. The authors acknowledge published papers that discuss spectrometry, to which the reader is referred (3, 12, 23, 29, 36, 41, 45, 57, 61).

## SPECTROMETRY

A discussion of NIR spectrometry requires both definitions and ground rules. Consider the following definitions. NIR spectroscopy is the art and science of interactions of NIR energy (760 nm) with matter. A spectroscopist is a person who studies spectroscopy. NIR spectrometry, on the other hand, is the art and science of building NIR instrumentation that a spectroscopist would use to pursue spectroscopy. NIR spectroscopy, requiring human interaction and interpretation, results in a report or paper. NIR spectrometry, also requiring human interaction and interpretation, culminates in an instrument—a spectrometer.

The question may be raised, "Which came first, spectroscopy or spectrometry?" It may be argued, with considerable confidence, that spectroscopy was performed with the human eye long before electricity and electronic spectrometers were invented. For example, one selects clothing based on spectroscopy performed with the eye. Individual dyes in cloth reflect specific colors or band of wavelengths. Historically, it appears that the limitations of "vision-spectrometry" stimulated researchers to develop instruments that would improve spectroscopy. The ground rule here is that spectrometers were developed to improve spectroscopy. Spectroscopy was improved by the contributions of researcher and spectrometers.

**Figure 4.1.1.** Generalized NIR spectrometer having three basic parts: lamp, wavelength isolator, and detector

## GENERALIZED SPECTROMETRY

A generalized spectrometer (Fig. 4.1.1) has three parts: 1) an NIR source, 2) a wavelength isolator, and 3) a detector. More recently, computers have been added to the combination. Although essential for efficiency, computers are not an integral part of the spectrometer. The quartz-halogen lamp is by far the most popular source of NIR energy.[1] Quartz-halogen lamps are quite popular now because of their widespread use for household lighting. Light-emitting diodes are the second most popular source of NIR energy. Wavelength isolators range in technology: photodiode arrays (PDA), diode array detectors (DAD), laser diodes (LD), fixed filters (FF), wedge interference filters (WIF), tilting filters (TF), acousto-optical tunable filters (AOTF), liquid crystal tunable filters (LCTF), prisms, gratings, Fourier transform NIR (FT-NIR), and Hadamard. Spectrometer designers have a number of detectors from which to choose, the most popular of which are lead sulfide (PbS)-, indium gallium arsenide (InGAS)-, and silicon-based devices (voltaic cell, diodes, and transistors for the Herschel region[2]). Silicon-based detectors are restricted by their spectral response to ultraviolet, visible, and short-wavelength NIR (200–1000 nm).

## NIR SPECTROMETERS

### Classified According to Wavelength Isolation Technology

Wavelength isolators fall into one of two groups: 1) Discrete-value and 2) full-spectrum devices. Full-spectrum spectrometers (sometimes called "scanning instruments") produce spectra with equally spaced data (*viz.* 2 nm apart) across the full range from 780 to 2500 nm. Discrete-value spectrometers may be further categorized by the technology used to produce narrow wavelength bands. Table 4.1.1 provides a listing of the major technologies for isolating wavelength bands for conducting NIR spectroscopy. In the following pages of this chapter filter and diode spectrometers (discrete-value instruments) are listed first, followed by full-spectrum spectrometers.

---

[1] The word "energy" is used here to avoid confusion with "light." NIR energy is not visible to the human eye and therefore is not light.
[2] The Herschel region of the NIR spectrum has been defined as 700–1000 nm, the region covered by silicon-based detectors.

**TABLE 4.1.1. NIR Spectrometers Classified According to the Wavelength-Isolating Technology**

| Technology | Comments |
|---|---|
| *Diodes* | |
| Emitting diode arrays (EDA) Fig. 4.1.2 | There is a limited number of light-emitting diodes available in the marketplace. These emitters make possible very compact spectrometers for specialized applications requiring only a few wavelengths. Therefore, NIREDS are nonscanning instruments. |
| Diode-array detectors (DAD) Fig. 4.1.4 | DAD, implemented in conjunction with a fixed grating, can produce a compact spectrometer with resolution limited only by the number of receptors in the array. Diode array spectrometers are full-spectrum instruments. |
| Laser diodes Fig. 4.1.5 | Production of many wavelengths from laser diodes is still lagging other technologies. Laser-based instruments are wavelength limited and therefore are not full-spectrum instruments. |
| *Filters* | Filters appear to be the first method for isolating wavelengths for performing spectrometry, albeit the first filters were liquid filters. |
| Fixed Fig. 4.1.6 | Fixing the plane of a narrow-band interference filter (NBIF) normal to an NIR beam provides a single wavelength band. The bandpass of these filters may be as narrow as 1 nm. Fixed-filter spectrometers are not capable of producing a full spectrum. |
| Wedge Fig. 4.1.7 Fig. 4.1.8 | A wedge-interference filter (WIF) consists of two quartz plates spaced by a dielectric wedge. Moving the wedge in front of a slit produces incremental wavelength bands. Paul Wilks has been its strongest supporter. Spectrometers that incorporate a WIF are full-spectrum instruments. |
| Tilting Fig. 4.1.9 | Tilting a NBIF in a beam of parallel NIR energy will produce incremental wavelength bands. Unlike the WIF, tilting results in an increased band width and diminished transmission. Tilting-filter spectrometers, though limited to a narrow range of wavelengths, are full-spectrum spectrometers. |
| AOTF Fig. 4.1.10 | An acousto-optical tunable filter is a specialized optic whose bandpass is determined by the radio frequency applied across the optic. AOTF spectrometers are full-spectrum instruments. |
| LCTF Fig. 4.1.11 | Liquid crystal tunable filter (LCTF) spectrometers may be designed to operate in the visible, NIR, MIR and FIR. However, the switching speed is much slower than the AOTF (maximum of 1 ms). |
| *Prism* Fig. 4.1.12A | Prisms, used in the early days of spectrometry, produce nonlinear dispersion of NIR spectrum, making it difficult to coordinate prism position with wavelength. Prism based spectrometers are full-spectrum instruments. |
| *Grating* Fig. 4.1.12B Fig. 4.1.13 | Gratings produce a near-linear dispersion with wavelength. The remaining nonlinearity can be removed with software. Spectrometers incorporating gratings are said to be full-spectrum instruments. |

**TABLE 4.1.1.** (*Continued*)

| Technology | Comments |
|---|---|
| *FT-NIR*<br>Fig. 4.1.14 | Fourier transform near-infrared (FT-NIR) spectrometers produce reflection spectra by moving mirrors. Once plagued by noise, modern FT-NIR spectrometers boast noise levels equivalent to grating-based instruments. FT-NIR spectrometers are full-spectrum instruments. |
| *Hadamard*<br>Fig. 4.1.15 | Originally implemented with a complex arrangement of shutters, Hadamard technology has never competed with dispersion-type instruments. This technology is capable of producing a full spectrum. Professor Bill Fateley has been one of its strongest supporters. |

AOTF and LCTF function both in discrete and full-spectrum spectrometers in that both can move randomly to individual wavelengths within the operating range. Finally, prism, grating, FT-NIR, and Hadamard are discussed as strictly full-spectrum analyzers.

### Diodes

***Emitting Diode Arrays (EDA).*** Several light-emitting diodes (LEDs) are commercially available, making possible the construction of multichannel (or multiband) NIR spectrometers that can be used for numerous applications in the food industry. A schematic of a EDA array spectrometer is shown in Figure 4.1.2A. Several emitting diodes are arranged systematically in a matrix array. As needed, the bandpass of individual diodes may be narrowed by adding narrow-band interference filters. A diffuser is interposed between the sample and the diode matrix to provide "uniform illumination" from all diodes simultaneously.

The Zeltex Model ZX101 (see Fig. 4.1.2B) is a typical example of EDA technology (51). The instrument, having approximately 11 emitting diodes, provides results within 10 s after the reading button is pressed. The ZX101 includes a RS-232 port for downloading spectral values. Values are easily read on the display before they are stored.

***Other Emitting Diode Developments.*** McClure and his associates produced a hand-held NIR meter that implemented three diode emitters for determining chlorophyll and moisture (35) (Fig. 4.1.3). Figure 4.1.3A is a picture of the so-called "TWmeter." Its size is 10 cm W by 20 cm L by 4 cm thick. Powered by 4 AA batteries, the TWmeter takes measurements at 700, 880, and 940 nm by sequentially exciting (Fig. 4.1.3B) the three emitting diodes. The diodes, in this case, are unfiltered. The three narrow-band diode emitters (see the circuit in Fig. 4.1.3C) have broadband LEDs mounted in parallel to indicate which diode emitter is excited. The big cost-reducing feature of this hand-held unit is the detector, a Texas Instrument voltage-to-frequency (V/F) silicon detector. The V/F detector eliminates the need for an external and costly A/D converter by allowing the microprocessor to count the

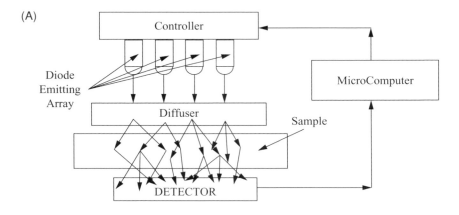

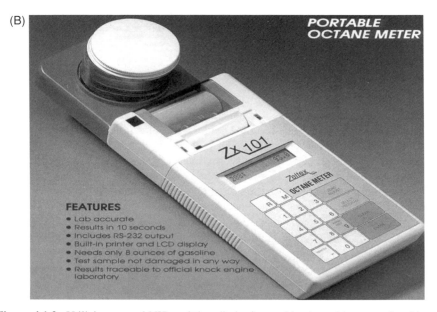

**Figure 4.1.2.** Utilizing several NIR emitting diodes in combination with narrow-band interference filters and a diffuser (A). The Zeltex ZX101 (B), originally designed for determining the octane rating of petroleum products, has found many applications in the food industry.

output of the detector for a fixed period of time. Software adjustments make it easy to account for sensitivity variations from one diode emitter to the other. The parts cost for this meter was less than $250.00 when it was built in 2002.

Massie demonstrated that NIR measurements may be made with thermally driven gallium arsenide emitters in which the temperature is changed by magnitude of the diode current. He used a single-diode emitter with a nominal wavelength of 900 nm.

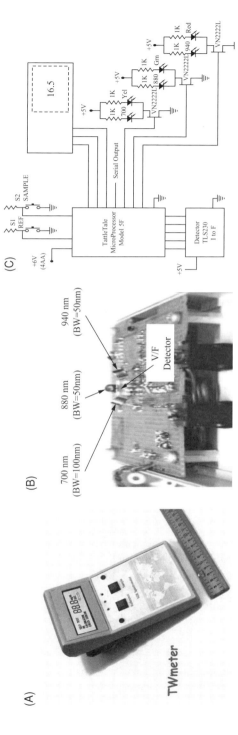

**Figure 4.1.3.** The TWmeter [McClure, 2002 #4049]: a hand-held NIR spectrometer (A) that consisting of three unfiltered NIR-emitting diodes, a silicon-V/F detector and a minimum-featured microprocessor (B) for determining chlorophyll and moisture. (C): complete circuit that controls to meter functions. Cost reductions were achieved by letting the microprocessor count the output of the V/F converter, thus avoiding the need for an external analog-to-digital converter.

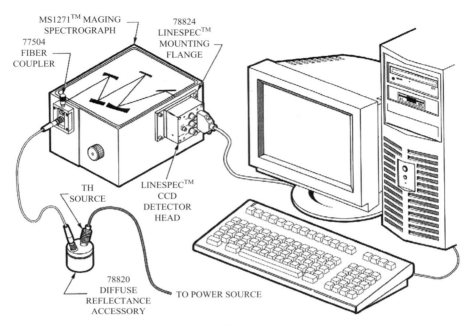

**Figure 4.1.4.** Basic setup of Oriel's MS127i® diode-array spectrometer utilizing a single fiber-optics cable to collect spectra from solid samples. Oriel's Linespec diode array detectors cover the range 200–1000 nm with 1048 pixels. The system includes 12-bit A/D converter. There are no moving parts in this system. (Used by permission of Oriel Instruments, 150 Long Branch Boulevard, Stratford, CT 06615-0872)

By changing the temperature from 82.8 to 103.6°C, the peak wavelength changed from 916.2 to 922 nm. The correlation coefficient between fat content determined by the infrared emitter instrument and by chemical analysis was 0.82, with a standard error of 1.98% fat. The standard deviation of replications for the chemical analysis was 0.35% fat (30).

**Photodiode Detector Arrays.** Photodiode detector arrays (PDA) offer a *no moving parts* technology that is attractive for certain applications. Figure 4.1.4 shows the basic setup of a diode array spectrometer (Oriel MS127i) that is configured with a fiber-optic bundle for conducting diffusely reflected light from the sample to the fixed grating monochromator. This instrument uses a linear diode array (Oriel Linespec® Series) for measuring the reflected energy. These arrays cover the range 200–1000 nm, although the signal becomes more noisy below 450 nm and above 900 nm.

**Other Developments: PDA.** Greensill and Walsh have developed procedures for standardizing the miniature Zeiss MMS1 PDA spectrometers involving six steps (14).

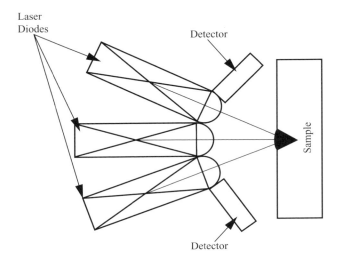

**Figure 4.1.5.** Schematic of a three-wavelength laser-diode array spectrometer for making reflection measurements.

Clancy demonstrates that a simple linear slope and bias correction can be effective in normalizing instruments and thus allowing a single master calibration to be used on all instruments (4). Morimoto developed two field-operable PDA instruments for determining fruit quality based on PDA technology (42).

***Laser Diodes.*** Laser diode spectrometers are still pretty much in the research stage. Although the cost of laser diodes has fallen over the last 20 years, this expense is still nontrivial. Where applicable, laser diode spectrometers offer definite advantages: 1) The bandwidths of lasers are very narrow, and 2) the output intensity can be very high compared to other spectrometers. A schematic of a laser diode spectrometer is given in Figure 4.1.5.

## Filters

***Fixed Filters.*** A schematic of a fixed-filter (FF) spectrometer (FFS) is shown in Figure 4.1.6A. The first commercial NIR analyzers were of this type and were used at grain elevators, where wet chemistry was not timely enough to be of use. Today several companies produce multiple FF instruments. The QUIK/20, incorporating 20 narrow-band interference filters (shown in Fig. 4.1.6B), is typical of the FF technology. It has a filter wheel with 20 filters and implements a noise reduction routine based on the fast Fourier transform (8, 20, 38).

FFS capable of producing a limited number of bands cannot do all things for all people. Yet, because the NIR absorptions of food and food products are broad and

(A)

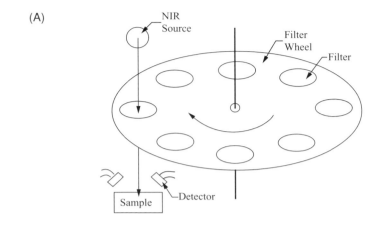

(B)

**Figure 4.1.6.** Fixed-filter spectrometer: (A) schematic and (B) the Leco QUIK/20(Leco Corporation, 3000 Lakeview Avenue, St. Joseph, MI 49085), typical of fixed-filter NIR spectrometers. It incorporates 19 narrowband interference filters (1445, 1680, 1722, 1734, 1759, 1778, 1818, 1940, 1982, 2100, 2139, 2180, 2190, 2208, 2230, 2270, 2310, 2336, 2348 nms).

overlapping, it is surprising how many things the FFS can do. Typical FFS include filters that are useful for calibrations in high-volume food applications, such as moisture, protein, and fat. As far back as 1976, Watson and his colleagues (56) made an evaluation of two FFS (the Dickey-john GAC and the Neotec GQA) for measuring protein. Although the nominal wavelength of the six filters in each instrument

**TABLE 4.1.2. Fix Filters Incorporated into Two Commercial Fixed-Filter Spectrometers**

| GAC (Dickey-john) | GAQ (Neotec) |
|---|---|
| 1680 | 1867 |
| 1940 | 1920 |
| 2100 | 2118 |
| 2180 | 2167 |
| 2230 | 2250 |
| 2310 | 2297 |

Performances of the instruments were similar, although the filter wavelengths differed.

differed (see Table 4.1.2), performances of the two instruments were similar. Now that Morimoto et al. have demonstrated that derivative calibrations can be developed with filters (43), there appears to be renewed interest in the FFS.

*Other Fixed-Filter Developments.* Worthington et al. (67) developed a four-filter spectrometer for determining the ripeness of intact tomatoes. Massie (31) and his colleagues designed and tested a high-intensity spectrometer, two filter wheels each with 11 fixed filters (one on either side of the tomato). They reported linear absorbance measurements to nine optical density units.

*Wedge-Interference Filters.* Figure 4.1.7 is a schematic of a wedge-interference filter (WIF) spectrometer. Wedge filters are constructed similarly to fixed filters, with

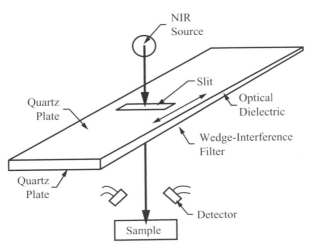

**Figure 4.1.7.** Schematic of a wedge-interference filter spectrometer. The optical dielectric is held statically by the two quartz plates.

the single exception that the optical dielectric between the plates is wedge shaped. That is, the dielectric at one end is thicker than at the other end, producing longer to shorter wavelengths, respectively. A slit interposed between the source and the sample allows the passing of a narrow band of wavelengths, with the band changing as the wedge is moved from one end to the other (62, 64). WIF are also available in circular form in which the thickness of the optical dielectric varies with filter rotation.

Paul Wilks has always been a proponent of the WIF. His most recent addition to his product line is the "variable filter array (VFA) spectrometer." It is a new-concept instrument that enables the user to obtain NIR spectra on a variety of materials wherever they occur, in the production plant or in the field. It consists of an ATR sample plate with an elongated pulseable source mounted close to one end and a linear variable filter attached to a detector array mounted close to the other (Fig. 4.1.8). The net result is a very compact spectrometer with no moving parts and no optical path exposed to air and capable of producing NIR spectra of powders, films, liquids, slurries, semisolids, and solid surfaces. The array has 64 elements giving an approximate 12 wavenumber resolution in the mid IR. Sample loading simply involves loading the ATR with a suitable thickness of material. Sample cups are not required.

(A)

(B)

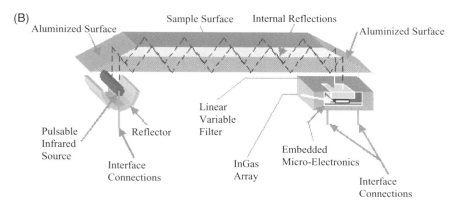

**Figure 4.1.8.** The Wilks wedge-interference filter spectrometer: (A) picture and (B) schematic. (Reproduced with permission from Wilks Enterprises, 140 Water Street, Norwalk, CT 06854).

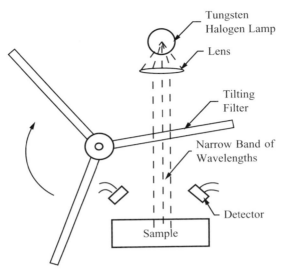

**Figure 4.1.9.** Schematic of a tilting-filter spectrometer. The three tilting filters provide for wavelength isolation in three regions of the NIR spectrum.

***Tilting Filter.*** Tilting-filter spectrometers (Fig. 4.1.9) have three inherent problems. First, much like a prism, the relationship between angle and wavelength is nonlinear. Second, the bandpass of the filter increases as the angle from the normal energy beam increases (clockwise or counterclockwise). Third, the peak transmission of the filter decreases as the angle of the filter from normal to the source beam increases (clockwise or counterclockwise). All three disadvantages make it difficult to reproduce instrument specifications from one instrument to another. Rosenthal (50) (CEO of Neotec, now FOSS NIR Systems), the first to produce a tilting-filter spectrometer, dropped production of tilting-filter spectrometers in 1993. Currently, no one produces a commercial version of this technology.

Filters have definite advantages over other wavelength-isolation methods. First, narrow-band NIR interference filters can be produced for any wavelength in the NIR region. Second, NIR filters characteristics can be reproduced, making it much easier to duplicate spectrometer characteristics. Third, the bandpass of a filter may be increased or decreased (increasing or decreasing the energy falling on the sample, respectively) depending on the application. The major disadvantage is that filter technology remains expensive ($300–$500 each). Nonetheless, spectrometers implementing narrow-band interference filters compete quite well for online and field (hand-held) applications where the objective is to measure a limited number of parameters.

***AOTF.*** An acousto-optical tunable filter (AOTF) (53) is a radio frequency (RF)-controlled narrow-band interference filter. Figure 4.1.10 is a schematic of a typical AOTF spectrometer. Most contemporary AOTF crystals are tellurium dioxide ($TeO_2$)

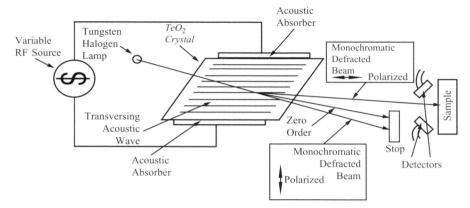

**Figure 4.1.10.** Schematic of a AOTF spectrometer. The major advantage of AOTF is the "snap-to feature" that allows the immediate and random isolation of specific wavelengths. This means that once the calibration wavelengths are determined the wavelengths can be read without scanning all wavelengths.

crystals. The emerging horizontally polarized beam is the analytical beam of choice, and the zero-order and vertically polarized beam are blocked (not used). For certain applications the horizontally polarized beam is blocked, allowing the vertically polarized beam to emerge as the analytical beam.

Application of an RF signal across the crystal sets up an acoustical pattern that functions very similarly to a transmission grating. Unlike gratings, the acoustical frequency allows only a narrow band of wavelengths to pass—thus, the crystal acts more like a narrow-band interference filter. The wavelength of energy emerging from the crystal is inversely related to the RF signal impressed on the crystal. The angle between the beams and the bandwidth of the emerging beam are a function of the design of the instrument but are usually a few degrees and as small as 1 nm, respectively. Varying the power of the applied RF signal determines the intensity of the analytical beam.

***Liquid Crystal Tunable Filters.*** The liquid crystal tunable filter (LCTF) has had considerable success in the visible region, with limited success in the near infrared. A LCTF element (shown in Fig. 4.1.11A) is a stack of retardation plates that becomes polarization sensitive (1, 11). Specific regions of the NIR spectrum are determined by the number and thicknesses of the retardation plates (usually crystal quartz plates).

Switching speeds from one wavelength band to another is dependent on the relaxation time of the crystal and can be as high as 50 ms. Although special crystals have been made with a switching time of 5 ms, this time far exceeds that of grating and AOTF technology and is restricted to a short sequence of wavelengths.

The spectral resolution is on the order of 10–20 nm, although special crystals can reduce the bandpass to 5–10 nm. Blocking filters are needed to block out-of-range

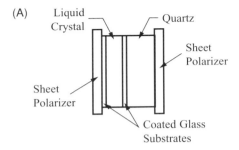

**Figure 4.1.11.** Schematic of a single LCTF element (A) and a working verion (B) for the NIR region. [Scientific Solutions, Inc. (SSI) 55 Middlesex Street, Unit 210, North Chelmsford, MA 01863-1561 USA]

transmission of the filter. A picture of Scientific Solution's LCTF is shown in Figure 4.1.11B.

***Other LCTF Developments.*** Tilotta et al. (54) utilized liquid crystals to modulate radiation in a Hadamard spectrometer. It was the first in the development of a no moving parts Hadamard spectrometer.

***Prisms.*** Prisms come in three forms: 1) dispersing, 2) reflecting, and 3) polarizing (22). Dispersing (or transmission) prisms were most popular in the early development of NIR technology. Reflecting prisms were designed to change the orientation or direction (or both) of a NIR beam. Polarizing prisms are made of birefringent materials that determine which polarization emerges (22).

Prism-based spectrometers (Fig. 4.1.12A) were used by NIR pioneers (23). In particular, the Beckman Model DU was modified by Wilbur Kaye (25) to record absorption spectra from 210 to 2700 nm automatically. His modifications included replacement of the photomultiplier tube with a lead sulfide (PbS) cell and inclusion of a chopper and an electronic recorder. Kaye later modified the same instrument for collecting transmission spectra in the NIR region of the electromagnetic spectrum (26).

(A)

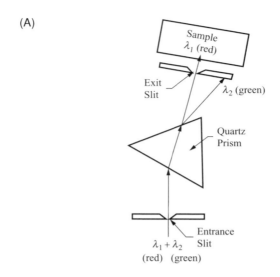

(B)

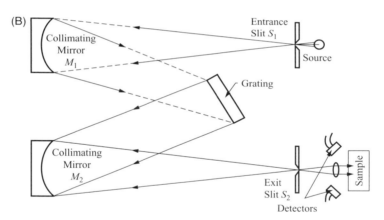

**Figure 4.1.12.** Two types of NIR spectrometers: (A) simple prism spectrometer and (B) Czerny–Turner plane grating spectrometer. As the prism or grating rotate, the exit slit isolates a particular wavelength interval.

McClure used the Perkin-Elmer (Model 450) spectrometer for studying the spectral characteristics of tobacco in the range from 600 to 2600 nm (32). This long scanning range was achieved in the dual-beam mode with slit-control electronics that varied the bandpass of the instrument from 12 to 88 nm at 600 and 2600 nm, respectively. Absence of a digitizer for the photometric signal was a major drawback of this instrument. Most researchers agree with Wilbur Kaye's assessment of NIR technology (24):

There are four aspects of (NIR) instrumentation that deserve mentioning: 1) An ordinary incandescent source radiates a maximum amount of energy in this region so that most instruments are not "energy limited." Also, stray light can be kept very low. 2) The lead sulfide detector useful throughout the near infrared region is very sensitive and rapid responding and yields a large output signal. 3) Quartz prisms and gratings of high dispersion permit high resolution. Commercial instruments are now available with resolving power of 5000 or more and custom built instruments have achieved values as high as 150,000. 4) Long, accurately measured and permanently sealed cells can be used, eliminating one of the troublesome variables in quantitative analysis.

***Gratings.*** The first commercial gratings were machine ruled (machine grooved) with a diamond ruling tool. They were called replica gratings because they were made from a pattern called "the master grating." Replica gratings made today are far superior to those the master gratings made 20 years ago. Holographic gratings are rapidly replacing the replica grating. They are made by a depositing photosensitive material onto a flat glass plate. Lasers are used to etch grooves in the material, and aluminum is vacuum deposited on the surface of the grooves to make them reflective. Holographic gratings are somewhat less efficient than replica gratings, but the precision of the grooves eliminates ghosts and reduces scattered light. Consequently, today NIR holographic-grating spectrometers perform much better than the older prism- or replica-grating types.

Figure 4.1.12B gives a schematic of a Czerny–Turner plane grating spectrometer. Grating spectrometers, especially with the development and enhancements of holographic gratings, are capable of much higher resolution than prisms and are much easier to implement in small packages. Thus we see a trend in the last few years toward smaller instrument volumes. Once only a lab bench instrument, grating-based spectrometers are finding their way into the field and on the processing line (46).

Figure 4.1.13A gives a schematic of a FOSS Model XDS grating monochromator with an attached fiber-optic multiplexer for performing on-line measurement of food products. The monochromator is a modified Czerny–Turner configuration with an oscillating concave holographic grating. Attached to the monochromator is a 10-channel fiber-optic multiplexer with 1 channel taken for the reference channel. This multiplexer attachment allows interactance and transmission measurements simultaneously or separately. Figure 4.1.13B is a picture of the XDS with an interactance probe lying in front.

***Fourier Transform-Near Infrared (FT-NIR).*** Only within the last 20 years has FT-NIR instrumentation (Fig. 4.1.14) become available. Even then, the first commercial instruments had a distinct disadvantage compared to grating-based scanning instruments. FT-NIR spectrometers employ an entirely different method for producing spectra. There is no dispersion involved. Energy patterns set up by an interaction with a sample and a reference and moving mirrors (or other optical components) produce sample and reference interferograms that are used to calculate the absorbance spectrum of the sample.

There are two advantages of FT-NIR spectrometers that make it attractive. The first is the throughput advantage. In the absence of dispersion, the energy at the output of an FT-NIR interferometer (similar to a monochromator) can be many times greater than that obtained from a grating monochromator. Nonetheless, the first commercial instruments literally starved for output energy, resulting in noise levels incommensurate

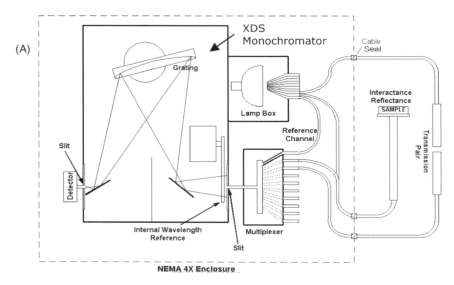

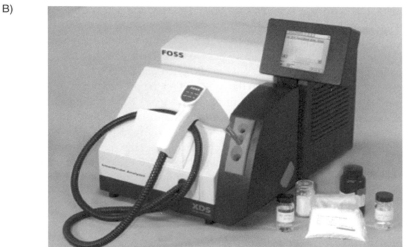

**Figure 4.1.13.** Schematic of the FOSS Model XDS spectrometer with multiplexing apparatus attached for on-line applications in the food industry (A) and a table top XDS with a Smart Probe (B).

with the standards set by dispersion-type spectrometers. More recently, manufacturers have improved the noise levels of FT-NIR spectrometers to the point where they readily compete with all other types of NIR technology.

**_Hadamard._**  Hadamard-transform NIR spectrometry (see Fig. 4.1.15) (HT-NIR) is based on the combination of multiplexing and dispersive spectrometers where the

(A)

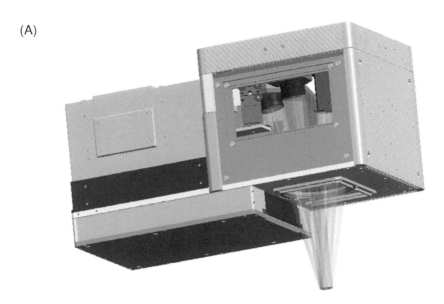

(B)

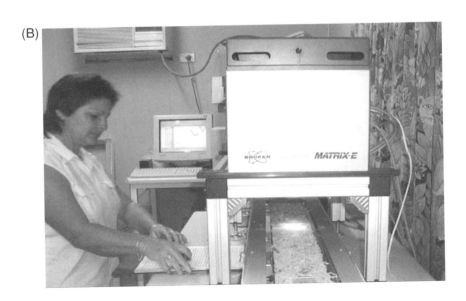

**Figure 4.1.14.** The Bruker Matrix-E, a Fourier transform near-infrared (FT-NIR) spectrometer (Bruker Optics Inc., 19 Fortune Drive, Manning Park, Billerica, MA 01821-3991): (A) illustrating the noncontact measuring concept and (B) mounted for analyzing sugarcane pulp passing underneath.

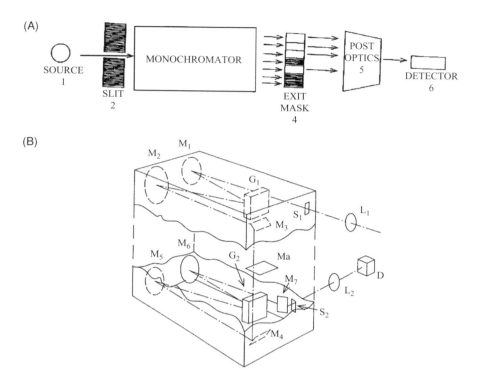

**Figure 4.1.15.** Simplified (A) and detailed (B) schematic of a dispersive Hadamard transform spectrometer. In A the source provides radiation for the monochromator where the energy is dispersed for focusing onto the exiting mask. The post optics collect the radiation from the transparent slots in the mask and focus this radiation onto the detector. The figure in B is given to illustrate the complexity of a Hadamard spectrometer.

choice of transparent or opaque elements of a multislit mask provides information that may be transformed into conventional NIR spectrum with Hadamard mathematics (16, 54). That is, the energy from an NIR source (usually tungsten halogen lamps) is first dispersed into spectral elements (wavelengths) that are collected and focused onto a focal plane. Unlike purely dispersive systems where there is only one exit slit, the focal plane of a Hadamard system implements a multislit array. Signals from this multiple-slit arrangement are collected by a single-element detector, resulting in an endcodegram that, with Hadamard mathematics, yields a spectrum (10). Hadamard technology fell on hard times because, in the beginning, mechanical masks were the only means of multiplexing the needed information. Later, liquid crystal modulators were developed that renewed hopes for Hadamard technology (54). Nonetheless, problems remain that render Hadamard spectrometry unattractive for spectrometer manufacturers. In a recent publication of Hammaker *et al.* (15), they discuss a digital micromirror array that "promises to be the best Hadamard encoding mask yet developed." Yet HT-NIR spectrometry is still weighted down with complex problems and, for those problems, remains in the research environment.

## TRENDS

The strongest trend in the NIR spectrometer industry today is a move toward smaller and more compact instruments (34). The original instruments (*viz.* the Neotec 6250) were roll-around instruments. The successor to the 6250 was the FOSS NIR Systems 6500, a laptop instrument that has a footprint less than 2 ft$^2$. McClure's roll-around instrument shown in Figure 4.1.16 had a 15-ft$^2$ footprint (37). Further work by McClure and others has produced hand-held instruments (called "NIR meters") for measuring nicotine and moisture in tobacco (40), vanillin and moisture in vanilla beans (39), moisture and nitrogen in grass tissue (43), and chlorophyll and moisture in plant leaves (33). A general consensus has emerged within the last 20 years calling for a full line of instruments that are portable or hand-held and easily tailored to perform specific "analytical tricks." Starting with the first hand-held instrument (35) and looking at the

**Figure 4.1.16.** The author is sitting with a 1981 custom built grating/prism (Cary 17) spectrometer that has a footprint of 15 ft$^2$; the FOSS XDS take less than 1.5 ft$^2$ of lab bench space.

emphases of others (2, 21), the trend toward compact portables and hand-held NIR instruments will continue well into the twenty-first century.

## FUTURE

With the Food and Drug Administration pushing for zero defects at the end of drug processing lines (27, 47, 49, 55), NIR spectrometry has a rosy future. Twenty years ago there were fewer than 15 manufacturers in the NIR spectrometer business. Today, there are more than (58) companies around the world producing NIR spectrometers (see Table 4.1.3). Economic indicators continue to lure small businesses into this fascinating market. From where these authors sit, the sky is the limit.

**TABLE 4.1.3. Partial List of NIR Instrument/Supply Vendors**

| Manufacturer/Address | WWW/E-mail/Phone Numbers | Type of Equipment |
|---|---|---|
| ABB BOMEM | 585 Charest Blvd East, Suite 300 Quebec G1K 9H4 CANADA TEL: (800) 858-3847 FAX: (418) 877-2834 Email: ftir@ca.abb.com URL: www.bomem.com | FT-IR and FT-NIR |
| Acton Research Corporation | 15 Discovery Way Acton, MA 01720 USA TEL: 978.263.3584 FAX: 978.263.5086 Email: mail@acton-research. com URL: www. acton-research.com | Monochromator Integrated spectrometers |
| Analect Instruments Applied Instrument Technologies | 2771 North Garey Avenue Pomona, CA 91767 TEL: 909-593-3581 Toll Free: 800-326-2328 FAX: 909-392-3207 Email: AIT@hs.utc.com URL: www.orbital-ait.com | FT-NIR |
| Analytical Spectral Devices | 5335 Sterling Drive Boulder, CO 80301 TEL: (303) 444-6522 FAX: (303) 444-6825 URL: www.asdi.com | Spectrometer LabSpec Pro (350-2500nm) |
| Axiom Analytical, Inc. 17751 Sky Park Circle, Suite B Irvine CA 92614 USA | TEL: 949-757-9300 FAX: 949-757-9306 Email: info@goaxiom.com URL: www.goaxiom.com | Probes, cells, fiber optics |

**TABLE 4.1.3.** *Continued*

| Manufacturer/Address | WWW/E-mail/Phone Numbers | Type of Equipment |
|---|---|---|
| Bio-Rad Laboratories Headquarters 1000 Alfred Nobel Drive Hercules, CA 94547 USA | TEL: (510) 724-7000 FAX: (510) 741-5817 Email: lsg_websupport@bio-rad .com | FT-NIR NIR database of organic compounds |
| Brimrose Corporation of America 19 Loveton Circle Baltimore, MD 21152-9201, USA | TEL: 410-472-7070 FAX: 410-472-7960 URL: www.brimrose.com | AOTF; fiber optics |
| Bruker Optics Inc 19 Fortune Drive, Manning Park Billerica, MA 01821-3991 | TEL: (978) 439-9899 FAX: (978) 663-9177 URL: optics@brukeroptics.com | FT-NIR spectrometer Fiber optics |
| Buchi Analytical Buchi Labortechnik AG Postfach, CH-930 Falwil 1 Switzerland | TEL: +41 71 394 6363 FAX: +41 71 394 6464 Email: buchi@buchi.com URL: www.buchi-analytical.com | FT-NIR Sponsor of Buchi NIR Award |
| ChemImage Corp. 7301 Penn Avenue Pittsburgh, PA 15208 USA | TEL: 1-412-241-7335 FAX: 1-412-241-7311 E-Mail: ChemImage@ chemimage.com | Chemical imaging systems |
| Control Development, Inc. 2633 Foundation Drive South Bend, IN 46613, USA | TEL: 574-288-7338 FAX: 574-288-7339 URL: www.controldevelopment .com | Compact/wireless PC board with optics |
| CVI Laser, LLC Corporate Headquarters 200 Dorado Place SE Albuquerque, New Mexico USA | TEL: 800-296-9541 FAX: 505-298-9908 Email: Sales@CVIlaser.com URL: www.cvilaser.com | Lasers, optics, filters |
| DHC Analysis Inc. 3645 Warrensville Center Road Suite 325 Cleveland, OH 44122 USA | TEL: 216-295-0755 FAX: 216-295-1887 Email: casal@dhcanalysis.com URL: www.dhcanalysis.com | Specialized NIR instruments |
| DICKEY-john Corporate Headquarters 5200 DICKEY-john Road Auburn, IL 62615 USA | TEL: (217) 438-3371 Toll Free: (800) 637-2952 FAX: (217) 438-6012 | Filter spectrometers |

*Continued*

**TABLE 4.1.3.** *Continued*

| Manufacturer/Address | WWW/E-mail/Phone Numbers | Type of Equipment |
|---|---|---|
| EG&G Princeton Applied Research Advanced Measurement Technology 1321 Teakwood Drive Fort Collins, CO 80525 USA | TEL: 970-482-1823 FAX: 970-484-0547 E-Mail: Frank.Heiliger@ pari-online.com URL: http://www.princeton appliedresearch.com/ | CCD and DA |
| Celetronix USA, Inc. 2125-B Madera Road Simi Valley, CA 93065 USA | TEL: 805-955-3600 FAX: 805-582-4431 | Optics |
| Foss NIRSystems 7682 Executive Dr. Eden Prairie, MN 55344 USA | TEL: 301-680-9600 URL: www.foss-nirsystems.com | Moving grating |
| FUTREX, Inc. 6 Montgomery Village Ave Gaithersburg, MD 20879, USA | TEL: 800-255-4206 FAX: 301-670-1103 Email: info@futrex.com URL: http://www.futrex.com/ | LED Body composition analyzer |
| Infrared Fiber Systems, Inc. 2301-A Broadbirch Dr. Silver Spring, MD 20904 USA | TEL: 301-622-9546 FAX: 301-622-7135 Email: info@infraredfiber systems.com | Optical fibers, laser windows, AOTF spectrometers |
| Int'l Equipment Trading, LTD 900 Woodlands Parkway Vernon Hill, IL 60061 USA | TEL:(800)913-0777 FAX:(847)913-0785 info@ietltd.com | New and refurbished instruments |
| HORIBA Jobin Yvon Inc 3880 Park Avenue Edison NJ 08820-3012 USA | TEL: 732-494-8660 FAX: 732-549-5125 Email: systems@jyhoriba.com URL: www.jyhoriba.com | InGaAS Det CCD Ar. Det Spectrometers Software |
| Integrated Spectronics 22 Delhi Road NORTH RYDE, NSW, 2113 Australia | TEL: +61 2 9878 0977 FAX: +61 2 9878 3615 Email: ispl@intspec.com URL: http://www.intspec.com http://members.ozemail.com .au/~tdc/products.htm | Airborne hyperspectral mapper Field portable mineral analyzer (1300–2500 nm) |
| JASCO Corporation 2967-5 Ishikawa-Cho Hachioji City Tokyo 192-8537, JAPAN | TEL: 800-333-5272 TEL: +81-426-46-4111 FAX: +81-426-46-4120 | Grating UV/VIS/NIR FT-NIR RAMAN |

**TABLE 4.1.3.** *Continued*

| Manufacturer/Address | WWW/E-mail/Phone Numbers | Type of Equipment |
|---|---|---|
| KES Analysis, Inc.<br>160 West End Ave.<br>New York, NY 10023<br>USA | TEL: 212-595-7046<br>FAX: 212-787-3858<br>Email: Starkdw@aol.com | Diode array<br>Software |
| Kett<br>PMB 504<br>17853 Santiago Blvd.,<br>Suite 107<br>Villa Park, CA<br>92861-2633 | TEL: (800)438-5388<br>FAX: (714)693-2923<br>Email: sales@kett.com<br>URL: www.kett.com | Filter<br>Hand-held<br>Moisture,cholesterol |
| LabX<br>P.O. Box 216<br>478 Bay Street<br>Midland, ON, Canada L4R<br>1K9 | Toll Free: 888-781-0328<br>TEL: 705-528-6888<br>FAX: 705-528-0270<br>Email: help@labx.com | Fiber-optic<br>spectrometers<br>Up to 2200 nm |
| Labsphere, Inc.<br>PO Box 70, 231 Shaker<br>Street<br>North Sutton, NH 03260<br>USA | TEL: 603-927-4266<br>FAX: 603-927-4694<br>Email: labsphere@labsphere.com<br>URL:http://www.labsphere.com/ | Standard spectral<br>materials |
| LECO<br>LECO Corporation<br>3000 Lakeview Avenue<br>St. Joseph, MI 49085 | TEL: 269-985-5496<br>FAX: 269-982-8977<br>Toll Free: 800-292-6141<br>URL: http://www.leco.com/ | Filter |
| LT Industries, Inc.<br>811 Russell Ave<br>Gaithersburg, MD 20879 | TEL: 301- 990 - 4050<br>FAX: 301- 990 - 7525<br>Email: Sales@ltindustries.com<br>URL: www.ltindustries.com | Moving grating, fiber<br>optics, processing |
| NIR Technology Australia<br>34 Clements Avenue<br>Bankstown NSW 2200<br>Australia | TEL: +61 (0) 2 9790 6450<br>FAX: +61 (0) 2 9790 1552<br>Email: nirtech@zipworld.com.au<br>URL: http://www.lineart.zip<br>.com.au/homepage.htm | Near-infrared<br>analyzers (Herschel<br>region) |
| Mattson<br>47131 Bayside Parkway<br>Fremont, CA 94538 | Toll Free: 800-3277433<br>TEL: 510-657-5900<br>FAX: 510-492-5911<br>Email: info@mattson.com<br>URL: http://www.mattson.com/ | FT-NIR |

*Continued*

**TABLE 4.1.3.** *Continued*

| Manufacturer/Address | WWW/E-mail/Phone Numbers | Type of Equipment |
|---|---|---|
| MIDAC Corp. 130 McCormick Avenue #111, Costa Mesa, CA 92626 | TEL: 714-546-4322 FAX: 714.546.4311 Email: info@midac.com URL: http://www.midac.com/ | FTIR, FTNIR, open path |
| Minarad Spectrum, Inc. 7 Cambridge Dr. Trumbull, CT 06611 | TEL: 203-371-6348 FAX: 203-371-6349 | Filter |
| Moisture Systems 117 South Street Hopkinton, MA 01748 | TEL: 508-435-6881 FAX: 508-435-6677 | Filter, noncontact NIR |
| NDC Infrared Engineering USA 5314 North Irwindale Avenue Irwindale California 91706 | TEL: 626-960 3300 FAX: 626-939-3870 E-mail: info@ndcinfrared.com | Filter, noncontact |
| Thermo Nicolet Process Instruments 2555 N. Interstate 35 Round Rock, TX United States | Toll Free: 877-843-7668 TEL: 561-688-8700 URL: http://www.thermo.com/ | FT-NIR |
| Ocean Optics, Inc. 380 Main Street Dunedin, FL 34698 | TEL: (727) 733-2447 FAX: (727) 733-3962 Email: info@OceanOptics.com | Compact spectrometer – PC mounted Wavelength calibration source Calibrated light source A/D converters / software |
| NIR Technology Australia Suite 1, 173 Canterbury Road Bankstown NSW 2200 AUSTRALIA | TEL: 02-9790-6450 FAX: 02-9790-1552 Email: lineart@zipworld.com.au URL: http://www.lineart.zip .com.au/ | Filter-based systems Fiber optics, spectrograph, Lin Arr Detector Grain analyzers, food, polymers Pharmaceuticals |
| Olis, Inc. 130 Conway Drive, Suites A/B Bogart, GA 30622 | TEL: (706) 353-6547 FAX: 800-852-3504 | Used/refurbished/ upgraded spectrometers Software Computerized instruments |
| Optical Solutions, Inc. 333 Sunrise Avenue, Suite 801 Roseville, CA 95661-3479 | TEL: 916-677-1671 FAX: 916-677-1675 Email: sales@optical-solutions .com URL: http://www. optical-solutions.com/ | DA Photometers Software Process analyzers |

**TABLE 4.1.3.** *Continued*

| Manufacturer/Address | WWW/E-mail/Phone Numbers | Type of Equipment |
|---|---|---|
| Optometrics, LLC. <br> 8 Nemco Way <br> Stony Brook Industrial Park <br> Ayer, MA 01432 USA | TEL: 978-772-1700 <br> FAX: 978-772-0017 <br> Email: sales@optometrics.com <br> URL: www.optometrics.com | Mini-monochromators <br> Lasers optics <br> Filters, gratings <br> Optical components |
| Optronic Laboratories, Inc. <br> 4632 36th Street <br> Orlando, FL, 32811 - USA | Toll Free: 800-899-3171 <br> TEL: 407-422-3171 <br> FAX: 407-648-5412 <br> URL: http://www.olinet.com/ | Spectroradiometers <br> Reflection standards <br> LED systems <br> Monochromators |
| Orbital Sciences Corporation <br> 21839 Atlantic Blvd <br> Dulles, VA 20166 | TEL: 703-406-5000 <br> URL: http://www.orbital.com/ | Satellite Optics <br> Spectrometers |
| Newport/Spectral Physica/Oriel <br> 1791 Deere Ave. <br> Irvine, CA 92606 | Toll Free: 800-222-6440 <br> TEL: 949-863-3144 <br> FAX: 949-253-1680 <br> URL: http://www.spectra-physics.com/ | Gratings/lamps/optical <br> Motion control <br> Spectroscopy instruments |
| PerkinElmer Life and Analytical Sciences, Inc. <br> 549 Albany Street <br> Boston, MA 02118-2512, USA | Toll Free: 800-762-4000 <br> TEL: 203-925-4602 <br> FAX: 203-944-4904 <br> Email: productinfo@perkinelmer.com <br> URL: www.perkinelmer.com | FT-NIR <br> Imaging systems <br> Software <br> Consumables |
| Perten Instruments Inc. USA <br> 6444 South 6th Street Road <br> SPRINGFIELD, IL 62707 | TEL: 217-585-9440 <br> FAX: 217-585-9441 <br> Email: perteninc@perten.com <br> URL: http://www.perten.com/ | Diode arrays <br> Filters |
| Process Sensors Corp. <br> 113 Cedar Street <br> Milford, MA 01757 USA | TEL: 508-473-9901 <br> FAX: 508-473-0715 <br> Email: info@processsensors.com <br> URL: www.processsensors.com | Filter/noncontact <br> Over-the-line <br> Laboratory/benchtop |
| Scientific Solutions, Inc. (SSI) <br> 55 Middlesex Street, Unit 210 <br> North Chelmsford, MA 01863-1561 <br> USA | TEL: 978-251-4554 <br> FAX: 978-251-8822 <br> Email: info@sci-sol.com <br> URL: http://www.sci-sol.com/ | Liquid crystal tunable filters (LCTF) |

*Continued*

**TABLE 4.1.3.** *Continued*

| Manufacturer/Address | WWW/E-mail/Phone Numbers | Type of Equipment |
|---|---|---|
| Sentronic, GmbH<br>Germany | | Laser Diode analyzer<br>Miniature NIR<br>spectrometers<br>Modular optical<br>spectroscopy |
| Spectral Dimensions, Inc.<br>3416 Olandwood Court,<br>Suite 210,<br>Olney, MD 20832 USA | TEL: 301-260-0290<br>FAX: 301-260-0292<br>E-mail: info@spectraldimensions<br>.com<br>URL: http://www.spectral<br>dimensions.com/ | Imaging spectromters<br>Chemical imaging |
| Shimadzu America<br>Shimadzu Scientific<br>Instruments, Inc.<br>7102 Riverwood Drive<br>Columbia, MD 21046 | Toll Free: 800-477-1227<br>TEL: 410-381-1227<br>FAX: 410-381-1222<br>URL: http://www.shimadzu.com/ | DM, FT |
| Spectral Instruments, Inc.<br>420 North Bonita Avenue<br>Tucson, Arizona 85745 | TEL: 520-884-8821<br>FAX: 520-884-8803<br>Email: info@specinst.com<br>URL: http://www.specinst.com/ | Chemical imaging<br>Cooled CCD cameras<br>16-bit digitization |
| UOP Guided Wave<br>Incorporated<br>BV 5190 Golden Foothill<br>Parkway<br>P.O. Box 427 El Dorado<br>Hills, CA<br>95762 7550 | TEL: 916-939-4300<br>FAX: 916-939-4307<br>Email: GWinfo@guided-wave<br>.com<br>URL: http://www.guided-wave<br>.com/ | Moving grating, fiber<br>optics, processing |
| Varian Instruments<br>2700 Mitchell Dr.<br>Walnut Creek, CA 94598 | TEL: 800-926-3000<br>FAX:.925-945-2102<br>E-mail: customer.service@<br>varianinc.com<br>URL: http://www.varianinc.com/ | Grating spectrometers<br><br>Software |
| Visionex Automation,<br>LLC<br>27650 Farmington Road,<br>Suite B3<br>Farmington Hills, MI<br>48334 | TEL: 248-613-1915<br>FAX: 248-848-1788<br>Email: info@visionexautomation<br>.com<br>URL: http://www.<br>visionexautomation.com/ | Lasers, machine<br>vision, optics |
| Wilks Enterprise, Inc.<br>140 Water Street<br>So. Norwalk, CT 06854<br>USA | TEL: 203-895-9136<br>FAX: 203-838-9868<br>URL: http://www.wilksir.com/ | DA, wedge<br>interference filters |

**TABLE 4.1.3.** *Continued*

| Manufacturer/Address | WWW/E-mail/Phone Numbers | Type of Equipment |
|---|---|---|
| Zeltex, Inc.<br>130 Western Maryland Parkway<br>Hagerstown, Maryland 21740 USA | Toll Free: 800-732-1950<br>TEL: 301-791-7080<br>FAX: 301-733-9398<br>Email: toddr@zeltex.com<br>URL: http://www.zeltex.com/ | DA, filters |
| Zeiss Optronics, GmbH<br>Carl Zeiss Str. 22<br>73447 Oberkochen<br>Germany | TEL: (+49) 73 64 20 01<br>Fax: (+49) 73 64 20 45 88<br>Email: zeo@zeiss.de<br>URL: www.zeiss-optronik.de | DA, fiber-optic spectrometers 195–2150 nm |

## REFERENCES

1. D. Bonaccini, L. Casini, P. Stefanini. Fabry-Perot tunable filter for the visible and near IR using nematic liquid crystals. In: *Current Developments in Optical Engineering IV*, Robert E. Fischer and Warren J. Smith, eds. SPIE—The International Society for Optical Engineering, San Diego, CA, 1990, 1334: 221–230.

2. D. A. Burns. Hand-held NIR spectrometry in medicine. Personal Communication, 2001.

3. E. W. Ciurczak. *Molecular Spectroscopy Workbench: Advances, Applications and Practical Advice on Modern Spectroscopic Analysis*. John Wiley & Sons, New York, NY, 1998.

4. P. J. Clancy. Transfer of calibration between on-farm whole grain analyzers. In: *Near Infrared Spectroscopy: Proceedings of the 10th International Conference on Near Infrared Spectroscopy, Kuonjgu, Korea*, R. K. Cho, A. M. C. Davies, eds. NIR Publications, Chichester, UK, 2002.

5. W. W. Coblentz. *Investigations of Infrared Spectra: Part 1.* (1905). (Reproduced in 1962 under the joint sponsorship of the Coblentz Society and the Perkin-Elmer Corporation), Carnegie Institute, Washington, DC, Publication No. 35.

6. J. R. Collins. Change in the infra-red absorption spectrum of water with temperature. *Phys Rev* **26**: 771–778, 1925.

7. A. M. C. Davies. Near infrared spectroscopy: time for the giant to wake up! *Eur Spectrosc News* **73:** 16 Oct, 1987.

8. A. M. C. Davies, S. M. Ring, J. Franklin, A. Grant, W. F. McClure. Prospects for process control using Fourier transformed near infrared data. *SPIE* 553: 330–331, 1985.

9. J. W. Ellis. *Trans Faraday Soc* **25**: 888, 1929.

10. R. D. B. Fraser. Interpretation of infrared dichroism in fibrous proteins—the 2u region. *J Chem Phys* **24**: 89–97, 1956.

11. N. Gat. Imaging spectroscopy using tunable filters: A review. In: *Wavelet Applications VII*, H. Szu Harold, Vetterli Martin, J. Campbell William, R. Buss James, eds. SPIE—The International Society for Optical Engineering, 4056: 50–64, 2000.

12. R. F. Goddu and D. Delker. Aids for the analyst: spectra-structure correlations for the near-infrared region. *Anal Chem* **32**: 140–141, 1960.

13. R. F. Goddu, D. A. Delker. Determination of terminal epoxides by near-infrared spectrophotometry. *Anal Chem* **30(12)**: 2013–2016, 1958.

14. C. V. Greensill, K. B. Walsh. Standardization of near infrared spectra across miniature photodiode array-based spectrometers in the near infrared assessment of citrus soluble solids content. In: *Near Infrared Spectroscopy: Proceedings of the 10th International Conference on Near Infrared Spectroscopy, Kuonjgu, Korea.* R. K. Cho, A. M. C. Davies. NIR Publications, Chichester, UK, 2002.

15. R. M. Hammaker, R. A. DeVerse, D. J. Asunskis, W. G. Fateley. Hadamard transform near-infrared spectrometers. In: *Handbook of Vibrational Spectroscopy*, J. M. Chalmers, P. R. Griffiths, eds. John Wiley & Sons Ltd., Chichester, West Susses, UK, **1**: 453–460, 2002.

16. R. M. Hammaker, J. A. Graham, D. C. Tilotta, W. G. Fateley. What is Hadamard transform spectroscopy. In: *Vibrational Spectra and Structure*, J. R. Durig, ed. Amsterdam: Elsevier, **15**: 401–485, 1986.

17. W. Herschel. Experiments on the refrangibility of the rays of the sun. *Phil Trans R Soc Lond* **90(XIV)**: 284–292, 1800.

18. W. Herschel. Experiments on the solar, and on the terrestrial rays that occasion heat; with a comparative view of the laws to which light and heat, or rather the rays which occasion them, are subject, in order to determine whether they are the same, or different. *Phil Trans R S Lond* **90(XV)**: 293–331, 1800.

19. W. Herschel. Investigation of the powers of the prismatic colours to heat and illuminate objects; with Remarks, that prove the different refrangibility of radiant heat. To which is added, and inquiry into the method of viewing the sun advantageously, with telescopes of large apertures and high magnifying powers. *Phil Trans R Soc Lond* **90(XIII)**: 255–283, 1800.

20. W. R. Hruschka. The Fourier transform. In: *Near-Infrared Technology in the Agricultural and Food Industries*, St. Paul, MN: American Assoc. of Cereal Chemists, **1**: 39–58, 1987.

21. T. Hyvarinen, P. Niemela. Rugged multiwavelength NIR and IR analyzer for industrial process measurements (This folder has patents on small spectrometers. Also, see 3412 for other portable (small) spectrometers.). *SPIE (Process Optical Measurements and Industrial Methods)*, **1266**: 99–104, 1990.

22. J. D. Ingle, Jr., S. R. Crouch. Spectrochemical Analysis. Englewood Cliffs, NJ Prentice-Hall, 1988.

23. W. Kaye. Near-infrared spectroscopy—a review: part 2. Instrumentation and techniques. *Spectrochimi Acta* **7**: 181–204, 1955.

24. W. Kaye. Near Infrared Spectroscopy. In: *Encyclopedia of Spectroscopy*, H. Clark ed. John Wiley & Sons, New York, NY, 494–505, 1960.

25. W. Kaye, C. Canon, R. G. Devaney. Modification of a Beckman model DU spectrophotometer for automatic operation at 210–2700 $\mu$m *J Opt Soc Am* **41(10)**: 658–664, 1951.

26. W. Kaye. R. G. Devaney. An automatic relative-transmission attachment for the Beckman Model DU Spectrophotometer. *J Opt Soc Am* **42(6)**: 567–571, 1952.

27. H. Mark, G. E. Ritchie, R. W. Roller, E. W. Ciurczak, C. Tso, S. A. MacDonald. Validation of a near-infrared transmission spectroscopic procedure. Part A: Validation protocols. *J Pharm Biomed Anal* **28**: 251–260, 2002.

28. G. C. Marten, J. S. Shenk, I. F. E. Barton. Near Infrared Reflectance Spectroscopy (NIRS): Analysis of Forage Quality (Handbook No. 643). Agriculture Handbook No. 643. Springfield, VA: National Technical Information Service. 1989.

29. K. A. Martin. Recent advances in near-infrared reflectance spectroscopy. In: *Applied Spectroscopy Reviews*. G. Brame Jr. Edward, ed. Marcel Dekker, Inc. **27**: 325–383, 1992.

30. D. R. Massie. Fat measurement of ground beef with a gallium arsenide infrared emitter. In: *Quality Detection of Foods*. J. J. Gaffney, ed. St. Joseph, MI: American Society of Agricultural Engineers (ASAE). **1**: 24–26, 1976.

31. D. R. Massie. A high-intensity spectrophotometer interfaced with a computer for food quality measurement. In: *Quality Detection in Foods*. J. J. Gaffney, ed. St. Joseph, MI: American Society of Agricultural Engineers. **1**: 12–15, 1976.

32. W. F. McClure. Spectral characteristics of tobacco in the near infrared region from 0.6 to 2.6 microns. *Tobacco Sci* **12**: 232–235, 1968.

33. W. F. McClure. Hand-held NIR spectrometry: Part 2. An economical no-moving parts meter for measuring chlorophyll and moisture. *Appl Spectrosc* **56(2)**: 32–34, 2001.

34. W. F. McClure. Hand-held near infrared spectrometry: Status, trends and futuristic concepts. In: *Near Infrared Spectroscopy: Proceedings of the 10th International Conference on Near Infrared Spectroscopy, Kyonjgu, Korea*. A. M. C. Davies ed. NIR Publications, Chichester, UK, pp. 131–136, 2002.

35. W. F. McClure. Hand-held near-infrared spectrometry: Status, trends and futuristic concepts. In: *Near Infrared Spectroscopy: Proceedings of the 10th International Conference*. A. M. C. Davies, R. K. Cho, eds. NIR Publications, Chichester, UK, p. 131, 2002.

36. W. F. McClure. 204 Years of near infrared technology: 1800–2003. *J Near Infared Spectrosc* **11**: 487–518, 2003.

37. W. F. McClure, A. Hamid. Rapid NIR measurement of the chemical composition of foods and food products: Part 1. *Hardware Am Lab* **12(9)**: 57–69, 1980.

38. W. F. McClure, A. Hamid, F. G. Giesbrecht, W. W. Weeks. Fourier analysis enhances NIR diffuse reflectance spectroscopy. *Appl Spectrosc* **38(3)**: 322–328, 1984.

39. W. F. McClure, C. M. Hargrove, M. Zapf, D. Stanfield. Hand-held NIR spectrometry: Part 3. An instrument for measuring vanillin and moisture and vanilla bean. Internal Research Report, NC State University, Department of Biological Engineering, Raleigh, NC 28695-7625, USA.

40. W. F. McClure, D. L. Stanfield. Hand-held NIR spectrometry: Part 4. A spectrometer for measuring nicotine and moisture in tobacco leaves and tobacco products. Internal Research Report, NC State University, Department of Biological Engineering, Raleigh, NC 28695-7625, USA.

41. C. E. Miller. Near-infrared spectroscopy of synthetic polymers. *Appl Spec Rev* **26(4)**: 277–339, 1991.

42. S. Morimoto. A nondestructive NIR spectrometer: Development of a portable fruit quality meter. In: *10th Intenational Conference on Near Infrared Spectroscopy, Kyongju, Korea*, R. K. Cho, A. M. C. Davies, eds. NIR Publication, UK, 2002.

43. S. Morimoto, W. F. McClure, D. L. Stanfield. Hand-held NIR spectrometry: Part 1. An instrument based upon gap-second derivative theory. *Appl Spectrosc* **55(1)**: 182–189, 2001.

44. K. H. Norris. A simple spectroradiometer for the 0.4- to 1.2-micron region. ASAE Paper No. 63-329, 1–5, 1963.

45. C. Pasquini. Near Infrared Spectroscopy: Fundamentals, Practical Aspects and Analytical Applications. *J Braz Chem Soc* **14(2)**: 198–219, 2003.

46. C. Paul. NIR climbs on the harvester and goes out to the field. *NIR Spectrum* **1(3)**: 8–9, 2003.

47. G. E. Ritchie. NIRS process analytics. *NIR Spectrum* **1(1)**: 4, 2003. (http://www.idrc-chambersburg.org).

48. G. E. Ritchie. NIRS Process Analytics: Back to the beginning again. *NIR Spectrum* **1(2)**: 9, 2003. (http://www.idrc-chambersburg.org).

49. G. E. Ritchie, R. W. Roller, E. W. Ciurczak, H. Mark, C. Tso, S. A. MacDonald. Validation of a near-infrared transmission spectroscopic procedure. Part B: Application to alternate content uniformity and release assay methods for pharmaceutical solid dosage forms. *J Pharm Biomed Anal* **29**: 159–171, 2002.

50. R. D. Rosenthal, R. Hayden, S. Engler. A highly flexible 9.0 OD light transmittance/reflectance research spectrophotometer. In: *Quality Detection in Foods*, J. J. Gaffney, ed. St. Joseph, MI: American Society of Agricultural Engineers, **23(1)**: 16–19, 1976.

51. T. Rosenthal. Infrared-emitting diodes for near-infrared spectrometry. In: *H andbook of Vibrational Spectroscopy*, J. M. Chalmers, P. R. Griffiths, eds. John Wiley & Sons, Chichester, West Sussex, UK, **1**: 461–465, 2002.

52. N. Sheppard, H. A. Willis, J. C. Rigg. International union of pure and applied chemistry agreements. *Pure and Applied Chemistry* **57**: 105–120, 1985.

53. Staff. Introduction to acousto-optics. Brimrose Corporation of America. Accessed 11–28, 2004. Internet: http://www.brimrose.com.

54. D. C. Tilotta, R. M. Hammaker, W. G. Fateley. A visible-near-infrared Hadamard transform spectrometer based on a liquid crystal spatial light modulator array: A new approach in spectrometry. *Appl Spectrosc* **41(6)**: 727–734, 1987.

55. C. Tso, G. E. Ritchie, L. Gehrlein, E. W. Ciurczak. A general test method for the development, validation and routine use of disposable near infrared spectroscopic libraries. *JNIRS* **9**: 165–184, 2001.

56. C. A. Watson, D. Carville, E. Dikeman, G. Daigger, G. D. Booth. Evaluation of two infrared instruments for determining protein content of hard red winter wheat. *Cer Chem* **53(2)**: 214–222, 1976.

57. L. G. Weyer. Near-Infrared Spectroscopy of organic substances. *Appl Spectrosc Rev* **21(1&2)**: 1–43, 1985.

58. O. H. Wheeler. Near infrared spectra of organic compounds. *Chem Rev* **59**: 629–666, 1959.

59. K. B. Whetsel. Near-infrared spectrophotometry. In: *Applied Spectroscopy Reviews*, G. Brame Jr. Edward ed. Marcel Dekker, New York, NY, 1–67, 1968.

60. K. B. Whetsel. American development in near infrared spectroscopy (1952–1970). *NIR News* **2(5)**: 12–13, 1991.

61. K. B. Whetsel. The first fifty years of near-infrared spectroscopy in America. *NIR News* **2(3)**: 4–5, 1991.

62. P. A. Wilks. The origins of commercial infrared spectrometers. In: *The History and Preservation of Chemical Instrumentation*, J. T. Stock, M. V. Orna, eds. New York: D. Reidel Publishing Company. **1**: 27–32, 1986.

63. P. A. Wilks. How the infrared spectrometer reached the bench chemist. *Spectroscopy* **16(12)**: 14–15, 2001.

64. P. A. J. Wilks. A practical approach to internal reflection spectroscopy. *Am Lab* **4**: 42–56, 1972.

65. H. A. Willis. Laboratory Methods in Vibrational Spectrosocpy. John Wiley & Sons, Chichester UK, 1987.

66. H. A. Willis, R. G. Miller. Quantitative analysis in the 2-u region applied to synthetic polymers. *J Appl Chem* **6**: 385–391, 1956.

67. J. T. Worthington, D. R. Massie, K. H. Norris. Light transmission technique for the predicting ripening time for intact green tomatoes. ASAE Paper No. 73-6526, 1973.

68. O. R. Wulf. 1935. *J Am Chem Soc* **57**: 1935.

# Time-of-Flight Spectroscopy

SATORU TSUCHIKAWA AND W. FRED MCCLURE

## INTRODUCTION

Consumer demands with respect to agricultural products are becoming increasingly diverse. The producer must not only supply them safely while maintaining freshness but also ensure taste and nutritional values are maintained. A nondestructive measurement system that can accurately monitor these indices simultaneously with minimum time and effort is therefore very desirable for marketing of agricultural commodities. On the other hand, biological materials like wood are commonly used in applications where both the cellular structure and the original physical shape are retained. Also in this case, a truly nondestructive measurement system, which can accurately monitor the physical conditions and chemical components of a wooden material at the same time, is very desirable for making an article with high quality and reliability, without waste of raw materials, time, and energy.

In the past few decades, many researchers have focused on the potential use of near-infrared (NIR) spectroscopy, a practical spectroscopic procedure, for detection of organic compounds in matter. In the fields of agriculture, food, medicine, paper, polymer, etc. considerable interest has been directed toward NIR spectroscopy because it is nondestructive and accurate, has a fast response, and is easy to operate. Detection of NIR light from a sample may be by either transmittance or reflectance. The transmittance method is very desirable for detecting internal information of matter with large volume, whereas the optical information from diffusely reflected spectra is confined to the subsurface layer of samples. It is especially important that progress be made in developing a NIR transmission device for detection of internal characteristics of high-moisture fruit and vegetable products or thick wood products (i.e., timber, lumber, or furniture). However, behavior of transmitted light from an agricultural or forest product is directly affected by both physical and chemical properties of the tissues, making it very difficult to examine in detail the optical characteristics of the tissue and the output origin of transmitted light for accurate evaluation of the sample constituents.

To resolve such problems, some researchers have given attention to time-of-flight (TOF) spectroscopy, which uses short pulses of illumination in which a portion of the light is scattered but most of the light propagates through the sample. A time-resolved measurement, or time domain system, could provide the TOF information of the detected light. Patterson et al. (1) theoretically proposed the usefulness of the time-resolved reflectance and transmittance for the noninvasive measurement of tissue optical properties. Profio (2) and Sevick et al. (3) reported on the properties of light scattering in various samples. Leonardi and Burns (4, 5) investigated quantitative measurements in scattering media on the basis of TOF spectroscopy with analytical descriptors. They revealed that experimental analysis of the time-resolved profile was very efficient for estimating the absorption and scattering coefficient.

Recently, Tsuchikawa and Tsutsumi (6, 7) have proposed a hyphenated technique between TOF and NIR spectroscopy, which is called TOF-NIRS. This system combines the best features of the spectrophotometer (i.e., tunability of wavelength) and the laser beam (i.e., permeability of incident light), and, more advantageously, the behavior of transmitted light could be observed by time-resolved state within very a short time domain. The system, made up of a parametric tunable laser and a NIR photoelectric multiplier, was constructed on the basis of a new concept, in which a time-resolved profile of transmitted output power was measured with nanosecond sensitivity. In the field of medical science, such a time-domain method is expected from developing optical tomography techniques, which can especially be used for the noninvasive detection of breast cancer (8) or hemoglobin concentration changes associated with the neural activation of the human brain (9). Therefore, TOF-NIRS should be also applicable for image construction concerning agricultural products. In this chapter, we introduce the principle of TOF-NIRS and some applications to agricultural and forest products.

## MEASURING APPARATUS

A typical schematic diagram of the measuring system for TOF-NIRS is illustrated in Figure 4.2.1. The system consists of an exciter laser (Q-switched Nd:YAG laser), an

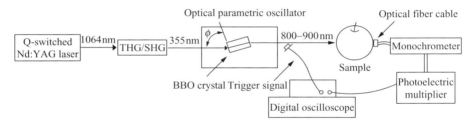

**Figure 4.2.1.** Schematic diagram of the measuring system (TOF-NIRS).

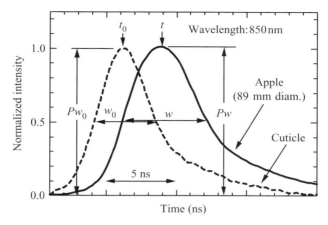

**Figure 4.2.2.** Normalized time-resolved profiles of an apple and its cuticle.

optical parametric oscillator, a monochromator, a near-infrared photoelectric multiplier, and a digital oscilloscope. The wavelength of the pulsed laser ($\lambda$) could be tuned from 410 nm to 2550 nm by the optical parametric oscillation of a BBO ($\beta$-BaB$_2$O$_4$) crystal (10). The monochromator is utilized to keep the spectral line width constant because it varies with the wavelength.

The transmitted output power from the sample is measured by a NIR photoelectric multiplier with a spectral response ranging from 300 nm to 1700 nm, which is cooled to $-80°$C, through an optical fiber cable. A Si pin-type photodiode is placed near the optical parametric oscillator to generate a trigger signal. The optical fiber cable is directly in contact with the sample.

## OUTLINE OF TIME-RESOLVED PROFILE

The time-resolved profile refers to the variation in intensity of the detected light beam with time. The normalized time-resolved profiles of an apple (89 mm in diameter) and its cuticle (1.0 mm in thickness) are illustrated in Figure 4.2.2. The sampling time and the number of times the transmitted output power was averaged were 100 ns and 300, respectively. The time-resolved profile of the cuticle was conveniently employed as the reference. It is important to focus on some typical parameters representing the variation of the time-resolved profile. For example, variations in peak maxima, the time delay of peak maxima, and the full width at half-maximum of the profile are fundamental optical factors. The measure of attenuance ($A_t$) is defined as follows:

$$A_t = \log\left(\frac{Pw_0}{Pw}\right) \tag{4.2.1}$$

where $Pw_0$ and $Pw$ indicate the peak maxima of the reference and the object, respectively. The measure of time delay of peak maxima ($\Delta t$) is expressed as follows:

$$\Delta t = t - t_0 \qquad (4.2.2)$$

here $t_0$ and $t$ indicate the time at peak maxima of the reference and the object, respectively. The variation of the full width at half-maximum value of the profile ($\Delta w$) is also expressed as follows:

$$\Delta w = w - w_0 \qquad (4.2.3)$$

where $w_0$ and $w$ indicate the full width at half-maximum value of the profile for the reference and the object, respectively. Leonardi and Burns (4) also referred to other informative optical parameters concerning the time-resolved profile.

## APPLICATION OF TOF-NIRS TO AGRICULTURAL AND FOREST PRODUCTS

The following are examples of application of TOF-NIRS to agricultural and forest products. These basic data can be helpful in realizing a measurement system for the agricultural and forest industries.

### 1.   Application of TOF-NIRS to Wood as Anisotropic Cellular Structure (6, 7)

TOF-NIRS was introduced to clarify the optical characteristics of wood having a cellular or porous structure as an anisotropic medium. The combined effects of the wood structure, the wavelength of the laser beam, and the detection position of transmitted light on the time-resolved profiles were investigated. In this study, the time-resolved profile of a sample with thickness of 1 mm was taken as a reference.

The optical penetrating length was strikingly improved as compared to the previous measuring system. Variations in the attenuance of peak maxima $\Delta t$, the time delay of peak maxima $\Delta t$, and the variation of full width at half-maximum $\Delta w$ were strongly dependent on the cellular structure of a sample and the wavelength of the laser beam (Fig. 4.2.3). The substantial optical path length was estimated from $\Delta t$. In the case of the detection of the transmitted output power at the opposite face, it became about 30 to 35 times as long as sample thickness, except for the absorption band of water. $\Delta t \times \Delta w$, representing the light scattering condition, increased exponentially with the sample thickness or the distance between the irradiation point and the end of the sample (Fig. 4.2.4). Around $\Delta = 900$–950 nm, there may be considerable light scattering in the lumen of the tracheid, which is multiple specular reflection and easy to propagate along the length of the wood fiber. Such a tendency was remarkable for soft wood (e.g., spruce in Fig. 4.2.4) with an aggregate of thin layers of cell walls. When we apply TOF-NIRS to cellular structural materials like wood, it is very important to give attention to the difference in the light scattering within the cell wall and the multiple specular-like reflections between cell walls.

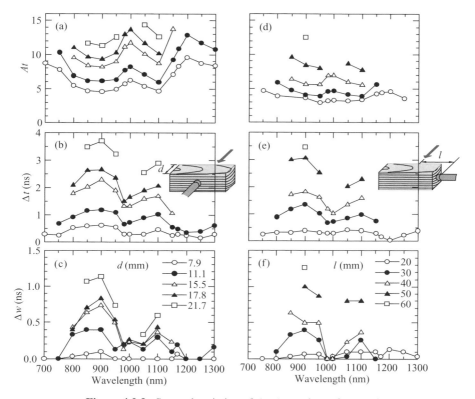

**Figure 4.2.3.** Spectral variation of $A_t$, $\Delta t$, and $\Delta w$ for wood.

## 2.    Application of TOF-NIRS for Detecting Water Core in Apples (11)

TOF-NIRS was applied to detection of the water core in apples. Figure 4.2.5 shows the cross sections of tested apples. The combined effects on the time-resolved profiles of water cores, the wavelength of the laser beam, and the detection position of transmitted light were investigated in detail. $A_t$, $\Delta t$, and $\Delta w$, where the time-resolved profile of the cuticle was employed as the reference, decreased gradually as the size of the water core area increased (Fig. 4.2.6). The water-cored tissue would be expected to transmit much more energy because of the filling of the intercellular spaces with liquid. This results in less light scattering, so that the light path time through a sample decreases.

The substantial optical path length was estimated from $\Delta t$, and the ratio of the substantial path length to the nominal optical path length $R_{S/N}$ was calculated. It could be estimated that the value of $R_{S/N}$ on the equator at $\Delta = 800$ nm was $\sim 10$ to 17 and that for $\Delta = 900$ nm varied from 6 to 11. Fluctuation of $\Delta t$ standardized by the distance between the irradiation position and the detection position $l$ is very small, and it differs clearly with the presence and size of a water core. One can reliably detect the water-cored condition by $\Delta t$, although the present classification of water core is generally visual and sensory (Fig. 4.2.7).

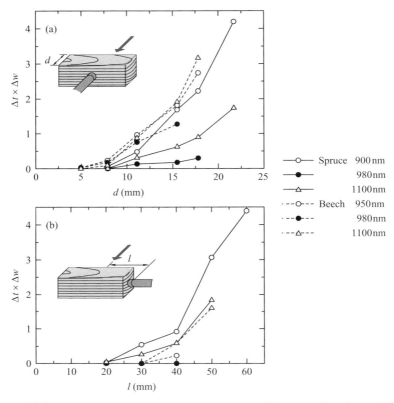

—○— Spruce 900 nm
—●— 980 nm
—△— 1100 nm
--○-- Beech 950 nm
--●-- 980 nm
--△-- 1100 nm

**Figure 4.2.4.** Variation of $\Delta t \times \Delta w$ with wood sample thickness or detection position.

|  Rank 0 | Rank 1 | Rank 2 | Rank 3 |

**Figure 4.2.5.** Cross sections of tested apples.

## 3. Application of TOF-NIRS for Detecting Sugar Content and Acidity in Fruits (12, 13)

The effect of sugar and acid contents in fruit on the transmitted output power was fully examined and used to estimate the optical parameter preferable for detecting internal quality. The performances of multiple linear regression (MLR), principal component regression (PCR), and partial least-squares regression (PLS) analysis by TOF-NIRS were also compared to those from normal methods using reflectance data.

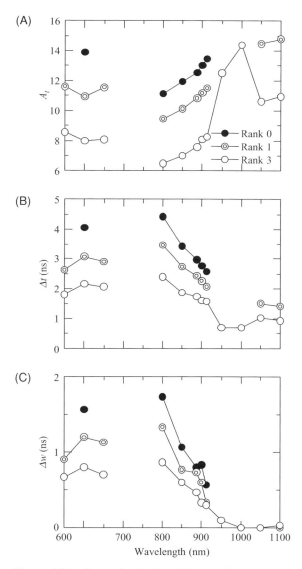

**Figure 4.2.6.** Spectral variation of $A_t$, $\Delta t$, $\Delta w$ for apple.

Figure 4.2.8 shows the variation of $A_t$, $\Delta t$, and $\Delta w$ standardized by dividing by sample thickness $d$ with the sugar content or acidity in apple. A pulsed laser beam of $\Delta = 800$ nm irradiated the sample, and the transmitted output power was detected at its opposite face. $A_t/d$, $\Delta t/d$, and $\Delta w/d$ decreased gradually as sugar content increased. These results mean that the intensity of transmitted light increased and the optical path length decreased incrementally with sugar content. To correctly interpret this phenomenon, we may consider that the difference of refractive index between the

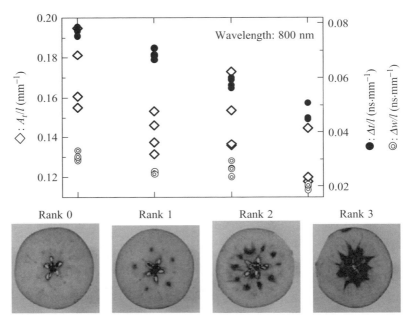

**Figure 4.2.7.** Relationship between the rank of water core and $A_t$, $\Delta t$, and $\Delta w$ standardized by dividing the distance between the irradiation position and the detection position $l$.

cellular structure and the intercellular spaces of an apple differs with the sugar content. The refractive index of the cellular structure may approach those of the intercellular spaces as sugar content increases. Therefore, the variations of refractive index with sugar content should relate directly to the time-resolved profiles. On the other hand, $A_t/d$ and $\Delta w/d$ increased as acidity increased. It may be also supposed that refractive index would vary with acidity.

An attempt was made to find robust calibration equations for sugar content and acidity by MLR, PCR, and PLS, where the optical parameters were employed as explanatory variables. Normally, chemometrics by NIR spectra employs the absorbance as the explanatory variable, where only wavelength-dependent characteristics of the materials can be considered. In this case, it is very difficult to precisely evaluate the small amount of a constituent such as acid content in a fruit. On the other hand, chemometrics by TOF-NIRS would be related to both wavelength- and time-dependent characteristics as the explanatory variables, where the light absorption and light scattering phenomena in a sample are included. It may therefore be possible to detect the acid content in a fruit on the basis of this new optical concept. The statistical results are summarized in Tables 4.2.1 and 4.2.2. Figure 4.2.9 shows the PLS analysis in a optimum model for acidity in apple. In the case of normal analysis by second-derivative NIR spectra, standard error of calibration (SEC) and correlation coefficient between measured and predicted acidity $r$ were limited to 0.048% and

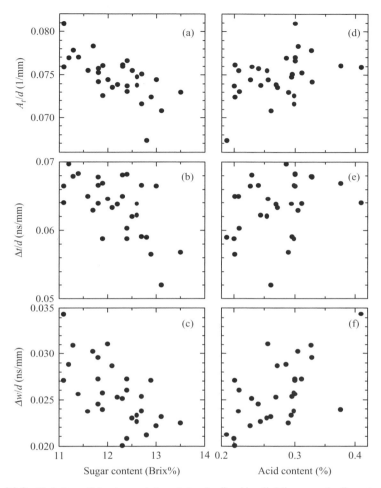

**Figure 4.2.8.** Variation of $A_t$, $\Delta t$, and $\Delta w$ at standardized by dividing sample diameter $d$ with the sugar or acid content in apple.

**TABLE 4.2.1. Statistical Parameters for Multiple Linear Regression (MLR) Analysis**

| Spectroscopic Method | Constituents | Selected Wavelength (nm) | $r$ | SEC (%) |
|---|---|---|---|---|
| Normal NIR | Sugar content | 840, 884, 892 | 0.70 | 0.47 |
|  | Acid content | 816, 846, 900 | 0.57 | 0.046 |
| TOF-NIRS | Sugar content | 810[a], 860[b], 800[c] | 0.85 | 0.35 |
|  | Acid content | 860[a], 890[b], 800[c] | 0.73 | 0.039 |

$r$: Regression coefficient between real value and evaluated value. SEC: Standard error of calibration. [a]$A t/d$; [b]$\Delta t/d$; [c]$\Delta w/d$.

**TABLE 4.2.2. Statistical Parameters for Principle Component Regression (PCR) and Partial Least-Squares (PLS) Analysis**

| Chemometrics | Spectroscopic Method | Constituents | Number of Factors | r | SEC (%) |
|---|---|---|---|---|---|
| PCR | Normal NIR | Sugar content | 9 | 0.63 | 0.58 |
| | | Acid content | 9 | 0.56 | 0.052 |
| | TOF-NIRS | Sugar content | 6 | 0.84 | 0.38 |
| | | Acid content | 6 | 0.68 | 0.043 |
| PLS | Normal NIR | Sugar content | 9 | 0.72 | 0.52 |
| | | Acid content | 9 | 0.66 | 0.048 |
| | TOF-NIRS | Sugar content | 6 | 0.94 | 0.23 |
| | | Acid content | 6 | 0.92 | 0.023 |

$r$: Regression coefficient between real value and evaluated value. SEC: Standard error of calibration.

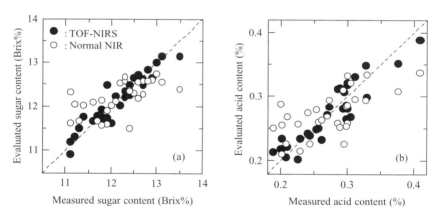

**Figure 4.2.9.** PLS analysis for sugar or acid content in apples.

0.66, respectively. However, in the case of the new PLS analysis by TOF-NIRS in which $A_t/d$, $\Delta t/d$, and $\Delta w/d$ were employed as explanatory variables, SEC and $r$ were improved up to 0.023% and 0.92, respectively. These results clearly indicate the usefulness of TOF-NIRS for the detection of a small amount of constituent in a fruit.

It is concluded that the evaluation of sugar content or acidity by TOF-NIRS provided high accuracy compared to the normal method independent of chemometric procedure. In particular, PLS analysis presented good calibration with the equation described above.

## ACKNOWLEDGMENTS

This study was partly supported by Grant-in-Aids from Agriculture, Forestry and Fisheries Research Council of Japan (Grant number 1521 to S. T.).

## REFERENCES

1. M. S. Patterson, B. Chance, B. C. Wilson. Time resolved reflectance and transmittance for non-invasive measurement of tissue optical properties. *Appl Optics* **28:** 2331–2336, 1989.

2. A. E. Profio. Light transport in tissue. *Appl Optics* **53:** 2216–2221, 1989.

3. E. M. Sevick, B. Chance, J. Leigh, S. Nioka, M. Maris. Quantitation of time- and frequency-resolved optical spectra for the determination of tissue oxygenation. *Anal Biochem* **195:** 330–351, 1991.

4. L. Leonardi, D. H. Burns. Quantitative measurements in scattering media: photon time-of-flight analysis with analytical descriptions. *Appl Spectrosc* **53:** 628–636, 1999.

5. L. Leonardi, D. H. Burns. Quantitative mutiwavelength constituent measurements using single-wavelength photon time-of-flight correction. *Appl Spectrosc* **53:** 637–646, 1999.

6. S. Tsuchikawa, S. Tsutsumi. Application of time-of-flight near infrared spectroscopy to wood with anisotropic cellular structure. *Appl Spectrosc* **56:** 1117–1124, 2002.

7. S. Tsuchikawa. Application of time-of-flight near infrared spectroscopy to wood. In: *Useful and Advanced Information in the Field of Near Infrared Spectroscopy.* S. Tsuchikawa, ed. Research Signpost, 2004, pp. 293–307.

8. J. C. Hebden, M. Tziraki, D. T. Delpy. Evaluation of the temporally extrapolated absorbance method for dual-wavelength imaging through tissuelike scattering media. *Appl Optics* **36:** 3802–3817, 1997.

9. J. P. Van Houten, D. A. Benaron, S. Spilman, D. K. Stevenson. Imaging brain injury using time-resolved near infrared light scanning. *Pediatr Res* **39:** 470–476, 1996.

10. P. G. Harper, B. S. Wherrett. *Nonlinear Optics.* Academic Press, New York, 1977.

11. S. Tsuchikawa, S. Kumada, K. Inoue, R. Cho. Application of time-of-flight near-infrared spectroscopy for detecting water core in apples. *J Am Soc Hort Sci* **127:** 303–308, 2002.

12. S. Tsuchikawa, T. Hamada. Application of time-of-flight near infrared spectroscopy for detecting sugar content and acidity in apples. *NIR News* **14:** 6–8, 2003.

13. S. Tsuchikawa, T. Hamada. Application of time-of-flight near infrared spectroscopy for detecting sugar and acid contents in apples. *J Agric Food Chem* **52:** 2434–2439, 2004.

# NIR Imaging and its Applications to Agricultural and Food Engineering

E. NEIL LEWIS, JANIE DUBOIS, and LINDA H. KIDDER

## INTRODUCTION

Chemical imaging spectroscopy is a maturing analytical technique that provides a qualitative and quantitative characterization of the spatial distribution of species of interest in complex, heterogeneous samples. Advances in array detectors and computing in the late 1980s and early 1990s enabled the coupling of macroscopic and microscopic imaging capabilities and spectroscopy for mid-infrared (MIR), near-infrared (NIR), and Raman spectroscopic techniques (1–3). NIR chemical imaging (NIR-CI), in particular, is a robust technique that can be used for the characterization of complex biological materials such as food and greatly extends the capabilities of NIR spectroscopy.

## METHODOLOGY AND INSTRUMENTATION

Near-infrared imaging can be considered as a massively parallel implementation of single-point, or bulk NIR spectroscopy. Indeed, the instrumental principles are largely identical: The sample is illuminated with a source of radiation in the NIR range, and after interaction with the sample the resulting absorbed or reflected radiation is directed toward an appropriately sensitive detector. However, the difference between single-point NIR spectroscopy and imaging is the simultaneous acquisition of tens of thousands of spatially resolved spectra as compared to a single spectrum. This is enabled by the use of a two-dimensional NIR detector, the focal plane array (FPA). The choice of FPA is based mostly on required spectral range, but the need for cryogenic cooling for some detectors can reduce their utility for industrial applications.

*Near-Infrared Spectroscopy in Food Science and Technology*, Edited by Yukihiro Ozaki, W. Fred McClure, and Alfred A. Christy.

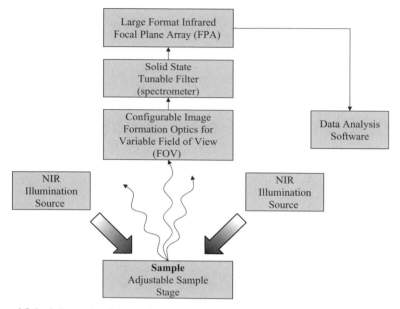

**Figure 4.3.1.** Schematic of the major optical components and mode of operation of a commercial NIR-CI instrument.

Wavelength selection is achieved either by modulating the source, as with Fourier transform (FT) imaging or through image filtering using tunable or fixed-bandpass filters. Depending on the optics employed, NIR imaging may be a microscopic or macroscopic method.

Although a number of experimental implementations exist for generating NIR spectroscopic imaging data, (4) the tunable filter approach using a liquid crystal tunable filter (LCTF) has some distinct advantages. Rapid tuning of discrete wavelengths through software control with no moving parts enables the collection of data sets comprising approximately 80,000 spectra in a matter of a couple of minutes. Figure 4.3.1 shows a schematic description of a commercial NIR-CI instrument incorporating an LCTF. In addition to tuning continuously throughout the entire NIR spectral tuning range, the LCTF is also able to collect data over a narrow spectral range or discretely using just a few analytically relevant wavelengths. A random-access tunable filter offers compelling advantages over FT methods, for instance, by minimizing the size of the data sets and dramatically speeding up the data collection process. Once a data collection method has been optimized and validated, it can readily be "bundled" such that collection and analysis is integrated and accomplished in "real time." These chemical imaging protocols can proceed with little or no operator intervention.

Another benefit of the tunable filter approach is that the FPA, LCTF, and imaging optics can be unified into a single wavelength-selectable chemical imaging module. This results in an instrument that is compact and has a highly configurable field of

view (FOV), suitable for installation in either the laboratory or an industrial environment. Further benefits are realized because these types of instruments also have no
moving parts and, with the exception of power, require no external utilities. Standard
commercial instruments of this type are normally mounted in a downward-looking
orientation, on an optical rail with an illumination source and a sampling stage. However, this basic design also permits the instrument to be mounted in side-looking or
inverted orientations for more specialized applications.

## DATA ANALYSIS

A typical commercially available NIR imaging system collects approximately 80,000
spatially resolved spectra simultaneously, providing orders of magnitude more spectral information than is collected with single-point spectrometers. A chemical imaging
data set is represented by a three-dimensional cube where two axes describe vertical and horizontal spatial dimensions and the third dimension represents the spectral
wavelength dimension (Fig. 4.3.2).

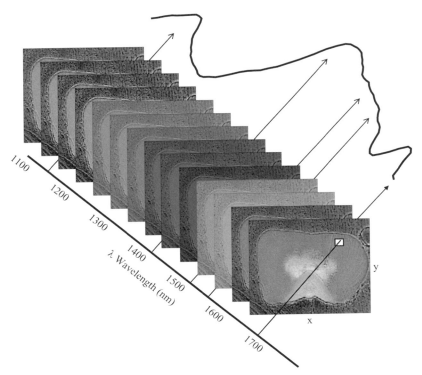

**Figure 4.3.2.** Schematic of an image cube collection with a NIR-CI instrument. The data
cube is constructed plane by plane, where each plane contains the spectral information at one
wavelength ($\lambda$) for all spatial positions ($x_i, y_i$) in the field of view.

A variety of data analyses can be applied to visualize and process chemical imaging data sets, the goal being to evaluate and quantify the heterogeneity (contrast) of a sample. However, the process can be generalized into four basic steps: spectral correction, spectral preprocessing, classification, and image analysis. The first two steps are customary for all comparative spectroscopic analyses, whether for single-point or imaging implementations. Classification techniques are also performed for both single-point and imaging data, although the imaging data sets enable unique and novel applications of the familiar methods. Image analysis of spectroscopic data is of course unique to imaging data and differentiates it from "spectral" thinking.

Spectral correction involves taking the ratio of the data to a background reference in order to remove the instrument response component. Because the filter-based NIR-CI instruments operate in a true "staring" or DC mode, it is also necessary to subtract the "dark" camera response from the image cube. The correction to reflectance ($R$) is therefore:

$$R = (\text{Sample} - \text{Dark}) \, / \, (\text{Background} - \text{Dark})$$

Further processing is also usually performed to transform the reflectance image cube to its $\log_{10} (1/R)$ form, which is effectively the sample "absorbance." The same set of operations can also be performed for transmittance data cubes. This results in chemical images in which brightness linearly maps to the analyte concentration and is generally more useful for comparative as well as quantitative purposes. The next step, spectral preprocessing, is typically used to remove physical effects from the data, such as, scattering effects due to particle sizes, etc.

In some instances a simple gray scale image, at a single wavelength, can highlight the spatial distribution of a particular chemical component within a sample, based on differences in the absorbance of all the sample components at that wavelength. In addition, these individual image planes can be superimposed by mapping onto red, green, and blue (RGB) channels to create composite color chemical images that summarize the distribution of multiple components. However, things are not usually that simple, and data sets must be at least normalized to account for pathlength effects or to further isolate components of interest. Other processing may also be warranted to account for scattering effects. For cases with significant spectral overlap, common in NIR, or for low signal-to-noise ratio data, statistical and chemometric approaches such as principal component analysis (PCA) or partial least-squares (PLS) regression can be utilized. Additionally, after these methods are employed to isolate spectral information unique to the component of interest, standard image processing approaches can be applied to further characterize the resolved spatial and chemical structure of the sample. These techniques can include particle sizing and particle counting, or quantitative estimates of sample heterogeneity and relative species abundance with other statistical approaches. In practice the value of NIR chemical imaging is highly dependent on the use and development of effective multivariate data processing and data mining strategies because the ability to analyze so many spectra by any other means is highly inefficient.

It is also possible to produce a four-dimensional data cube by adding a third spatial coordinate. This is achieved by physically sectioning a sample and acquiring multiple image cubes along the depth of the sample. An example of such an approach is shown later in this chapter where the three-dimensional chemical image of a corn kernel is examined.

## THE "VALUE PROPOSITION" FOR NIR CHEMICAL IMAGING

Spectroscopic (chemical) imaging is principally designed for the investigation of systems with a complex spatial distribution of chemical components and is therefore ideal for the study of biological materials and, by analogy, raw and processed foods. The wavelength dimension carries chemical information, and the two-dimensional image acquisition provides the spatial distribution of these chemical moieties. This type of spatial and chemical information is highly specific and can be used to characterize the physiology of hybrid varieties of grains, for example. What chemical species, how much is present, and where they are located in a sample may be answered in a single experiment using NIR-CI.

The technique can also be used as a parallel spectroscopic tool. By spatially isolating multiple samples and selecting an appropriate FOV, tens to hundreds of samples can be simultaneously imaged onto the detector and investigated in a single measurement. For this type of high-throughput application, the intent is typically to perform intersample characterization, quickly screening samples for outliers, or looking to characterize multiple samples with specific spectroscopic criteria. The capability is operative over a range of size scales and sample shapes that are difficult or impossible for current single-point instrumentation to address.

A further strength of an imaging instrument is the ability to view both samples and pure components, or reference materials, simultaneously by choosing a FOV and an arrangement of samples that enable this. In effect the instrument extends the double-beam spectrometer paradigm to a multiple-beam spectrometer where the number of beam paths is a function of the number of internal references simultaneously viewed by the instrument. Creating single data sets that contain both the sample and the reference materials simplifies many aspects of the experiment. The data set is "internally" consistent because the reference and sample data are measured simultaneously by the same instrument—temporal variations in instrument response are automatically accounted for. Also, because the relevant data always remain together in a single image data file, the data and reference samples are always processed together in identical ways and cannot be separated. The reference samples may be left in the instrument when new "unknown" samples are introduced, and therefore a new experiment simultaneously results in new reference spectra. Any changes that affect instrument response or data collection parameters result in those changes being reflected in the reference sample spectra. In principle, this can help obviate the need to apply transfer of calibration procedures to analyze samples on different instruments.

Additionally, because the spectra collected are spatially independent, a localized particle as small as a single pixel can be identified among hundreds or thousands of others without the "dilution" effects encountered in bulk NIR. The question of detection limits becomes dependent on particle statistics rather than signal-to-noise ratio and concentration, implying that NIR imaging can detect components at significantly lower concentrations than is possible with bulk NIR. The main assumption in this statement is that the component of interest is heterogeneously dispersed and spatially localized.

NIR chemical imaging also inherits the capabilities that make single-point or bulk NIR spectroscopy useful in so many different applications. Sample preparation is limited or nonexistent, high signal-to-noise ratio data are acquired rapidly, and data analysis can be automated. Despite the range of potential applications, the use of this technology is still in its infancy in the food sciences. Most of the methods developed so far involve machine vision for on-line or at-line quality control, using the visible and short wavelength NIR spectral ranges (5–7). This is due, in part, to the relative low cost and easy availability of cameras operating in this spectral range. However, as the technology becomes more mainstream and longer-wavelength cameras are more readily available, the greater chemical specificity afforded in the 1000- to 2500-nm spectral region will help promote the adoption of NIR chemical imaging. To highlight the utility of the technique for the analysis of food and food products, we present two examples, an examination of the three-dimensional distribution and localization of components (oil, starch) in a corn kernel and a high-throughput characterization of cattle feed.

## APPLICATIONS

### Three-Dimensional NIR Chemical Imaging of a Corn Kernel

NIR spectroscopic quantitative analysis to characterize the nutritional and commercial value of grains is widespread following pioneering work by Norris and collaborators (8) (for reviews of the principles see Refs. 9 and 10). Typically, the bulk quantification of sample components like oil, protein, starch, and water is the information of interest and determinable with NIR spectroscopy. By adding the spatial dimension, the chemistry of the sample can be probed for purposes such as comparing the physiology of different breeds and, by extension, their economic and nutritional value. It is anticipated that this information could enable the development of tighter control standards for breeding programs and commensurate improvements in the quality of the product.

In a typical NIR spectroscopic measurement, spectral contribution from the entire corn kernel is averaged in a single spectrum. In the following example, the chemical information contained in the NIR spectrum and the spatial resolution provided by NIR imaging instrumentation were used to investigate the three-dimensional biochemistry and morphology of a corn kernel.

Reflectance images were acquired at a magnification of $30 \times 30$ μm per pixel with a MatrixNIR™ NIR system (Spectral Dimensions Inc., Olney, MD) equipped with an InGaAs focal-plane array detector with $320 \times 256$ pixels. A wavelength increment of 7 nm was used to collect data over the spectral range 1100–1750 nm. The first data cube was acquired from the intact kernel, and subsequent reflectance image cubes were obtained after shaving a 250-μm longitudinal slice off the kernel. The complete four-dimensional data set is built from 43 consecutive three-dimensional data cubes acquired from adjacent slices of the corn kernel, each three-dimensional image cube requiring about 2 minutes of acquisition, for a total of 83 minutes to image the entire kernel. The final data set consists of 3,522,560 NIR spectra. Dark and background image cubes were also collected and used to convert each separate image cube to reflectance before being converted to absorbance with $\log_{10} (1/R)$.

Figure 4.3.3, A and B, shows a typical NIR chemical image and a reflectance spectrum acquired in the hard endosperm structure of the kernel. Although each image cube visualizes the biochemical composition of specific morphological structures at each physical depth into the sample, all of the data sets can also be assembled to construct an image of the three-dimensional distribution of the chemical components of interest. The individual image cubes are first processed, using a commercially available software package (ISys 3.0, Spectral Dimensions, Inc.), by first subtracting

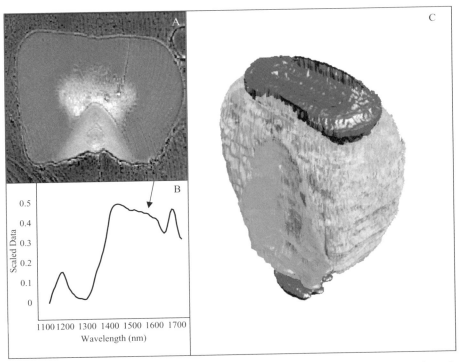

**Figure 4.3.3.** A, gray scale NIR-CI acquired from a longitudinal section of a corn kernel. B, NIR reflectance spectrum acquired at a single pixel position corresponding to the hard endosperm of the kernel. C, three-dimensional isosurface plot calculated from principal components of the NIR spectra of the kernel arising predominantly from oil (green) and different starches (red, yellow).

the mean spectrum and scaling each spectrum to unit variance. The resulting data sets are then further processed to obtain PCA scores and loadings. PCA factors are selected that correspond most closely to the chemical components of interest, and their corresponding score images are used to represent the two-dimensional distribution of that chemical component in each slice. The corresponding score images for all slices representing a single component are then assembled into a new image cube from which an isosurface plot is created with Matlab 6.0 (The Mathworks, Natick, MA). The isosurface plot is a three-dimensional contour plot in which the points lying on the surface are of the same value, that is, the isovalue. The effect is to highlight contiguous spatial regions in three dimensions that have similar chemical composition. Isovalues are determined by examining the score images and a histogram of the score image intensities for the chemical species of interest (ISys 3.0). Separate isovalue plots are created in a similar manner for the other chemical species in the sample, and finally the complete three-dimensional isosurface plot for all components is rendered again with Matlab 6.0. Figure 4.3.3C shows the three-dimensional reconstruction of the corn kernel based on the isovalues calculated for principal components indicative of oil and starch. Like most grains, the corn kernel has a relatively simple anatomy with radically different protein, lipid, and carbohydrate contents in its individual structures. NIR spectra of the corn kernel clearly differentiate between the pericarp (primarily cellulose and hemicellulose), the endosperm (90% starch), and the germ (30% oil) (11). Indeed, although chemically similar, the endosperms differ in the packing of the starch and the protein bodies encased in the matrix (11), which results in differences in the NIR spectra of the two types of endosperm. These are clearly visible in the volumetric image. Among other things, it is possible to use these data to calculate the relative volumes of these two structures and assess hardness of the corn kernel. This is an attribute of significant economic importance.

Finally, although this is a relatively straightforward example, it illustrates the value of a NIR imaging measurement in providing not only a biochemical and morphological description of the different components in a spatially and chemically complex sample but also the possibility to use simple pixel statistics based on the spatial "chemical content" to obtain quantitative information as well. Interestingly, FTIR microscopic imaging has also shown potential as a tool for the investigation of corn and wheat kernel physiology and could be used as complementary technique (12, 13).

## High-Throughput Screening of Cattle Feed

Outbreaks of bovine spongiform encephalopathy (BSE) in Europe, and sporadic cases in North America and Japan, have prompted changes throughout the cattle industry. One of the first lines of defense against the spread of the disease has been tighter control on the content of animal feed. It is generally accepted that BSE spreads through feed containing meat and bone meal from contaminated ruminants. As a result, proteins of mammalian origin, with some exceptions, may not be used to feed ruminant farm animals in the European Union and the United States(14–16).

Although the only testing method currently approved for the identification and estimation of constituents of animal origin in feed is light microscopy, a variety of additional methods for the detection of proteins of animal origin are also being developed. These include polymerase chain reaction, ELISA, and NIR spectroscopy, microscopy, and imaging (17–19). The detection limits vary, from 0.01% attained by light microscopy, which measures the presence of bone fragments (18, 19), to 1% by NIR spectroscopy (18). However, according to results obtained in the European STRATFEED project (17, 18) 0.05% has been attained by coupling NIR microscopy with a sedimentation step. A NIR imaging method employing an infrared array detector is being developed that improves the throughput of the NIR microscopic method while retaining its specificity (20).

Although the modified light microscopy method has the desired sensitivity, it imposes a heavy workload, involving multiple steps of grinding, sieving, solvent extraction, and subjective visual microscopic examination. Particles of animal origin are identified by the presence of animal bone, recognized by its morphology. The typical lacunae of land animal bones (i.e., mammals and birds) distinguish them from fish bones. A quantitative result is calculated on the basis of the estimated proportion by weight of animal bone fragments to the total weight, taking into account the proportion of bone in the constituent animal of origin. In contrast, the NIR imaging technique requires little sample preparation. The subjective qualification of bone in a feed sample is replaced by discrimination based on the direct measurement of the specific chemical signature of particles of animal origin from the other ingredients in the unprocessed feed. The following example outlines how the NIR imaging method using a FPA detector may be used as a high-throughput method to screen for the presence of animal protein particles in the presence of other feed particle types.

A sample containing a randomly distributed array of particles of cattle feed (8.5 mg) contaminated by "spiking" with approximately 1% poultry meal is imaged with a MatrixNIR$^{TM}$ NIR chemical imaging system (Spectral Dimensions Inc.). Data are simultaneously acquired over a FOV of approximately $13 \times 10$ mm with an InGaAs FPA detector containing $320 \times 240$ pixels. The resulting image cube consists of 76,800 spatially resolved spectra where each spectrum spans the wavelength range 1350–1650 nm with a 5-nm data increment and corresponds to approximately $40 \times 40$ µm of the sample surface. The data acquisition time for a single image cube is approximately 5 minutes. A background data set, collected from a reflectance standard, is used to calculate reflectance data before being further transformed to absorbance with $\log_{10} (1/R)$.

Figure 4.3.4A shows a NIR bright-field image of the feed sample showing no discernible difference between the individual particles. This image essentially corresponds to what visual inspection or visible digital imaging would provide. However, the NIR spectra (Fig. 4.3.4B) of the two principal fractions of this feed, namely cattle feed component and poultry-based contaminant, show significant differences. This difference can be exploited to generate image contrast and identify the individual components within the "unknown" sample shown in Figure 4.3.4A. A PLS calibration set was created from single image cubes of the cattle feed and poultry meal 'pure' components (not shown), where each calibration image cube contains 76,800 spectra from hundreds of different "pure" component particles. In principle these two data sets statistically span the chemical variation found in each "pure" component. To

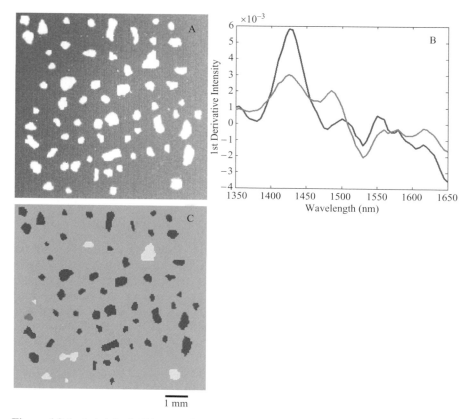

**Figure 4.3.4.** A, bright-field image of cattle feed contaminated with 1% poultry meal. B, NIR spectra of cattle feed and poultry meal. C, results of a PLS prediction showing the pixels assigned to cattle feed (blue) and the pixels assigned to poultry meal (red).

identify regions of the unknown sample corresponding to these two major fractions, a PLS calibration model is developed by assigning a unit value for the concentration of each pure component in the calibration data set and applying this model to the contaminated feed image cube. The results are shown in Figure 4.3.4C, where the pixels assigned to poultry meal contaminants are colored red and the pixels assigned to the cattle feed base are blue. The yellow color identifies pixels that are classified as "mixtures" and are estimated to contain a low concentration of poultry meal. A quantitative estimate of the contamination level can be obtained from these results by measuring the PLS weighted area of the pixels classified by PLS relative to the total area of all particles. This yields a contamination level estimate of 1.5%, in relatively good agreement with the actual content of 1%.

The benefits of NIR imaging, especially as compared to the established microscopic method, for identifying animal protein contaminant in cattle feed are severalfold. It is an objective classification and, once developed can be automated, not needing specially trained personnel to perform the analysis. In addition, it is significantly faster because it works on the raw materials without sample preparation and requires only 5 minutes to acquire a complete chemical image cube. This single image cube may contain hundreds or even several thousand particles.

There is an obvious need for high-throughput detection of land animal proteins in feedstuffs, and the accepted method cannot meet this requirement. Near-infrared imaging is a promising alternative because it shows sensitivity comparable to the currently accepted method while the optical and data processing characteristics of the technique make it amenable to automation and significantly higher throughput. With some modification of the laboratory-based implementation, NIR images could be acquired and statistically analyzed in a near-real-time measurement at a site of feed transfer or loading.

## FUTURE PERSPECTIVES

The penetration of this technique in the field of food technology will be driven by a number of factors, not least of which is a recent renewed interest in the protection of the food supply. The ability of the technique to rapidly, noninvasively, and nondestructively screen relatively large quantities of food for foreign entities, contaminants, or intentional adulteration sets it apart from many other analytical methods. In addition the technique is highly robust and transportable, and as such is also amenable to "field" or process applications, giving it additional value for the food industry. In conclusion, the future of the technique looks bright and the utilization of two-dimensional infrared array detectors not only as imaging devices but also as high-throughput spectrometers is having, and will continue to have, a major impact on the field of NIR spectroscopy.

## ACKNOWLEDGMENTS

The authors would like to thank Dr. Frederick Koehler IV for providing the data processing expertise necessary to process the corn data and produce the volumetric corn image.

## REFERENCES

1. P. J. Treado, I. W. Levin, E. N. Lewis. High fidelity Raman imaging spectroscopy: a rapid method using an acousto-optic tunable filter. *Appl Spectrosc* **46**: 1211–1216, 1992.

2. P. J. Treado, I. W. Levin, E. N. Lewis. Indium antimonide (InSb) focal plane array (FPA) detection for near-infrared imaging microscopy. *Appl Spectrosc* **48**: 607–615, 1994.

3. E. N. Lewis, P. J. Treado, R. C. Reeder, G. M. Story, A. E. Dowrey, C. Marcott, I. W. Levin. Fourier transform spectroscopic imaging using an infrared focal-plane array detector. *Anal Chem* **67**: 3377–3381, 1995.

4. P. Colarusso, L. H. Kidder, I. W. Levin, J. C. Fraser, J. F. Arens, E. N. Lewis. Infrared spectroscopic imaging: from planetary to cellular systems. *Appl Spectrosc* **52(3)**: 107A–120A, 1998.

5. B. Park, Y. R. Chen, R. W. Huffman. Integration of visible/NIR spectroscopy and multi-spectral imaging for poultry carcass inspection. *J Food Eng* **30**: 197–207, 1996.

6. D. Guyer, X. Yang. Use of genetic artificial neural networks and spectral imaging for defect detection on cherries. *Comput Electron Agric* **29**: 179–194, 2000.

7. P. M. Mehl, Y.-R. Chen, M. S. Kim, D. E. Chan. Development of hyperspectral imaging technique for the detection of apple surface defects and contaminations. *J Food Eng* **61**: 67–81, 2004.

8. K. H. Norris, J. R. Hart. Direct spectrophotometric determination of moisture content of grain and seeds. In: *Principles and Methods for of Measuring Moisture in Liquids and Solids*, A. Wexler, ed. Reinhold, New York, **4**: 19–25, 1965.

9. J. S. Shenk, J. J. Jr Workman, M. O. Westerhaus. Application of NIR spectroscopy to agricultural products. In: *Handbook of Near-Infrared Analysis*. D. A. Burns, E. W. Ciurczak, eds. Marcel Dekker, New York, pp. 383–431, 1992.

10. P. C. Williams. Application of near infrared reflectance spectroscopy to analysis of cereal grains and oilseeds. *Cereal Chem* **52**: 561–576, 1975.

11. A. R. Hallauer, ed. Specialty Corns, 2nd ed. CRC Press, Boca Raton, FL, 2001, p. 479.

12. B. O. Budevska, S. T. Sum, T. J. Jones. Application of multivariate curve resolution for analysis of FT-IR microscopic images of in situ plant tissue. *Appl Spectrosc* **57(2)**: 124–131, 2003.

13. C. Marcott, R. C. Reeder, J. A. Sweat, D. D. Panzer, D. L. Wetzel. FT-IR spectroscopic imaging microsocopy of wheat kernels using a mercury-cadmium-telluride focal-plane array detector. *Vibrat Spectrosc* **19**: 123–129, 1999.

14. European Commission, Commission of the European Communities. Regulation (EC) No 999/2001 of the European Parliament and of the Council of 22 May 2001 laying down rules for the prevention, control and eradication of certain transmissible spongiform encephalopathies. *Off J Eur Communities* **L 147**: 1–40, 22 May, 2001.

15. European Commission, Commission of the European Communities. Regulation (EC) No 1774/2002 of the European Parliament and of the Council of 3 October 2002 laying down health rules concerning animal by-products not intended for human consumption. *Off J Eur Communities* **L 273**: 1–95, 3 October, 2002.

16. Department of Health and Human Services, US Food and Drug Administration. Rule 21 CFR Part 589.2000: Substances Prohibited from Use in Animal Food or Feed: Animal Proteins Prohibited in Ruminant Feed. *Federal Register* **62(108)**: 30936–30978, 1997.

17. http://stratfeed.cra.wallonie.be/Public_Web_site/Home_page/index.cfm

18. G. Gizzi, L. W. D. van Raamsdonk, V. Baeten, I. Murray, G. Berben, G. Brambilla, C. von Holst. An overview of tests for animal tissues in feeds applied in response to public health concerns regarding bovine spongiform encephalopathy. *Rev Sci Tech Off Int Epiz* **22(1)**: 311–331, 2003.

19. European Commission. Commission Decision 98/88/EC of November 1998 establishing guidelines for microscopic identification and estimation of constituents of animal origin for the official control of feedingstuffs. *Off J Eur Communities*, **L 318**: 45–50, 27 November, 1998.

20. Baeten V. and Dardenne P. (2005). Applications of near-infrared imaging for monitoring agricultural food and feed products, In: *Spectrochemical Analysis using Infrared Multichannel Detectors*, R. Bhargava, I. W. Levin eds. Blackwell Publishing, pp. 283–301.

# CHAPTER 5

# Sampling Techniques

SATORU TSUCHIKAWA

## INTRODUCTION

Effective sampling technique applicable to various optical systems is an essential requirement for any chemical method. It is important to achieve a useful analysis that properly represents the composition of the larger sample. Increasingly accurate and precise analytical method in the laboratory will not improve on poor sampling technique, nor will it give more accurate numbers for estimating actual sample quality. Needless to say, sampling is the most important technique in making a good calibration equation with highly predictable performance. The calibration sample set that is applied to make a calibration equation should contain representative characteristics of the population for unknown samples that will be predicted by NIR spectroscopy. When we arrange to make a calibration equation, it is important to consider real sample conditions.

In the case of agricultural products, which present various features as natural materials, the sample set should be sufficiently variable with respect to variety, producing area, producing year, and maturity stage to meet the conditions. The distribution of the constituent being calibrated for over the calibration samples is also important (see Fig. 5.1). As some researchers have pointed out, from the viewpoint of statistical analysis the range of the variability should be as large as that expected in any future sample, and it is better to keep a uniform spread of values over the whole range (1, 2). This has the advantage of providing more precise estimates of the calibration constants. It would certainly not be a good idea to take a calibration set that is more concentrated in the center of the range than the natural distribution. Nor is it, in general, a good idea to concentrate too strongly on the extremes of the range.

*Near-Infrared Spectroscopy in Food Science and Technology*, Edited by Yukihiro Ozaki,
W. Fred McClure, and Alfred A. Christy.
Copyright © 2007 John Wiley & Sons, Inc.

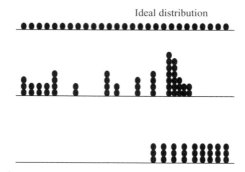

**Figure 5.1.** Model of some varying distributions of the content of a constituent.

## SAMPLE PREPARATION

Quite often, samples may require certain modifications in order to obtain satisfactory results. Such preparations may include grinding, slicing or cutting, shredding or juicing, homogenizing, or mixing and packing.

### Grinding

Grinding is the most basic process to make a uniform powder shape. In the early stages of NIR development, ground samples were commonly used to avoid variation of NIR spectra with difference of physical shape of the sample. The mean and the distribution of the particle size directly depend on the type of commercial grinder. Even if the same machine is used to make powder samples, their properties will be affected by grinding conditions such as warming-up time of the machine, sample loading, or moisture conditions. In the case of agricultural products, dried samples are normally chopped in a commercial blender for 30s to 1 min (1). When the transmission or diffuse reflected spectra are measured in slurry form, samples may be pulverized or homogenized with various cell disrupter devices or high-speed blenders. The solid particle material is suspended by vigorous mixing and shaking, and spectral measurements are taken immediately.

Some areas of the agricultural products industry recommend a specific grinder to make a suitable powder. According to the *Grain Inspection Handbook—Book V*, the Udy-Cyclone mill with a 1-mm round-hole screen and equipped with an automatic feeder is the only grinder approved by FGIS for official wheat protein testing (3).

### Slicing or Cutting

Some researchers examine to get good NIR spectra from sliced or cut samples. Slicing or cutting technique is used for basic NIR research, which is concerned with the light absorbing or scattering problem. Foe example, Birth measured the distribution of laser light (632 nm) transmitted through potatoes in high-moisture, nonpigmented slices of raw plant tissue (4). The results were used to develop equations that describe the boundary conditions for the geometric aspects of radiation interacting with plant tissue.

Dull et al. examined the NIR analysis of soluble solids in cantaloupe (5). They used intact and sliced samples to evaluate the effect of sample presentation on NIR analysis. A 3.8-cm-wide transverse section was removed from each fruit by slicing in planes perpendicular to the stem-blossom axis at 1.9 cm to each side of the axis midpoint. A core 2.0 cm in diameter taken from the transverse section was used for NIR transmittance measurements. NIR spectra of the sliced samples were much better than those of intact samples in this experiment.

Tsuchikawa et al. examined the effects of physical conditions found in wood on the absorption of NIR radiation by using sliced wood (6). In these experiments, conifers (Sitka spruce) that had various degrees of surface roughness and orientation of fibers to the direction of incident light were used. Results of these measurements showed that the orientation of fibers and the surface roughness of wood were directly related to absorbance.

## Shredding and Juicing

When some chemical constituent in a nonuniform solid sample must be predicted with high accuracy, shredding and juicing may be an effective technique for measuring the mean spectrum of the sample. In the work on NIR spectroscopic quality inspection of sugar cane by Sekiguchi et al., shredding and juicing were skillfully adopted as sample preparations for NIR analysis (7). We must, however, consider the effect of residuum of squeezing on NIR spectra, where light scattering conditions vary with them.

## Homogenizing

For the NIR measurement of a nonuniform liquid sample, such as untreated milk, homogenization is the preceding process. As untreated milk has difference size of fat globules, it gives fluctuations in NIR spectra due to high scattering. For example, the diameter of casein in cow milk varies from 40 to 200 μm. Therefore, it is important to eliminate variation in scattering conditions for a good calibration equation. In work on NIR analysis of milk constituents, a Miklo-tester Mk. II homogenizer (Foss Electric, Denmark) was used to make fat globules smaller and more uniform (1). As Kawano pointed out, NIR spectra for homogenized and untreated milk were successfully taken in the longer-wavelength region from 1100 nm to 2500 nm by using a sample cell with a sample thickness of 0.25 mm (2). If the shorter wavelength region from 700 nm to 1100 nm is used for NIR measurement, homogenization is not necessarily needed. However, also in this case, we must consider the scattering condition of the sample.

## Packing and Mixing

In the case of powder samples, a mixing process is required to minimize the effects of repack error due to variation in packing density. When there are differences

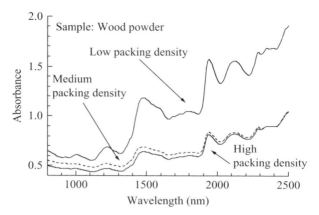

**Figure 5.2.** Effect of packing density on NIR spectra.

between NIR spectra for sample powder, especially within the longer-wavelength region, we may have some packing variation as shown in Figure 5.2. If this effect is not minimized, the correlation between optical data such NIR spectra and the chemical information of the sample is reduced.

Needless to say, repetition or reproducibility is a fundamental requirement for sample preparation for NIR measurement. Shenk et al. explained the devices currently available from the various instrument manufacturers for sample presentations (8). These variations of packing techniques may be applicable in several approaches to the sample quality analysis problem. The measurement of diffuse reflectance spectra of dried and ground samples in a closed-cup device is the most commonly used procedure. On the other hand, transflectance and transmission techniques are available to analyze wet or liquid samples. Details of sample presentation devices are described below in this chapter.

## TEMPERATURE AND MOISTURE CONTROL

It is well known that NIR spectra are easily affected by sample temperature. When samples that have temperatures different from that of the calibration sample set are predicted, a bias necessarily occurs because of the variation of the free water content. Particularly in the case of liquid, the effect of temperature is serious. Therefore it may be better to control the sample temperature before NIR measurements. In the case of liquids, a water bath is commonly used for this purpose. In the automatic chemical composition analyzer for soy sauce, a water bath is used as a temperature controller (2). As the sample goes through the glass coil in the water, the sample temperature is maintained at a constant temperature of 20°C.

However, NIR measurement is required under more severe conditions. The temperature at the on-line measurement system in packing houses for satsuma mandarins varies from 6°C to 24°C. Of course, it is very difficult to maintain constant

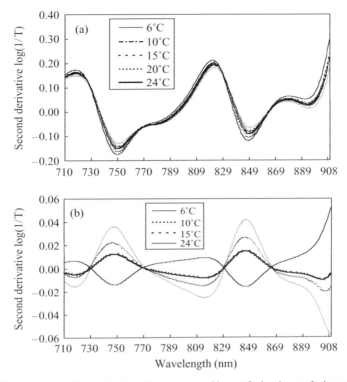

**Figure 5.3.** A. change of second-derivative spectra of intact fruits due to fruit temperatures. B. difference spectra on the basis of 20.

temperatures in such a fruit processing system. Figure 5.3 shows second-derivative spectra when the temperature of satsuma mandarins changed from 6–24°C and the difference spectra deducted a spectrum at 20°C from these spectra (9). Thus the changes in NIR spectra of intact fruits almost all depended on water, which is 80% or more of a fruit. Kawano et al. reported that they could automatically get a regression equation to compensate for the change of temperature if the regression equation is made from the sample set of various fruit temperatures (10). When temperature control of samples is not easy, the technique of sample temperature compensation is very useful.

In the case of quantitative analysis, variation in moisture content may be accepted if the variation is within the moisture range of the calibration samples. However, it should be checked that the moisture content of the sample submitted to NIR analysis is in that range, because the moisture content of samples changes easily when the samples are not kept in a closed environment. Sometimes the sample loses moisture and the moisture content goes below the moisture range. When grains are ground under high-moisture conditions, such as on a rainy day, the ground sample absorbs moisture and the moisture content becomes higher than the moisture range. It is important to keep the moisture content of samples within the range.

In the case of qualitative analysis, the moisture content of the samples is the most important factor that influences the NIR spectra. If there are differences in moisture content among the groups of samples, each group can be easily classified. To avoid this possibility, it is better to perform a moisture control of the samples. In a study on the classification of normal and aged soybean seeds by NIR discriminant analysis, the moisture content of each single kernel was controlled in a desiccator with silica gel at room temperature until each one had a constant moisture content of 14.6% (11).

## SAMPLE PRESENTATION DEVICES

The NIR researcher, user, and engineer must give attention to sample presentation devices that directly govern the quality of the spectra themselves. The type of sample presentations are mainly divided into "transmission," "reflection," "transflection," and "interaction" modes as shown in Figure 5.4.

In the case of the transmission mode, the incident light illuminates perpendicular to one side of the sample and the transmitted light is detected from the opposite side. This optical system is widely applicable for liquids without scattering or in low-scattering conditions where a cuvette is normally used. Lambert–Beer's law is almost available for this mode; therefore, we may pay attention to only the absorption coefficient, whereas some scattering effect due to small particles in the liquid varies with the medium.

In the case of the reflection mode, the incident light also illuminates perpendicular to the sample surface. Then NIR lights propagate in a sample with a series of absorption, scattering, diffraction, and transmission. And, finally, diffuse reflected light radiates from the sample surface. In this case, the sample should be opaque (for example, a powdered sample). The Kubelka–Munk equation is applicable for this mode where both absorption and scattering coefficients are important factors to explain the variation of NIR spectra. Normally, the incident light could not reach a

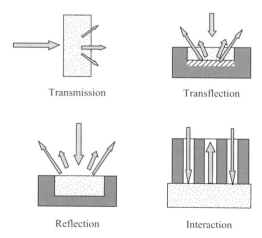

Transmission       Transflection

Reflection       Interaction

**Figure 5.4.** Type of sample presentation.

deeper position in a sample because of high absorption or multiple scattering. If the samples have sufficient thickness, the optical sample thickness should be regarded as "infinite," in which case we may pay attention to only the absorption coefficient in the Kubelka–Munk equation. Such a situation is very useful for analyzing NIR spectra of powdered and solid samples. Therefore, it is recommended to prepare a sample of more than 1-cm depth.

The transflection mode is the combination of transmission and reflection, which was originally developed by Technicon from the InfraAlyzer. Incident light is transmitted through the sample and then scattered back from a reflector, which is made of ceramic or aluminium to be compatible with the diffuse reflection characteristics of the instrument. The NIR spectra of not only liquid but also turbid material can be detected by the integrating sphere or the detector closely settled to the sample. NIR spectra from small volumes of the sample can be clearly measured.

In the case of the interaction mode, an interaction probe with a concentric outer ring of illuminator and an inner portion of receptor is usually used. Some companies produce specified NIR devices. The end of the probe is in contact with the surface of the sample. Therefore only the light transmitted through the sample can be detected. The interaction mode is used to determine the Brix value of fruits or vegetables.

## SAMPLE CELL OR SAMPLE HOLDER

It is important to match a proper sample cell/holder to the sample in order to obtain good NIR spectra. Various kinds of sample cells or sample holders have been proposed by researchers and manufacturers. Various types of fiber-optic probes have been developed to complement the type of sample being measured.

### Sample Cells for Whole Grains

Figure 5.5 shows sample cells for whole grains. The sample cell developed by NIR Systems has a required sample loading portion that can be used for NIR measurements of whole grains and seeds. The cell, which is mounted vertically with the NIR instrument, mechanically moves up or down slowly during NIR measurement to compensate for the heterogeneity of the samples; 30–50 scans are usually performed for acquiring a good NIR spectrum. Another type of the cell was designed by Bran and Luebbe. The cell is rotated quickly during NIR measurement. Consequently, NIR radiation illuminates a ring portion of the cell. An important point in measuring good and stable spectra by using such cells is to maintain consistency in the sample loading.

### Sample Cells for Powdered Sample

The construction of the cell for powdered samples is mostly similar, independent of the instruments. In any case, the most important matter is to ensure constant and reproducible packing density. The scattering conditions of each sample should be as identical as possible for a series of NIR experiments. Therefore, tight packing is strongly recommended. Figure 5.6 shows a typical sample cell, which has a circular

**Figure 5.5.** Sample cell for grain.

**Figure 5.6.** Sample cell for powder.

**Figure 5.7.** Measurement of paste sample with dish and beaker.

quartz window 3.5 cm in diameter and a black plastic vessel 1 cm deep. The powder sample is scooped into the cell and leveled by a specific spoon called a "scoopula."

## Sample Cells for Pastes

A sample cell for pastes is applied for high-moisture samples such as dough, ground meat, and mayonnaise. Both transmission and reflection techniques can be employed for the paste sample. In the case of the reflection mode, the surface of the sample in the open cup should be smoothed by using a paddle or mounting the glass on the sample. A laboratory dish or beaker including the paste sample can be measured as shown in Figure 5.7 (12).

## Sample Cells for Liquids

Transmittance and transflectance methods are used to obtain the NIR spectra of liquids. When the transmittance method is used, the detector has a simple structure and the sample must be strictly prepared so as not to contain light-scattering particles, with centrifugation or filtration if necessary. Measurement error increases with the distance between the sample cell and the detector when the sample is contaminated with light-scattering particles. It is also essential to minimize the losses of transmitted radiant energy, and hence the detector is placed immediately behind or beneath the sample. A cuvette cell made of quartz for the transmission mode (Fig. 5.8) is popular as a sample cell not only in IR spectroscopy but also in NIR spectroscopy for use with various liquids. The width of the cell depends on the wavelength region and samples used. In the case of water, the following conditions are recommended, taking into account the intensity of absorption: cells 2–3 cm in width at 970 nm, cells 2–3 mm in width at 1450 nm, and cells less than 1 mm in width at 1930 nm. Selection of a suitable cell width is essential for measuring good NIR spectra.

**Figure 5.8.** Sample cell for transmission mode.

In transflectance measurement, the transmitted light is reflected from a diffusely reflecting plate such as a ceramic reference, after which it acts like diffusely reflected light. The usual path length of the cell is 0.1 mm (see Fig. 5.8).

## FIBER-OPTIC PROBES

Recently, the fiber-optic probe has been employed frequently in NIR measurement. Many kinds of fiber-optic probes for NIR measurements are commercially available that can be used to perform on-line monitoring in food industries (Fig. 5.9; Ref. 12). However, reasonably priced glass fibers attenuate energy, particularly in the combination band region (1800–2500 nm) because the relatively high OH content of the glass.

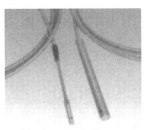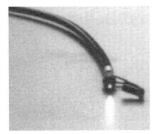

**Figure 5.9.** Various fiber-optic probes.

Therefore, further improvements in fiber technology will allow greater uses for glass and other materials as fiber-optic components for quantitative as well as qualitative analytical work.

## ACKNOWLEDGMENTS

The author sincerely thanks for Dr. Sumio Kawano (National Food Research Institute of Japan), BUCHI Co., and Soma Kogaku Co. for providing photographs.

## REFERENCES

1. B. G. Osborne, T. Fearn. *Near Infrared Spectroscopy in Food Analysis.* Longman Scientific & Technical, Harlow, Essex, UK. 1988.

2. S. Kawano. Sampling and sample presentation, In: *Near Infrared Spectroscopy—Principle, Instrumentation, and Applications.* H. W. Siesler, Y. Ozaki, S. Kawata, M. Heise eds. John Wiley-VCH, Weinheim, Germany, 2002, p. 115–124.

3. U. S. Department of Agriculture. *Grain Inspection Handbook—Book V.* 2004.

4. G. S. Birth. *J Food Sci* **43**: 916–925, 1978.

5. G. G. Dull, G. S. Birth, D. A. Smittle *J. Food Sci* **54**: 393–395, 1989.

6. S. Tsuchikawa, S. Tsutsumi. *Appl Spectrosc* **56**: 869–876, 2002.

7. R. Sekiguchi, K. Fuchigami, S. Hara, C. Tsutsumi. In: *Near Infrared Spectroscopy. The Future Waves*, A. M. C. Davies, P. Williams, eds. NIR Publication, Chichester, UK, p. 632–637, 1996.

8. J. S. Shenk, J. J. Workman, Jr., M. O. Westerhaus. Application of NIR spectroscopy to agricultural products, In: *Handbook of Near-Infrared Analysis*, D. A. Burns, E. W., Ciurczak, eds. Marcel Dekker, New York, p. 419–474, 2001.

9. K. Miyamoto. Internal quality control of citrus fruit in a packing house using near infrared spectroscopy, In: *Useful and Advanced Information in the Field of Near Infrared Spectroscopy*, S. Tsuchikawa, ed. Research Signpost, Trivandrum, India, p. 139–165, 2003.

10. S. Kawano, H. Abe, M. Iwamoto. *J Near Infrared Spectrosc* **3**: 211–218, 1995.

11. T. Kusama, H. Abe, S. Kawano, M. Iwamoto. *Nippon Shokuhin Kogyo Gakkaishi* **44**: 569–578, 1997.

12. Catalog from BUCHI Co., 2003.

# CHAPTER 6

# Latent-Variable Analysis of Multivariate Data in Infrared Spectrometry

ALFRED A. CHRISTY AND OLAV M. KVALHEIM

## INTRODUCTION

### Analytical Data Profiling in Spectroscopy

Spectroscopy deals with the interaction of electromagnetic radiation with matter. Depending on the energy of the electromagnetic radiation, different oscillations are excited. This excitation involves absorption of the corresponding energy of the oscillation involved. For example, microwave radiation excites rotational motions of the molecules, infrared radiation excites vibrational modes, near-infrared radiation excites overtones and combination frequencies, UV and visible radiation excite electronic transitions. When a polychromatic radiation source is used to excite a particular type of oscillation in a substance, the intensity of the absorbed electromagnetic radiation depends on the absorptivities and the concentrations of the different oscillators in the sample. The absorption peak of a pure oscillator is Gaussian type instead of a sharp single line because of the uncertainty associated with the frequency of absorption. However, when there is no overlap between the peaks arising from different oscillators, this absorption (for dilute solutions) obeys Beer's law: $A = abc$, where $a$ is absorptivity, $b$ is path length, and $c$ is concentration. When the absorptions arising from different oscillators overlap, and if there is no interaction between the oscillators, Beer's law can still be applicable and the absorptions arising from the oscillators can be resolved and can be used as the basis for quantitative analysis involving the oscillator. If not, the interactions between different oscillators in a sample affect the linearity between the absorption arising from a particular oscillator and its concentration.

Similar situations arise when measurements are made in the infrared region. In mid-infrared spectrometry, the infrared spectrum maps the internal vibrational frequencies

*Near-Infrared Spectroscopy in Food Science and Technology*, Edited by Yukihiro Ozaki,
W. Fred McClure, and Alfred A. Christy.

of a substance versus the intensity of the interaction of infrared radiation with vibrational motions. Absorption at each wavenumber is characteristic of that substance. In several cases it is possible to identify a particular vibration that can be used for quantification of a component in the system. The situation may not be the same when the measurements are made in the near-infrared region. This region gives rise to overtones and combination frequencies of the fundamental vibrations, and the profiles are usually broad and overlapping. These overlappings arise because of the absorption contribution of the other components present in the sample matrix. For example, food samples, meat, or any sample of biological origin may contain carbohydrates, protein, fat, and water in the matrix. A near-infrared spectrum measured on a sample of the above types may contain contributions from these components in the matrix. There is no selectivity for any one variable approach in the spectrum, and quantification of mixtures using one variable approach is difficult and sometimes impossible. In these cases we lack selectivity in using a single absorption for quantitative purposes. The alternative is to use absorption values at several variables and calibrate against one response variable.

In certain applications spectral profiles are measured on samples, for example, raw samples and samples that are exposed to an external condition or factor that can exert certain changes in the system. These spectral profiles should reflect the effect of the external factors. However, these changes can be small, and visual inspection of the profiles may be difficult. What we need here is a tool that can classify the spectral profiles or group the spectral profiles according to the external factors involved. A methodology that uses the whole profiles collectively is needed.

Approaches that exploited the information in the whole analytical profiles or multivariate data of samples had to wait because of the lack of computing facilities and proper statistical programs that could handle multivariate data produced by modern analytical instruments. The past two decades have witnessed rapid growth in both computer hardware capacity and new methodologies for solving problems involving, for example, resolution of analytical profiles and quantification of mixtures. The success of multivariate classification and calibration using NIR spectroscopic data profiles is evident from the enormous number of industrial applications reported in the literature.

Most of the applications with infrared spectroscopic data involve classification, calibration, and prediction. This is because the infrared profiles describe chemical structure that is directly related to the physical and/or chemical properties of the systems analyzed. Changes in the infrared profile of a system imply variation in the chemical structure and hence variations in the properties of the system. These data profiles contain information, and extraction of this information requires multivariate data analysis.

This book contains several different NIR applications in food analysis, and many of them use multivariate data handling. Our aim in this chapter is to discuss the aspects of latent variable decomposition in principal component analysis and partial least squares regression and to illustrate their use by an application in the NIR region.

## THEORY

Many books are available in the market that explains the philosophy (1, 2) and mathematical frame of methods (2–5) for latent-variable analysis. This theory section only

$$\begin{bmatrix} x_{11} & x_{12} & x_{13} & x_{14} & \cdot & \cdot & \cdot & \cdot & \cdot & x_{1m} \\ x_{21} & x_{22} & x_{23} & x_{24} & \cdot & \cdot & \cdot & \cdot & \cdot & x_{2m} \\ x_{31} & \cdot & \cdot & \cdot & \cdot & \cdot & \cdot & \cdot & \cdot & x_{3m} \\ \cdot & & & & & & & & & \\ \cdot & & & & & & & & & \\ \cdot & & & & & & & & & \\ \cdot & & & & & & & & & \\ \cdot & & & & & & & & & \\ \cdot & & & & & & & & & \\ x_{n1} & x_{n2} & x_{n3} & x_{n4} & \cdot & \cdot & \cdot & \cdot & \cdot & x_{nm} \end{bmatrix} \qquad \begin{bmatrix} y_1 \\ y_2 \\ y_3 \\ y_4 \\ \cdot \\ \cdot \\ \cdot \\ \cdot \\ y_n \end{bmatrix}$$

Infrared spectral profiles in rows     Measured<br>dependent variable

**Figure 6.1.** The spectral data arrangement and the measured property in matrix arrangement.

outlines aspects of latent-variable analysis that are necessary for an understanding of the PCA (principal component analysis) and PLS (partial least squares) techniques.

## Data Profiling in Infrared Spectrometry

As mentioned above a spectral profile in infrared spectrometry contains absorbencies at each wavenumber or wavelength of the absorbed radiation. This can be represented by an array of numbers that represent the absorbencies at different wavenumbers or wavelengths. This spectral profile may contain information that reflects the nature of the sample or can be correlated to a physical or chemical property of the system. When measurements are made with this intention, the spectral profiles can be represented by a matrix containing rows of numbers representing the spectral profiles and a column representing the external property (see Fig. 6.1). The relationship between the samples, that is, the relationship between the rows of the matrix, requires some clever mathematical decomposition of the data.

## Latent-Variable Decompositions

As shown by Kvalheim (6), approaches to decomposition of multivariate data in terms of latent variables can be developed within the frame of a generalized NIPAL algorithm (7), for instance, decomposition into principal components (8) and decomposition using the PLS approach (9, 10).

A latent-variable decomposition can be expressed as

$$\mathbf{X} = \sum_{a=1}^{A} \mathbf{t}_a \mathbf{p}_a^{t} + \mathbf{E} \qquad (6.1)$$

The matrix $\mathbf{X}$ represents the collected spectral data after optional procedures for pre-processing, such as normalization and column-centering. Each row of $\mathbf{X}$ corresponds to an object (sample), and each column represents a variable. Superscript t is used to imply transposition. Vectors are defined as columns so that transposition defines row vectors.

Decomposition into principal components is carried out by obtaining an ordinary least-squares solution that minimizes the residuals $\mathbf{E}$. This is the most common decomposition procedure and provides $A$ columns of orthogonal score vectors $\{\mathbf{t}_a\}$ and $A$ rows of orthonormal loading vectors $\{\mathbf{p}_a^t\}$.

Partial least-squares calibration uses the covariance between the data matrix $\mathbf{X}$ and another matrix $\mathbf{Y}$ (property matrix). Here, the score vectors are orthogonal, but the loading vectors become oblique.

The following is a general algorithm for the decomposition of $\mathbf{X}$ by the method of successive projections (6):

$$\text{Define } \mathbf{X}_1 = \mathbf{X}.$$

Repeat for $a = 1, 2, \ldots, A$ ($A$ is the number of latent variables extracted).

The procedure is carried out in four steps.

In the first step, a latent variable in the so-called variable space is selected, that is, a coordinate vector:

$$\mathbf{w}_a^t; \|\mathbf{w}_a\| = 1.$$

In the second step, the objects are projected on $\mathbf{w}_a$, whereby the column vector of scores $\mathbf{t}_a$ is calculated:

$$\mathbf{t}_a = \mathbf{X}_a \mathbf{w}_a.$$

The third step involves the calculation of the row vector of loadings $\mathbf{p}_a^t$ that expresses the covariances between the latent variable and the original variables:

$$\mathbf{p}_a^t = \mathbf{t}_a^t \mathbf{X}_a / \|\mathbf{t}_a\|$$

The final step involves the removal of the dimension of $\mathbf{X}_a$ associated with $\mathbf{w}_a$:

$$\mathbf{X}_{a+1} = \mathbf{X}_a - \mathbf{t}_a \mathbf{p}_a^t$$

The final step ensures orthogonal score vectors. The decomposition formula, Equation 6.1, holds only for orthogonal scores. The final step also guarantees orthonormal $\{\mathbf{w}_a\}$.

Different decompositions of $\mathbf{X}$ can now be selected by different choices of $\{\mathbf{w}_a\}$. The methods for direct decomposition for PCA and PLS techniques are outlined below.

1. Principal-component (PC) decomposition:

$$\mathbf{w}_a^t = \mathbf{p}_a^t / \|\mathbf{p}_a\|$$

where the objects are projected on to the (converged) loading vector $\mathbf{p}_a^t$ for each dimension.

2. Partial least-squares (PLS) decomposition (9, 10):

$$\mathbf{w}_a^t = \mathbf{u}_a^t \mathbf{X}_a / \|\mathbf{u}_a^t \mathbf{X}_a\|$$

where the objects are projected on to the (converged) latent variable vector $\mathbf{u}_a^t$ for an external (dependent, predicted) variable block $\mathbf{Y}$ ($\mathbf{u}_a^t$ is equal to $\mathbf{y}^t$ with one predicted variable only).

From the projection algorithm above note that the vectors $\{\mathbf{w}_a^t\}$ are coordinates of the latent variables on the original variable axes. The choice of these "weightings" defines the various methods. Principal components represent the special case of the general projection method defined above where $\mathbf{w}_a^t$ and normalized loadings $\mathbf{p}_a^t$ are identical. Principal components can thus be found by iterating steps 1–3 until $\mathbf{w}_a^t = \mathbf{p}_a^t$. For this special case, the row loading vectors defined in Equation 6.1 are orthonormal.

## Some Aspects Related to PCA and/or PLS

*Visualization of Latent Projections.* Visualization is of prime importance for interpreting multivariate data. Many plots are available. The most important are score and loading plots and their combination into biplots.

The relationships among samples are revealed by their projections (scores) on the latent variables. This information is displayed in bivariate scoreplots. Similar samples group together in the score plots. The orthonormality between latent variable $\mathbf{w}_a$ vectors means that the distance between samples represents a quantitative measure of relatedness. Standardized scores can be used. In most instances, however, scores weighted in accordance with the "size" (proportional to $\|\mathbf{t}_a\|$) of the latent variables are preferred. These are the scores used in Equation 6.1. Score plots are very useful for visual detection of atypical samples, that is, outliers.

Interpretation of a single latent variable and of the features of a latent variable model is possible through connection to the original variables. This information is best displayed in loading plots. The coordinates of a variable in a loading plot are its loadings on the PCs. Standardized loadings may be used. However, by weighting each latent variable after its importance (proportional to the norm of the score vector defined by Equation 6.1), a quantitative display of the relative importance of the variables is obtained, Such a loading plot displays the contribution of a variable to a model directly, that is, proportional to the square of the distance from the origin. A variable near the origin carries little or no information, whereas a large distance from the origin means that the variable is important in separating samples. Finally, the mutual locations of the variables reflect their relationships. Variables located in the same or an opposite domain of the loading plot carry similar information.

Scores and loading can be simultaneously displayed in biplots (11, 12). This may be extremely useful, especially when the number of samples and variables are not too large.

***Validation of the Calibration Model.***   To evaluate the prediction ability of the calibration model in a PLS analysis, the cross validation technique is used (13). Prediction ability is then tested by predicting samples with known dependent variable. Samples in the calibration set are grouped into smaller sets, and the dependent variables of these sets can be predicted from the calibration models of the remaining samples. Cross validation of a calibration model makes it possible to select the optimum number of latent variables or factors, that is, the number giving the minimum prediction error for the calibration set. When the number of latent variables exceeds the optimum number, the error of prediction increases again. By including too many latent variables, *overfitting* of the calibration model results. The standard error of prediction (SEP) of a cross-validated calibration model is calculated by the following equation

$$\text{SEP} = \left[ (1/I_{\text{p}}) \sum_{m=1}^{n_{\text{v}}} \sum_{i=1}^{I_{\text{v}_m}} f_i^2 \right] \tag{6.2}$$

Here $I_{\text{v}_m}$ is the number of predicted samples in the cross validation of group $m$, $n_{\text{v}}$ is the number of groups used in the cross validation, and $I_{\text{p}}$ is the total number of prediction samples; $f_{\text{v}}$ is the residual variance of the sample $i$ after $A$ factors.

***Outlier Detection.***   To obtain a good calibration model one has to remove outlying samples, that is, those which are extreme compared to the others in the calibration set. The outlying property may be due to interferences in the spectral data or to measurement error in the dependent variable. Interferences in spectral data extract additional latent variables, thus increasing the complexity and reducing the predictive ability of the calibration model. Such samples are outliers in the spectral data. On the other hand, some samples are well described by the calibration model but the predicted data are far away from the experimental value. These samples are outliers in the dependent variable. As mentioned above, score plots are very helpful in identifying outliers.

***Pretreatment of Spectroscopic Data.***   When spectra are measured on a set of samples containing several components, it is often necessary to pretreat the data before they are subjected to multivariate data analysis. The selection of this pretreatment must be based on the knowledge of the instrument, the nature of the samples, and the type of model that is intended to the particular application. For example, whether the data are intended for calibration, for classification, or just for investigation purposes. If a linear model is used for the calibration, the pretreatment should preserve the linearity between the data and response, or the data should be transformed such that the linearity between the data and response is achieved. Furthermore, the quantity effect on the data of the sample should be removed if one is only interested in the relative quantities of the components in the samples. This type of treatment is also necessary when the instruments give varying responses for samples over time.

In certain applications one must reduce the number of variables in order to speed up the analysis. Multivariate calibration problems require handling and inverting matrices with many entries. It takes a lot of computer memory and time. By means of variable reduction the number of spectral variables in the calibration sample set may be reduced by a factor of 10 without losing information contained in the data. This provides smaller matrices for calibration modeling. However, the reduction that can be performed without a significant loss of information is data dependent, and the reduction must be done in a trial-and-error manner. A procedure called *maximum entropy* (14, 15) may be used with infrared data profiles. However, the computers that are available in the market now can handle several thousands of variables and samples and will serve the purpose of the people who look at the variable reduction with skepticism.

## Regression and Target Projections in PLS

Regression models calculated with least-squares procedures under the constraint of linear models are usually written as

$$\mathbf{y} = \mathbf{Xb} + \mathbf{E} \tag{6.3a}$$

or

$$\hat{\mathbf{y}} = \mathbf{Xb} \tag{6.3b}$$

where $\mathbf{X}$ is a matrix containing the predictor variables as shown earlier and $\mathbf{y}$ is a column matrix containing the dependent variable. The parameter $\mathbf{b}$ is determined by least-squares regression. Once $\mathbf{b}$ is determined, the property $\mathbf{y}$ can be determined (in multivariate calibration) from the spectroscopic profiles.

The parameters $\mathbf{b}$ are calculated by constructing a generalized inverse $\mathbf{X}^+$.

$$\mathbf{b} = \mathbf{X}^+\mathbf{y} \tag{6.4}$$

The generalized inverse $\mathbf{X}^+$ is

$$\mathbf{X}^+ = \mathbf{W}^t(\mathbf{PW}^t)^{-1}\mathbf{G}^{-1}\mathbf{T}^t \tag{6.5a}$$

where $\mathbf{W}$, $\mathbf{P}$, and $\mathbf{T}$ are the matrices of coordinate vectors, loading vectors, and score vectors, respectively. The matrix $\mathbf{G}$ is a diagonal matrix with elements $\mathbf{t}_a^t\mathbf{t}_a$, $a = 1,2,\ldots, A$ on the diagonal. The parameters $\mathbf{b}$ are then obtained by Equation 6.4.

$$\mathbf{b} = \mathbf{W}^t(\mathbf{PW}^t)^{-1}\mathbf{G}^{-1}\mathbf{T}^t\mathbf{y} \tag{6.5b}$$

The parameters $\mathbf{b}$ are estimated so as to predict $\mathbf{y}$ as well as possible, for instance, by means of cross validation (10). The predicted values of the dependent variables $\mathbf{y}$ for the training set samples is obtained by inserting Equation 6.4 into Equation 6.3b:

$$\hat{\mathbf{y}} = \mathbf{XX}^+\mathbf{y} \tag{6.6}$$

In PLS calibrations, the measured property $\mathbf{y}$ does not need to be related to all the spectral variables. If the property measured is related to a particular functional group then the correlation will be limited to a portion of the spectral variables. If this relationship is strong between $y$ and some $x$-variables, predicted and measured $y$ will be similar and the correlation coefficient between measured and predicted $y$ will approach 1. In many applications, for interpretive purposes, it may be of interest to find the $X$ variables that are most closely correlated to $y$. This can be done directly by calculating the covariance between measured y and the intensities at each wavelength, that is, by calculating the covariances $\mathbf{r}_{y,x}$ as

$$\mathbf{r}_{y,x} = \mathbf{y}^t\mathbf{X} \tag{6.7a}$$

These covariances can be plotted as a graph (16). However, a better approach is to calculate the covariance between the predicted values of $y$ and the intensities at each wavelength:

$$\mathbf{r}_{\hat{y},x} = \hat{\mathbf{y}}^t\mathbf{X} \tag{6.7b}$$

The result of this so-called target projection can be plotted as a graph to show the connection between a spectral profile and $y$ (16, 17). The use of predicted $y$ ensures that the wavelengths with both high predictive ability and high correlation with $y$ are given increased importance compared to those that has only a good correlation. This can be shown by inserting Equation 6.3b in Equation 6.7b:

$$\mathbf{r}_{\hat{y},x} = \mathbf{b}^t\mathbf{X}^t\mathbf{X} \tag{6.8a}$$

The product $\mathbf{X}^t\mathbf{X}$ is the variance-covariance matrix for the spectral profiles. This matrix can be expanded as shown in Equation 6.8b.

$$r_{\hat{y},xi} = \sum_{j=1}^{M} bj \sum_{k=1}^{N} x_{ki}\, x_{kj} = \sum_{j=1}^{M} b_j r_{xi,xj}\,; \; i = 1,2,\ldots, M \tag{6.8b}$$

Here, $M$ and $N$ are the number of wavelength and samples, respectively. The co-variation $r_{\hat{y},xi}$ between the predicted values of $y$ and the intensities at a wavelength $i$ is calculated as a weighted sum of covariances $r_{xi,xj}$ between intensities at two wavelengths $i$ and $j$ (Equation 6.8b). The use of regression coefficients $b_j$ as weights shows that the target projection procedure weights the importance of prediction $(b_j)$ and correlation $(r_{xi,xj})$ so that wavelengths that are important for predicting the property $y$ have large variance (sensitivity) and are well correlated with other predictive wavelengths that will be highlighted in the target projection plots. These are exactly the wavelengths that are important for the interpretation of the descriptor in relation to $y$ (16, 17).

The target projection plots are easy to interpret and can be used to name factors influencing a system. Furthermore, for systems in which variation in the multivariate profile with a given dependent variable is small, target projection can amplify the changes by means of the target projection plots. The NIR spectral profiles of food samples are generally broad, and the changes in the spectra with external properties maybe small and sometimes undetectable by visual inspection. Target projection analysis is very beneficial in such systems for interpretative purposes. The last few years have seen a keen interest in the application of target projection because we showed the theoretical basis and application of the technique more than a decade ago.

## APPLICATION OF PCA AND PLS IN THE DETECTION OF ADULTERATION IN OLIVE OILS

### Introduction

The quality of foods prepared for the consumer industry is dependent on the quality of the components used to make the final products. These components are sometimes adulterated with similar cheap components by small, greedy businesspeople and suppliers. Consumers are not in a position to detect these inexpensive components in the foods for which they pay a high price.

Olive oil is one of the oils that have been subjected to adulteration by seed oils for a long time. Although the adulteration is done for economic reasons, the action affects the quality of the foods in which olive oil is a component. Furthermore, the health implications of adulteration of olive oil with seed oils are a concern.

There are several different methods such as the iodine value, the saponification value, colorimetric reactions, etc. as well as refractive index, density, and viscosity measurements (18). Furthermore, UV and fluorescence spectrometry have also been used (18). However, greedy suppliers can adulterate pure olive oil to a higher degree to give acceptable analytical values that can give an OK quality to the adulterated oil as pure olive oil. Techniques such as high-performance liquid chromatography (HPLC) and gas chromatography (GC) have also been used in identifying adulterants (19). These involve wet chemical methods and are expensive and time consuming. There have been several attempts to find more rapid and efficient methods for the detection of olive oil adulteration. This includes the use of infrared spectrometry, both in the mid- and near-infrared regions (20).

The overtones and combinations of the functional groups CH, OH, and NH present in food samples give rise to absorptions in the NIR region. The low absorbance of the overtone and combination NIR bands allows the use of NIR spectroscopy for the nondestructive analysis of raw samples.

This application includes the task of identification and quantification of the adulterant. To demonstrate the different options in multivariate data analysis and achieve the aim of the study, we investigated the adulteration of olive oil by corn oil, soya oil, sunflower oil, walnut oil, and hazelnut oil. The NIR profiles of the samples of olive oil and the adulterated mixtures were used as the analytical data in this application. The measured infrared spectra were subjected to several different spectral pretreatment procedures, including double-derivative techniques, to identify the best procedure for the classification of unknown oil and for the quantification of the degree of adulteration in olive oil.

## Experimental

***Samples and Measurements.*** Pure olive oil purchased from from WAKO Pure Chemical Industries, Ltd., Japan was quantitatively weighed in a 4-mm quartz cell, and its NIR spectrum was measured. A small amount of the chosen adulterant was then quantitatively added, and the NIR spectrum was again measured. The addition and NIR spectral measurements continued until the cell was nearly filled with the oil mixture. The cell was then washed and dried before starting the next series of measurements, in which an adulterant was put in the cell first, and then olive oil was added dropwise. In this way we were able to cover the whole range of adulteration from 0 to 100% weight/weight. The same procedure was repeated with walnut oil, hazelnut oil, corn oil, soya oil, and sunflower oil as adulterants.

The NIR transmission measurements were made in the region of 12,000–4000 $cm^{-1}$ by using a Bruker Vector 22/N FT-NIR spectrometer equipped with a Ge diode detector. A total of 32 scans were collected on each sample with a spectral resolution of 4 $cm^{-1}$. The quartz cell was thermostatted at 25°C during the measurements. A total of 120 NIR spectra were acquired for each adulterant.

## Data Analysis

The Unscrambler (version 7.08, CAMO AS, Trondheim, Norway) software program was used in the PLS calibration of the NIR profiles with the adulterant percentage as the dependent variable in limited spectral regions and different data pretreatment methods. All of these different procedures were evaluated to find the spectral region and the data pretreatment method that give the best prediction and classification. The following procedure was found to serve the purpose of quantification and classification of an adulterant in olive oil in an acceptable manner.

First, each set of NIR spectra of olive oil containing an adulterant was modeled by a PLS algorithm for one dependent variable (PLS1). The NIR spectra (in the region of 9000–4550 $cm^{-1}$) were subjected to Savitzky–Golay smoothing (18 points, 2 orders) before multivariate data analysis. Each set of spectra of olive oil containing the adulterants soya oil, corn oil, walnut oil, and hazelnut oil was subjected to a multiplicative signal correction using the mean spectrum of the set. The spectra of olive oil containing the adulterant sunflower oil were subjected to mean normalization. The spectra were then subjected batchwise to PLS calibration modeling. In this procedure, a total of 75 (for each adulterant) and 30 spectra from each set were selected for calibration and validation, respectively. The remaining 15 spectra from each set were set aside for classification. In the second step, the calibration spectra of all the five sets were subjected to only an MSC correction before applying a principal component analysis (PCA). Then, in the third step, the classification samples were classified with the PCA model. In the final step, the classified samples were quantified for the identified adulterant in the oil mixtures.

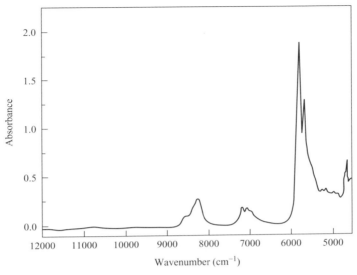

**Figure 6.2.** Some NIR spectra of walnut oil-adulterated olive oil mixtures.

## Results and Discussion

Some NIR spectra of olive oil adulterated by walnut oil in the region of 12,000–4500 cm$^{-1}$ are shown in Figure 6.2. The changes in the spectra are not very prominent. There are small differences around 5260 cm$^{-1}$. These small differences can easily be exploited by multivariate data analysis. The NIR band assignments for oils and fats are given in Table 6.1 (22–24). When an oil sample is adultered, the fatty acid composition of the oil is changed. The changes in the spectral profiles reflect little regarding the change in the composition. However, whether the changes in the profiles are small or large, the whole spectrum of the pure sample is affected.

**TABLE 6.1. Band Assignment**

| Wavenumber, cm$^{-1}$ | Functional Group | Assignment |
|---|---|---|
| 8560 | CH$_3$- | C-H stretch 2nd overtone |
| 8260 | -CH$_2$- | C-H stretch 2nd overtone |
| 7187 | CH$_3$- | 2 C-H str.+C-Hdef |
| 7074 | -CH$_2$- | 2 C-H str.+C-Hdef |
| 6010 | *cis* R$_1$CH=CHR$_2$CH$_3$- | *cis* CH |
| 5791 | -CH$_2$- | C-H 1st overtone |
| 5677 | | C-H 1st overtone |
| 5260 | C=O str. | 2nd overtone |
| 5179 | C=O str. | 2nd overtone esther |
| 4707 | -COOR | C-H str.+C=O str. |
| 4662 | -HC=CH- | =C-H str.+C=C str. |
| 4595 | -HC=CH- | CH assym. str.+C=C str. |

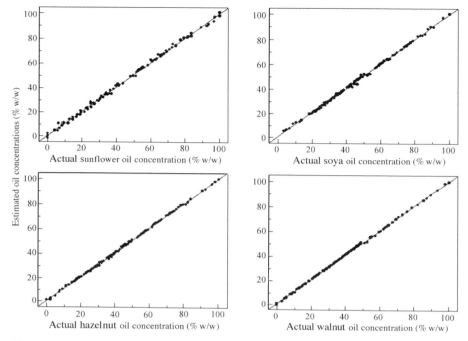

**Figure 6.3.** Calibration plots showing the relationships between estimated and actual concentrations of adulterants in olive oil samples (shown only for four oils).

Partial least-squares calibration plots of olive oil adulterated by four adulterants (sunflower oil, soya oil, hazelnut oil, and walnut oil) are shown in Figure 6.3. The statistics of these calibrations are given in Table 6.2. The figures and the table reveal excellent correlations between the measured and predicted values of the adulterants in olive oil.

**TABLE 6.2. Statistics of the PLS Models**

|  | Corn Oil | Sunflower Oil | Soya Oil | Walnut Oil | Hazelnut Oil |
|---|---|---|---|---|---|
| No. of spectra | 105 | 105 | 105 | 105 | 105 |
| No. of PLS comp. | 4 | 5 | 5 | 3 | 4 |
| Corr. coef. calibration | 0.9997 | 0.9978 | 0.9985 | 0.9994 | 0.9994 |
| Error of calc. (% w/w) | 0.53 | 1.26 | 0.87 | 0.55 | 0.66 |
| Corr. coef. prediction | 0.9998 | 0.9989 | 0.9995 | 0.9998 | 0.9998 |
| Error of Prediction (% w/w) | 0.57 | 1.32 | 0.96 | 0.56 | 0.57 |

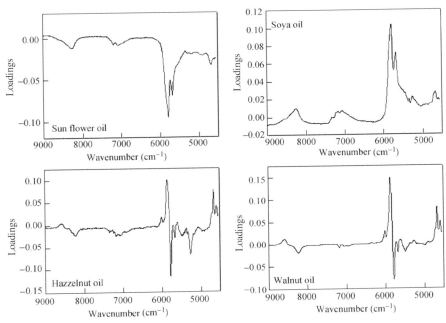

**Figure 6.4.** Loading plots related to the calibration models shown in Fig 6.3.

Loading plots for the same oil mixtures given in Figure 6.3 are shown in Figure 6.4. The loading plots give information regarding the spectral profiles that have been used in the PLS calibration. The small differences around 5260 cm$^{-1}$ and other parts of the spectral regions that are important for the calibration model are clearly presented in the loading plots. These differences come from the overtones and combinations of the -CH$_2$ and CH$_3$ stretchings in the oil glycerides. The compositions of the vegetable oils vary with the different fatty acids in the glyceride molecules. Different oils have different fatty acid compositions, and there should be differences in the intensities of bands arising from the overtones and combinations of the CH$_3$- and -CH$_2$-stretching vibrations. However, in reality, these differences appear to be smaller than expected, and the smaller percentage of the variances is responsible for explaining the variances in the adulterant concentrations. The loading plots reflect the smaller changes in the NIR spectra arising from the addition of adulterants in olive oil. It appears that the spectral profile changes throughout the whole spectral region. This is an indication that using the whole spectra with suitable data handling may give a better prediction than using a part of the spectra, or absorptions at single wavelengths, or wavenumber positions. The PLS calibrations of the NIR spectral profiles against adulterant concentrations gave excellent calibration models. These gave excellent predictions for the percentages of adulterants when predicted by the relevant models. These calibrations and predictions are restricted to the olive oil and the adulterants used in this work.

The prediction error plots of the calibration models for the four oils shown in Figure 6.3 are shown in Figure 6.5. The plots clearly show that the first five components

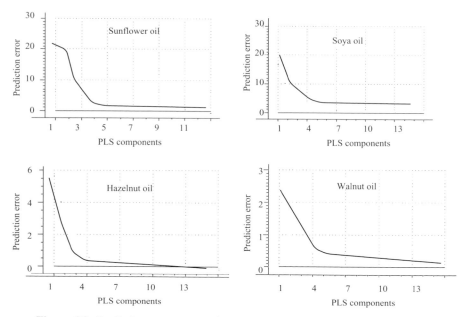

**Figure 6.5.** Prediction errors versus the number of PLS factors used in the model.

were responsible for explaining the variances in the data, and the minimum prediction errors are achieved in all four cases with 3–5 components. After this optimum number of components the prediction will not be improved. Use of several factors beyond this number will overfit the data.

A target projection plot for the calibration of hazelnut oil is shown in Figure 6.6. Because almost all variation in the spectroscopic data is explained by the model, the

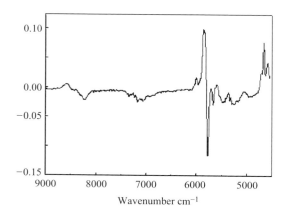

**Figure 6.6.** Target projection plot showing the profiles that give maximum correlation to the adulterant concentration.

## Calibration

*Scores*

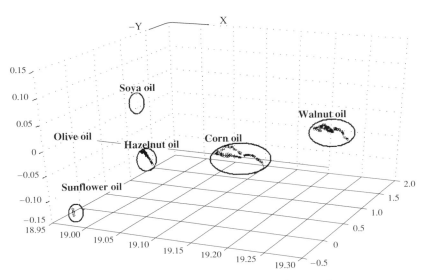

**Figure 6.7.** PCA classification of calibration samples.

plot is almost identical to the loadings. However, the plot shows only the predictive parts of the spectrum correlating with the adulterant concentration. The vibrational absorptions that are connected to the target projections profiles are given above.

The results of the PCA carried out on the spectral profiles of the mixtures used in the PLS calibration and validation are given in Figures 6.7 and 6.8, respectively. Three principal components explain 99% of the variance in the spectral data. The plot shows that the oils are grouped according to their adulterants. The groups are well separated and relatively far away from each other. This relative closeness between the groups gives an idea as to whether it is easy to detect the identity of an adulterant in an adulterated olive oil sample. The plot reveals that the multivariate profiles representing the oils are clearly separated in space and form five different classes in multivariate space.

The PCA models developed with the training samples (calibration samples in PLS analysis) classify the validation samples with almost 100% certainty. This again illustrates the effectiveness of the PCA model in classifying the samples. The data profiles of the 75 oil mixtures (15 from each adulterant) that were set aside were classified by the PCA model developed above.

The axes in Figures 6.7 and 6.8 represent the principal components used in classifying the calibration and validation samples. The olive oil spectrum lies at the center, and the other samples of adulterated olive oil spread out in the three-dimensional latent variable space, depending on their variation relative to the pure olive oil profile.

## Validation

*Scores*

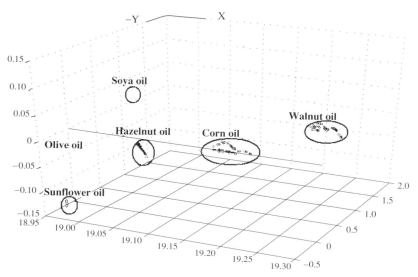

**Figure 6.8.** PCA classification of validation samples.

The closeness of the group to the olive oil sample indicates the closeness of the composition of the pure olive oil and the adulterant. Obviously, the fatty acid profile of the walnut oil adulterant seems to differ more than the fatty acid profiles of the other adulterants. Apart from one of the walnut oil-adulterated olive oil samples, all samples were correctly classified. An almost 100% prediction of the adulterants was achieved.

The adulteration problem presented here seems apparently simple. However, in practice the problem must be faced in a different manner. This is because the identities of the olive oil and the adulterant are not known when a sample of olive oil is bought from a retailer. Furthermore, olive oils are produced by several different countries and their fatty acids compositions vary. If a particular producer is interested in finding certain adulterants in his product, this can be easily solved by establishing calibration models with the olive oil and the adulterants concerned.

## CONCLUSION

We have shown that the procedure adopted above can classify and quantify adulterants in olive oil samples. A classification of oil mixtures according to the adulterants was achieved with almost 100% certainty in this case, and the degree of adulterant in the mixtures was achieved within acceptable error limits. A similar approach may be developed for other adulteration problems in the edible oil industry. However, the potential adulterants should be identified first. Once the adulterant is identified, it is easy to establish calibration models with the olive oil and adulterant concerned.

The problem becomes difficult when the identity of the adulterant is not known. In such circumstances some groundwork is needed to identify the adulterant. Obviously, gas chromatographic techniques will be chosen for such purposes. Once the adulterant is identified, further analysis involving quantification of the adulterant can be effectively carried out with a procedure described in this chapter.

In this chapter, we have explained latent-variable projection techniques for PCA and PLS analysis handling large amounts of correlated data as obtained, for example, by infrared profiling. These techniques can in several cases replace routine wet analytical methods and save a lot of money and time for the industrial laboratories.

## ACKNOWLEDGMENTS

The Japan Analytical Society is thanked for their kind permission to use parts of text and figures from *Analytical Sciences*, May 2004, Vol. 20, 1–6.

## REFERENCES

1. L. L. Thurstone. *Multiple-Factor Analysis*. University of Chicago Press, Chicago, London, 1947.

2. P. Horst. *Factor Analysis of Data Matrices*. Holt, Rinehart and Winston, New York, Chicago, London, 1964.

3. R. J. Rummel. *Applied Factor Analysis*. Northwestern University Press, Evanston, IL, 1970.

4. K. G. Jöreskog, J. E. Klovan, R. A. Reyment. *Geological Factor Analysis*. Elsevier, Amsterdam, 1976.

5. E. R. Malinowski. *Factor Analysis in Chemistry*. Wiley, New York, 1991.

6. O. M. Kvalheim. *Chemometrics Intell Lab Syst* **2**: 283, 1987.

7. S. Wold, C. Albano, W. J. Dunn III, U. Edlund, K. Esbensen, P. S. Geladi, Hellberg, E. Johansson, W. Lindbergand, M. Sjöström. Multivariate data analysis in chemistry. In: *Chemometrics – Mathematics and Statistic in Chemistry*. B.R. Kowalski, ed, 17–95, Reidel, Dordrecht, 1984.

8. J. Joliffe. *Principal Component Analysis*. Springer Verlag, Berlin, 1986.

9. S. Wold, A. Ruhe, H. Wold, W. J. Dunn III. *SIAM J Sci Stat Comput* **5**: 735, 1984.

10. H. Martens, T. Næs. *Multivariate Calibration*. Wiley, Chichester, 1989.

11. K. R. Gabriel. *Biometrika* **58**: 953, 1971.

12. O. M. Kvalheim. *Chemometrics Intell Lab Syst* **4**: 11, 1988.

13. H. A. Martens. Multivariate Calibration. Dr. Techn. Thesis. Technical University of Norway, Trondheim, 1985.

14. W. E. Full, T. Ehrlich, S. K. Kennedy. *J. Sedim. Petr.* **54**: 117, 1984.

15. T. V. Karstang, R. J. Eastgate. *Chemometrics Intell Lab Syst* **2**: 209, 1987.

16. O. M. Kvalheim, T. V. Karstang. *Chemometrics Intell Lab Syst* **7**: 39, 1989.

17. A. A. Christy, O. M. Kvalheim, F. O. Libnau, G. Aksnes, J. Toft. *Vibrational Spectrosc* **6**: 1, 1993.

18. G. Tous. In: *Analysis and Characterisation of Oils, Fats and Fat products*, H. A. Boekenoogen, ed, **2**: 315. Interscience, London, 1968.

19. N. K. Andrikopoulos, I. G. Giannakis, V. Tzamtzis. *J Chromatogr Sci* **39**: 4, 137, 2001.

20. M. D. Guillen, N. Cabo. *J Sci Food Agr* **75**: 1, 1997.

21. G. D. Batten, P. C. Flinn, L. A. Welsh, A. B. Blakeney. Leaping ahead with near infrared spectroscopy. In: *Proceedings of the 6th International Conference on Near Infrared Spectroscopy*, John Crainger of Johnzart Pty Ltd, Victoria, Australia. 1995.

22. G. Downey, P. McIntyre, A. N. Davies. *J Agr Food Chem* **50**: 5520, 2002.

23. B. G. Osborne, T. Fearn, P. H. Hindle. Practical NIR spectroscopy with applications in food and beverage analysis, 2nd ed, Longman Group UK Ltd., Essex, England, 1993.

24. P. Hourant, V. Baeten, M. T. Morales, M. Meurens, R. Aparicio. *Appl Spectrosc* **54**: 150, 2000.

# APPLICATIONS TO AGRICULTURAL AND MARINE PRODUCTS

# Grains and Seeds

PHIL WILLIAMS

## INTRODUCTION

There is nothing more difficult in its introduction, more hazardous in its undertaking and more uncertain in its success than to take the lead in the development of a new way of things

—(N. Machiavelli. Il Principe, 1532)

### The Bounty of Near-Infrared Spectroscopy

This chapter will be about applications. It will avoid the specifics of the chemistry of near-infrared reflectance spectroscopy (NIRS), and of chemometrics. It will attempt to illustrate in one locus the variables that can affect application of NIRS to grains and seeds and their derived products. Over 50 factors associated with the grains and seeds themselves have been identified (1), any of which can interact to complicate the operation. To confuse the picture yet further is the fact that they may interact in different ways, to different degrees, and under different conditions (of temperature, etc.) and may not even interact at all. Application of NIRS to grains and seeds is complicated by the wide-ranging variance in their size, shape, color, density, composition and texture.

Near-infrared spectroscopy has changed the world of commercial grain handling forever.

The first official "real-world" application of NIRS in 1975 (when the Canadian Grain Commission changed their massive wheat protein-marketing system from Kjeldahl to NIRS) alerted the world to a rapid, safe, reliable, and inexpensive method for composition analysis. Other industries began to capitalize on the technology, at

*Near-Infrared Spectroscopy in Food Science and Technology*, Edited by Yukihiro Ozaki,
W. Fred McClure, and Alfred A. Christy.
Copyright © 2007 John Wiley & Sons, Inc.

first slowly, then with almost exponential alacrity, until NIRS became a household word. The worldwide market for NIRS instruments has been estimated at between $US 100 and $US 200 million annually. Were the instrument companies to focus on marketing the technology (instead of just marketing the instruments) sales would likely increase even faster. This chapter discusses the intricacies and complexity of applying a new technique to an assortment of very diverse materials.

Near-infrared spectroscopy provides the client with ability to test things that could not be economically or practically tested before NIRS, because of the time and cost of testing. The technology provides the same accuracy (attainable by bias adjustment, where needed) and at least the same and usually superior reproducibility of reference methods, at a fraction of the cost, and in a fraction of the time.

The economics are very favorable. Testing farmers' grain at delivery has generated millions of dollars for farmers and grain companies. Careful use of NIRS in blending wheat by protein content for shipping has also generated generous revenue. There are big practical advantages and hidden economic benefits in obtaining instant information on the composition of crops, feed, and food ingredients at the time of delivery and processing, and of changes in the progression of processing operations. Feed mills can benefit by making more efficient use of ingredients in formulating feed mixes and ensuring efficient mixing. Food and feed plants can monitor their products progressively during processing. Plant-breeding institutions can avoid the frustrations and heavy expenses of advancing lines of unsuitable end-use quality by applying NIRS to selection in early generations. And the new concept of NIR precision agriculture will realize further substantial benefits for farmers by enabling them to improve the efficiency of grain storage (on the basis of protein content) and of fertilizer deployment.

## The Advantages of NIRS

What are the advantages of NIRS? They include:

1. Speed of analysis: commercial and industrial operations get their results in seconds, rather than hours (or days). This has great advantages. On-line NIRS analysis during processing enables close control of operations. Because of its speed and dramatically lower costs per test the technique enables testing of large numbers of samples in operations where the limitations of time and expense previously precluded large-volume testing.
2. Accuracy equal to reference testing
3. Reproducibility equal to (and often better than) reference testing
4. Low cost per test
5. Flexibility and efficiency—many constituents (up to 32) can be tested simultaneously
6. Environmentally clean—no chemicals needed (and therefore none to dispose of after analysis)

7. Easy and cheap to install (no requirements for fume exhaust or drainage)
8. Little or no sample preparation
9. Simple and totally safe to operate
10. Stand-alone instruments (no peripherals)
11. Small instrument size
12. Durable—instruments work well for many years
13. Networking capability of instruments within a plant or among a few or many plants, all under the control of a single bench-top computer (remote from the operation if necessary)

## The Disadvantages of NIRS

But there are some disadvantages as well. All of them affect the acceptability of the technology, but they can all be overcome:

1. Separate calibration for each commodity and constituent. This always has been, and remains, the most apparent impediment to NIR application. Although modern methods including "local" calibration and artificial neural networks (ANN) have streamlined large-scale NIRS and near-infrared transmittance (NITS) analysis of the most widely analyzed grains, to be most reliable these methods require large databases. The assembly of a large number (1000 or preferably more) of spectra of samples with accompanying reliable reference data for a constituent such as neutral detergent fiber (NDF) or metabolizable energy is a time-consuming and very expensive affair (for example, NDF by wet chemistry costs about $120.00 US per sample). As a result most calibrations for constituents other than protein, moisture, fat/oil, and starch are based on relatively small sample sets (200–300 is considered a small sample set for modern commercial or industrial NIR applications).
2. The need to monitor accuracy and precision. This is necessary, but it is also essential for laboratories to monitor the accuracy and precision of their reference "wet chemistry" methods.
3. Instruments are expensive to purchase. But the cost of a NIR instrument should be assessed by comparison with the value of testing the commodities or materials for the analysis of which the instrument has been purchased, and with the consequences of not testing them. The instruments cost thousands of dollars but are usually used to analyze commodities worth, or potentially worth millions.
4. Skepticism: People don't trust NIR (yet). Management still tends to adhere to the "tried and trusted" methods, such as forms of the Kjeldahl test for nitrogen or protein. Even the extremely efficient Dumas (combustion) method for nitrogen has been slow in widespread adoption, although in good hands it is more efficient and precise than any Kjeldahl method.

**TABLE 7.1.1. Items for Full Description of the Efficiency of a NIRS Calibration Model**

| No. | Item | Narrative |
|---|---|---|
| 1,2 | Number of samples | Calibration and validation |
| 3 | Source of samples | What are the samples and how were they assembled |
| 4 | Sample preparation | E.g., was the work performed on whole or processed samples? |
| 5,6 | Mean reference data | Calibration and validation |
| 7 | Std. Dev. Reference data | Validation samples |
| 8 | Cross-validation system | e.g., "one-out" cross-validation or number of subgroups |
| 9 | SECV | Standard error of cross-validation |
| 10 | SEP | Standard error of prediction |
| 11 | $r^2$ | Coefficient of determination |
| 12 | $b$ | Slope of validation computation |
| 13 | a | Intercept of validation computation |
| 14 | Bias | Mean difference between reference and NIR predicted data |
| 15,16 | RPD(2) or RER(3) | Ratio of SEP to SD of reference data of validation sample set (RPD) or to range of reference data of validation sample set (RER) |

5. Lack of knowledge of how to operate instruments really efficiently. Education in the application and interpretation of NIRS technology is the key to its proliferation. Better education in the use of NIR instruments would enhance confidence in the technology. Companies and other operations, such as plant-breeding stations, spend a lot of money on an instrument, then expect to get excellent results from an operator who has had about 2 days of training. Because of this the productivity is often substandard, exasperating to management, and embarrassing to the operators, who would really like to know more about their instrument and about NIR technology, and thereby do a better job.

## The Philosophy of Reporting NIR Results

Writing or reviewing articles on NIRS involves researching the literature. To communicate the results of a NIRS exercise effectively up to 16 pieces of information should be included. These are summarized in Table 7.1.1. Very few papers contain all of this information, or even enough for a full evaluation of the efficiency of the calibration exercise. Items 11, 14, and 15 (RPD) can give a good approximate evaluation of a calibration. Minimum acceptable values for the RPD and RER are 3.0 and 10.0, respectively. Some of the papers that are cited in Tables 7.1.4 and 7.1.5 have been found to have RPD values as low as 1.05. If the RPD is 1.0 this means that the SEP is equal to the SD of the reference data of the validation samples set (whether cross-validation or a test set is used), and the calibration model is of no practical value in analysis.

## Grains and Seeds: Their Role in Commerce and Industry and NIRS

Grains and seeds were the progenitors of NIRS application in commerce and industry. The first nonacademic applications were made to grain, specifically to wheat. Since those days, the embryo has grown into a lusty child, and NIR technology has become a household word, to the extent that most of the world's wheat of commerce is tested rapidly and accurately for protein and moisture contents by NIRS instruments.

First a word about grains and seeds. What are grains and seeds? A grain *is* a seed. A *seed* is defined as "a mature fertilized plant ovule, consisting of an embryo and its food store surrounded by a protective seed coat" (4). A *grain* is defined as "the small hard seedlike fruit of a grass, especially a cereal plant" (5).

There are practical differences between grains and seeds. Most grains are preprocessed into flours, malt, etc. before use in preparing a very wide variety of foods and beverages. Seeds are used directly for animal feed and for food (with or without cooking) and for extraction of nutritionally or industrially valuable constituents such as oil and, more recently, pharmaceutical substances. Analysis determines the amounts of these present in deliveries of the seeds as a means of assigning a value and a price. Seeds are also used as seeds, to propagate the next generations, but these seeds are usually not subjected to analysis. In plant-breeding institutions seeds of early generations are analyzed to determine the degree to which sought-after traits have been transmitted genetically.

By comparison with commercial and industrial operations, there are relatively few plant-breeding institutions in the world. But the on-going unique question that these institutions have to address is how to select the best genetic material to advance in the most economical way. First of all, why the need for plant breeding? There are several reasons, the most important of which is the need to ensure a constant supply of seed of varieties of grains and seeds that can overcome the constant encroachments of fungal disease and insect pests on yield. Grains and seeds are the most important sustainable food source for the world's burgeoning population. The approximately 2 billion metric tons of grain produced annually feed the 6.5 billion or so people, together with several billion domestic animals. Without farmers to grow it there would be no grain.

Plant breeders must also consider the functionality of grains and seeds. There is increasing interest in prediction of functionality in both grains and seeds. "Functionality" or application to grains and seeds is defined here as "the capability of a grain or seed to accomplish the intended purpose of the material for a specific role." Functionality in grains can also be defined as the end-use potential of the grain, be it for bread baking, noodle making, cake baking, beer making, animal feed manufacture, and a host of other functions. Functionality is strongly affected by physico-chemical properties that are conveyed to the grains by genetically introduced differences in (mainly) the structures of protein, oil, and starch, but partly by differences in the composition of the seed coat. Differences in the functionality of proteins are associated with both differences in the amino acid composition and the juxtaposition of these amino acids on the protein molecules. Changes in the functionality of starches are associated with differences in the ratio of amylose to amylopectin, whereas functionality in oils is linked to differences in fatty acid composition. The functionality of wheat transcends that of any other grain, about which more later.

Determination of functionality in the thousands of lines produced every year has always posed serious financial and logistics problems for plant breeders. Although pathologists and entomologists can assist in screening for the most resistant lines simply by walking through the plots and observing, determination of functionality involves chemical and/or physico-chemical analysis and actual processing, all of which are expensive. Probably even more important is the time restriction. Even the simpler physico-chemical tests, such as the Mixograph for estimating wheat bread baking potential, are restricted to about 50 tests per day, and there simply is not enough time to test every line. It is very frustrating for a breeder to advance lines to the stage at which the numbers are reduced, and the volume available increased sufficiently for testing, only to find that the chemists reject it because of inferior functionality, but it is a waste of time to proceed further with such material because the consumer will not be able to use it. An excellent example lies in triticale, once hailed as the savior of the world's potential food problems but now practically outmoded, because of its very poor functionality.

Near-infrared technology offers the plant breeder a method for screening thousands of lines quickly and cheaply, as well as enabling the determination of genetic/growing environment interactions, which have hitherto compounded the problem of large-scale testing of new genetic material.

Grains differ from seeds in that most grains are processed before final use or consumption, by some form of milling or by malting. The NIRS technique is also applied to commodities derived from grains, such as wheat flour and barley malt, and even to the final items accruing from these products (breads, pastas, noodles, beverages, and others). Grains are also used for animal feed directly. NIRS is applied to grains in the feed industry mainly to determine their composition and digestibility attributes, again, as a means of assessing their value.

But wheat represents less than 30% of total annual world cereal production. Table 7.1.2 gives the trends in grain production for 1994–2003. The grains most commonly used for food have shown a gradual increase in production, while the coarse grains have generally decreased. Maize can be considered as a food grain because, although only a small percentage is actually eaten as such by people, most of the crop is fed to domestic animals, which in turn feed people. Data on soybean are included for comparison. The increase of over 40% in world soybean production over the past 10 years (and 76% over the last 15 years) is remarkable. Soybean is the food legume equivalent of wheat, in that it has been widely recognized in Asia and elsewhere as a food grain with remarkable flexibility in its capacity as a raw material in production of an array of foods and flavorings.

Since its introduction to North America in the mid-eighteenth century, soybeans have become a major contributor to agricultural and industrial income. Soybeans contain about 20% oil, which is widely used in cooking. The oil-free meal, which contains over 40% of protein, is used mainly as a protein supplement for animal feeds and as a source of soy milk, from which an assortment of foods are made. On the average only about 6% of the world's soybeans is used in food production, although as much as 30% of Canada's annual crop of 2.5–3 million tons is used for food.

**TABLE 7.1.2. World Production of Most Common Cereal Grains 1994–2003 (000 Tonnes)**

| Year | Wheat | Durum | Maize | Rice | Barley | Sorghum | Oats | Millet | Rye | Soybean |
|------|-------|-------|-------|------|--------|---------|------|--------|-----|---------|
| 1994 | 527,202 | 26,423 | 570,750 | 537,338 | 159,925 | 60,718 | 33,251 | 28,688 | 22,303 | 136,000 |
| 1995 | 541,120 | 24,203 | 574,506 | 550,183 | 142,746 | 54,145 | 28,934 | 26,406 | 22,610 | 125,000 |
| 1996 | 582,689 | 30,405 | 587,716 | 569,236 | 155,622 | 73,685 | 30,940 | 28,875 | 22,944 | 130,000 |
| 1997 | 612,380 | 25,649 | 584,935 | 580,202 | 155,011 | 62,620 | 31,834 | 28,200 | 24,920 | 143,000 |
| 1998 | 588,842 | 31,128 | 604,013 | 563,188 | 138,820 | 63,451 | 23,802 | 29,204 | 20,977 | 158,000 |
| 1999 | 584,687 | 23,756 | 608,029 | 611,341 | 126,302 | 63,250 | 25,105 | 27,303 | 20,335 | 157,802 |
| 2000 | 582,276 | 26,666 | 591,695 | 599,051 | 133,505 | 56,235 | 26,971 | 27,754 | 20,493 | 161,412 |
| 2001 | 581,316 | 24,221 | 596,480 | 598,173 | 144,012 | 61,757 | 28,034 | 29,024 | 24,126 | 176,793 |
| 2002 | 566,151 | 26,829 | 604,636 | 569,527 | 134,962 | 53,497 | 26,734 | 24,142 | 21,477 | 180,729 |
| 2003 | 553,628 | 28,629 | 619,855 | 589,126 | 139,956 | 60,065 | 26,954 | 29,805 | 14,890 | 189,233 |
| Mean | 572,029 | 26,791 | 594,262 | 576,742 | 143,086 | 60,942 | 28,256 | 27,940 | 21,508 | 155,797 |
| Trend* | +4.8% | +8.4% | +6.9% | +6.5% | −9.2% | −1.1% | −13.7% | −2.1% | −19.0% | +41.7% |
| SD | 25,197 | 2565 | 15,390 | 23,251 | 108,08 | 5794 | 2998 | 1666 | 2767 | 22,086 |
| CV % | 4.4 | 9.6 | 2.6 | 4.0 | 7.6 | 9.5 | 10.6 | 6.0 | 12.9 | 14.2 |

*(((2003 + 2002)/2) − ((1994 + 1995)/2))/((1994 + 1995)/2) × 100

**TABLE 7.1.3. Grains and Seeds of Commerce and Industry**

| Cereals | Pulses | Oil-Bearing Pulses | Oilseeds |
|---------|--------|--------------------|----------|
| Wheat | *Phaseolus* beans | Soybean | Rapeseed |
| Maize | Dry peas | Ground nut | Canola |
| Rice | Chickpea | | Sunflower |
| Barley | Lentil | | Flax |
| Sorghum | Lupin | | Cottonseed |
| Oats | Faba bean (*Vicia*) | | Safflower |
| Millet | Mung bean | | Oil palm |
| Rye | *Lathyrus* (Grass pea) | | Industrial hemp |
| Eragrostis tef* | Cowpea | | Crambe seed |
| | | | Castor bean |

*Eragrostis abyssinica*

Table 7.1.3 summarizes the most important grains and seeds of industry. The main grain crops include cereals, food legumes (pulses), oil-bearing legumes, and oilseeds. Nutritional grasses, such as the millets, and Eragrostis tef, a millet-type grass that is a staple in Ethiopia, can be included among the cereals, all of which belong to the Gramineæ, the "Grass family".

Amaranth and buckwheat are two other commodities that are used as a main ingredient of some types of foods. Both are broad-leafed plants, and not true cereals. Amaranth belongs to the Amaranthaceæ family and buckwheat to the Polygonaceæ.

## WHAT HAS BEEN ACHIEVED TO DATE BY NIRS?

### Achievements in Composition Analysis

Since the first-ever application of NIR by Hart et al. in 1962 to the determination of moisture in seeds (6), the bibliography of NIRS technology has proliferated until it now numbers over 35,000 entries, many of which describe a very diverse assortment of applications to grains and seeds. The main areas have been composition analysis, analysis for prediction of functionality, and classification by NIR discriminant or classification analysis (NIRCA). Near-infrared spectroscopy has been applied to the analysis of many of the above commodities. Over 30 factors have been successfully predicted in cereals and pulses, and over 20 factors in oleaceous seeds. These applications have recently been comprehensively reviewed by Delwiche (7) and Dyer (8).

Only a few applications of NIRS to the analysis of lower-volume cereals, such as sorghum, the millets, and Eragrostis tef, have been reported. Stermer et al. reported on the determination of moisture content in sorghum (9) and Stevenson and Williams (10) were able to predict protein content in pearl, foxtail, proso and finger millets, sorghum, and Eragrostis tef. These workers also observed significant effects of growing location on NIRS sorghum analysis, and of species on millet analysis.

**TABLE 7.1.4. Applications of NIR Technology to Determination of Composition of Pulse Crops**

| Crop | Application |
|------|-------------|
| Soybean | Protein (11–16), moisture (13, 16), oil (13–16), crude fiber (12), trypsin inhibitor (17), P, Na, Ca, Mg, K (18), 7S/11S globulins (19), protein denaturation (20), 11S globulin (21), N solubility index (22), stachyose (23) |
| Dry Peas | Protein (24–28), fat (28), starch (28), stage of ripeness (29), sensory quality (30, 31) |
| Lupin | Protein (32), amino acids (33) |
| Chickpea | Protein (26) |
| Lentil* | Protein (26), tannins (34) |
| Faba Bean** | Protein (26), tannins (34) |
| Pigeon Pea | Protein (26) |
| Wild Vetch*** | Protein (26) |
| Groundnut | Protein, moisture, crude fiber, oil (35, 36) |
| Lathyrus | $\beta$-oxalyl amino-alanine (BOAA) (37) |

*Lens esculentum, L. culinaris*; ** *Vicia faba*; *** *V. narbonensis, V. gallilea*.

With the exception of soybeans (the pulse commodity that is most widely traded) the many advantages of NIR spectroscopy do not appear to have attracted the same amount of attention from pulse crop breeders and processors. Some compositional and functionality factors of pulse crops to which NIR technology has been applied are summarized in Table 7.1.4.

## Achievements in Functionality Analysis

Functionality means the ability of a commodity to perform the functions for which it is intended. The use of NIRS to predict functionality in grains and seeds is encumbered by the fact that the chemistry is essentially unknown. Functionality is conferred upon the commodity by a combination of chemical and physico-chemical factors that interact with each other to cause the commodity to behave in a certain manner during end-product processing. Functionality in wheat includes prediction of wheat "strength" factors, such as kernel texture, Farinograph, Extensigraph, Alveograph, and Mixograph parameters, and loaf volume. Most of the strength parameters are functions of the mystifying, and still unresolved, physico-chemical attributes of the wheat gluten protein complex.

The most consistent values are determined on flour-water doughs with instruments such as the instruments named above. These are unyeasted and unfermented, and consequently cannot truly relate to the actual bakery performance. But different cereal laboratories employ baking methods that differ widely, so that parameters such as loaf volume are not reproducible among laboratories. These laboratories can usually relate the physical dough figures to baking performance (and particularly to mixing behavior) within their own system. Baking is an expensive, time-consuming, and technique-demanding process. Different baking methods are developed to relate to commercial bakery practice in different countries and regions, so that the baking tests are very useful to the individual laboratory. Because physico-chemical characteristics are generally more reproducible among laboratories, and relate fairly well to end-use functionality, most of the world's wheat of commerce is traded on the basis of Farinograph, Alveograph, and Extensigraph values, rather than the results of baking tests.

Tests such as the Zeleny and sodium dodecyl sulfate (SDS) sedimentation tests have been developed by wheat breeding stations to predict "strength". They are cheaper in chemicals and equipment, and much faster, so that more samples can be processed in a day (and in the time between harvest and selection for planting). The ability to predict these results by NIRS would make it even cheaper to screen large numbers of plant breeders' lines for functionality, but the prediction of what is already a prediction test introduces an extra source of error, which is the error due to discrepancies in the relationship between the prediction test and what it is aimed at predicting, such as Farinograph stability time or loaf volume. Because of this "extraneous," but significant, source of error it is more practical to develop NIRS calibration models to predict such "strength" factors directly. It takes the same time to predict a set of Farinograph values as it does to predict Zeleny or SDS volume.

Functionality factors in barley include fermentability, digestibility, and true and apparent metabolizable energy. Functionality in oilseeds is based on differences in fatty acid composition. In pulse crops the most important marketing factors relate to the appearance of the grain. For food processing, soybeans with white hylum (the "joint" between the bean seed and the pod) are preferred. Varieties with dark hylum affect the color of tofu and other derived foods. Physical factors are those that can be measured without laboratory analysis, and can usually be measured very quickly. They include seed color, size, and uniformity and the tenuous "hard-to-cook" factor in peas. The presence of foreign material, broken kernels, and kernels damaged by factors such as weather damage and damage in storage also affect the value of the commodity. Seed size and shape are further physical factors. Seed shape affects dehulling or decorticating loss. Most of these factors do not lend themselves to accurate measurement by NIRS, although Edney et al. were able to predict plump kernel count in barley (62). Applications of NIRS to functionality are summarized in Table 7.1.5 Only a few applications to prediction of functionality in pulse grains have been reported.

Functionality factors include water absorption/hydration capacity, fat absorption, cooking time, protein functionality (amino acid and protein subunit composition), and starch functionality, which is associated with the amylose-to-amylopectin ratio. The amylose-to-amylopectin ratio affects the texture of cooked products. The texture of foods prepared from varieties with low amylose content tends to be stiffer than that of foods cooked from varieties with high amylose content.

## Achievements in Discriminant Analysis

Discriminant analysis is a way of classifying of substances on the basis of their spectra. As this stimulating application of NIR technology develops and becomes more

**TABLE 7.1.5. Application of NIRS to Prediction of Functionality in Grains and Seeds**

| Grain | References |
| --- | --- |
| Wheat | Kernel texture (38–47), Farinograph parameters (48–50), Alveograph parameters (48), Extensigraph parameters (50), Mixograph parameters (50), Zeleny sedimentation (38–40, 51), SDS sedimentation (52, 53), bread loaf volume (48, 49, 51, 54), water absorption (48, 51) |
| Barley | Fermentability, hot water extract, specific gravity (in beer) (55–58), metabolizeable energy (59–62) |
| Canola | Fatty acids (63, 64) |
| Rapeseed | Fatty acids (65–68) |
| Flaxseed | Iodine value (68) |
| Sunflower | Fatty acids (69, 70) |
| Soybean | Fatty acids (71–73), Nitrogen solubility index (NSI) (74), protein digestibility (74, 75), germination (76) |
| Mustard | Fatty acids (77–78), density (79) |
| Veg. Oils | Acid value, peroxide value, iodine value, phospholipid, color, viscosity (80) |
| Sesame oil | Fatty acid composition (81) |
| Peas | Sensory quality (30, 31), hardness, tenderometer value (27) |

refined it is destined to occupy a very significant role in agriculture, It is conceivable that the "traditional" system of development of calibration models for quantitative analysis could be replaced in many applications by systems based on discriminant analysis that identifies samples to be kept and those to be discarded, or degraded. This could apply to electronic grading of grains and seeds, where factors such as damage incurred by weather, harvest, or storage conditions could result in lowering of the value of the commodity. Most NIRS software ensembles offer a discriminant analysis or classification option, and calibration models can be developed that include equations for classification and analysis for composition and functionality.

The principle is based on the premise that grains and seeds that differ in spectral characteristics are indeed different, and belong to different classes, which can therefore be distinguished from one another. The reverse also applies. Differences in spectral signatures can be caused (and modified) by genetics, by growing conditions and seasons, and by other factors, and can even be modified during the day, because of changes in temperature and relative humidity. Spectral signals from some NIR instruments are more sensitive than others to these agencies. For effective discrimination with NIRS, such differences must be sufficiently consistent that the degree of discrimination or classification will persist and will not be significantly disturbed by factors that affect spectral characteristics. The technique has also been used to select sets of grains and seeds for use in calibration model development. For a more comprehensive understanding of the concept of discriminant analysis the reader is directed to references 84–86. Following are a selection of applications to grains and seeds, the results of which have been reported during the past two decades.

The concept was first introduced to modern NIRS technology by Rose et al. (82), for identification and differentiation among pharmaceuticals. The first application to agricultural products was documented by Bertrand et al. (83), who used it to identify wheat varieties suitable for bread baking. They achieved efficiency of 87% for correct identification of varieties. In a later paper the same group of workers extended their concept and described their application of pattern recognition to identification of French bread wheat varieties (87). In the same year Dardenne and Biston studied the application of discriminant analysis to Belgian wheat, and concluded that the second derivative of the log $1/R$ signal provided the most effective of four methods for variety identification (88). These workers used the Zeleny sedimentation volume as an indication of baking potential.

Williams and Sobering (89) outlined the parameters that could be used to categorize the western Canadian wheat classes (of which there are eight). Osborne and co-workers described a method for discrimination between Basmati and other long-grained rice types with NITS. They used Fisher linear discrimination calculated from the first five principal components of the NIT spectral data (90). Manley and her co-workers (91) attempted classification of UK wheat using NIRS and NITS methods. In all cases they found that the discrimination was affected by growing season and location.

Soybeans harvested at different times vary in processing quality, with autumn-harvested beans being of superior quality to beans harvested at other times. Differences in processing quality are based on protein recovery and gel yield during processing. Using these criteria, Tsai and colleagues (92) achieved discrimination of soybeans, with a Foss/NIRSystems Model 6500 scanning spectrophotometer, over a wavelength range of 400–2500 nm. Bewig and her co-workers developed a four-wavelength equation to classify vegetable oils. Their equation was developed using Mahalanobis distances between the various groups (93). Bertrand and co-workers (94) extended their studies on discriminant analysis to durum wheat products. They were able to differentiate among different types of semolina and flours, as well as ground whole wheat, using a stepwise discriminant analysis method based on principal components (95).

Using a SIMCA model developed in the Unscrambler on samples from one season, Turza and co-workers (96) were able to classify samples of soybeans from different years. Ootake and Kokot compared NIRS, photoacoustic spectroscopy, and Fourier transform Raman spectroscopy for differentiation between glutinous and nonglutinous rice (97). They concluded that FT-Raman gave the most promising results, and that the efficiency of classification could possibly be improved by using wavelengths in the mid-IR region between 280–1480 cm$^{-1}$. Hong et al., using canonical discriminant analysis over 400–2500 nm, were able to identify soybeans of superior processing quality, based on tannin content, degree of lipid oxidation, harvest season, and water absorption rate (98). High tannin content and high rate of lipid oxidation and low water absorption rate have adverse effects on soybean quality for tofu preparation.

Kwon and Cho (99) found that they could distinguish among Korean, Japanese, and Chinese whole sesame seeds with either a NIR filter or a monochromator-type instrument. They could not achieve the same accuracy of discrimination on either the roasted oil cake or the oil extracted from the seeds. Seregely (100) used the

Hungarian-developed polar qualification system (101) to distinguish between barley varieties suitable for malting and those more suitable for animal feed. With a Spectralyzer 1025 spectrometer operating between 1000–2500 nm that had been calibrated with samples of known composition, discrimination was achieved on the spectral data from whole barley seeds with no known analytical data.

## GRAINS AND SEEDS AND THEIR EMPATHY WITH DISPERSIVE SPECTROSCOPY—THE ANOMALIES OF NIR SPECTROSCOPY

The NIRS technique is logically based on the premise that whatever it is supposed to measure (sometimes called the "analyte") has wavelengths that appear to be appropriate to the analyte. Figure 7.1.1 and subsequent figures were developed with WINISI (InfraSoft International, State College, Pennsylvania). With this comprehensive software, mathematical pretreatment of spectral data is expressed in four values. Pretreatment includes the raw data (log $1/R$) and first or second derivatives thereof. First and second derivatives have the notations 1 and 2, respectively, whereas log $1/R$ has no derivative and has the notation 0. The "gap," or dimension of the derivative, is expressed in the second value, in terms of wavelength points. The degree of smoothing, or "segment," is expressed in the third value. The fourth value is sometimes referred to as "second smoothing," and is very rarely used in practice, the value is left at 1. A first derivative with a gap of 4 and a segment of 4 wavelength points would be described (in WINISI) as 1 4 4 1. A log of $1/R$ pretreatment with smoothing by four wavelength points would be described as 0 0 4 1, and so on. Figure 7.1.1 shows the second derivative (2 4 4 1) of wheat gluten, the main functional protein in wheat.

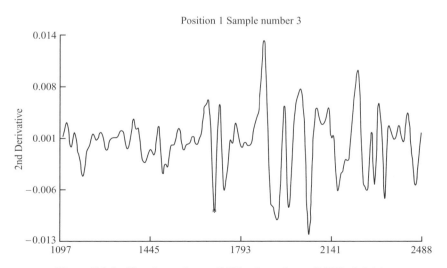

**Figure 7.1.1.** Absorbers of pure (96%) wheat gluten. D2OD: 2 4 4 1.

   The following sections are included to illustrate the complexity of the application of NIRS to a plethora of materials that contain widely variable amounts of constituents, most of which contain many different absorbers that interact with each other in different ways, depending on conditions and on their own composition, particularly with respect to moisture content.

   Although it is quite rational to associate absorbers such as the strong absorbers in the area of 2054 nm to N-H stretching/amide II or amide III, the strong "protein" absorber at 1696 nm is attributed to the first overtone of C-H stretching (102, 103). C-H stretching and bending frequencies are "traditionally" attributed to oil/fat, but there are a host of $CH/CH_2/CH_3$ groupings on protein, starch, and cellulose. The second derivative spectrum of wheat gluten (Fig. 7.1.1) shows 31 bands. Of these some are very small but often get selected in model development in favor of the "classic" protein absorbers at 1978, 2054, and 2180 nm.

   It is important to remember that there is really no such thing as an absolutely discrete absorber, and all of the so-called "classic" absorbers for water, protein, starch, cellulose, and oil/fat can occur over a range of several nanometers (104). In a system as complicated as grain the wavelengths selected during calibration model development may not be directly associated with the constituent for which the model has been intended. For example, in a series of models developed to test the influence of growing environment on the efficiency of testing whole-grain wheat for protein content, using three distinctly different growing regions, the wavelengths selected most consistently for all three models ranged from 1196 to 1220 nm [both assigned to C-H 2nd overtone (102)]. The 1196-nm band is close to a significant band at 1200 nm that occurs in starch. An absorber at 1260 nm [not assigned to anything in particular, but possibly associated with the 2nd overtone of C-H stretching, or P-OH stretching (105)] was also prominent in model development in this series. Figure 7.1.2 shows the second derivative spectra (2 4 4 1) of starch and dry starch (the most clear-cut), with the band at 1200 nm identified. Note also that, although the water band at 1928 nm was virtually eliminated on drying, the "water" band at 1428 nm *increased* in intensity.

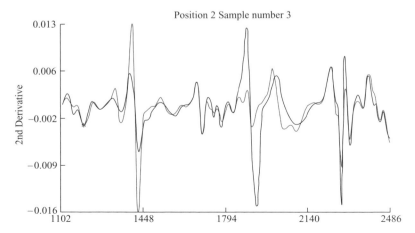

**Figure 7.1.2.** Dry wheat starch (bold) and starch (D2OD 2 4 4 1).

This implies that the O-H stretching/deformation absorber at 1428 nm is clearly not the same absorber as that associated with water in the 1920- to 1940-nm region.

A substance such as grain, dough, fresh grass, or manure should be visualized as a dynamic object, consisting mainly of space in which molecules and atomic combinations are constantly in motion. Conditions such as temperature change the velocity with which the atoms/molecules are in motion. Changes in relative humidity add extra water molecules to the milieu. In agricultural (and other) materials at any given time and at any given wavelength in the NIR region there are at least 3 and up to as many as 18 possible absorbers, all of which can influence wavelength selection in calibration development (105). This makes it difficult and rather ineffective to identify a single wavelength as being responsible for the success of a calibration model for these agricultural materials.

Partial least-squares (PLS) regression tends to use variance in spectral data in the same wavelength areas as MLR for model development (106), and both MLR and PLS use more than one wavelength. Assuming that one or more of the wavelengths used are the most strongly associated with known absorbers (based on F-values), usually variance in the areas of up to eight other wavelengths is also important in development of successful models, so the questions remain as to what the relative contributions of the variance at the other wavelengths are, and why are they used.

## FACTORS AFFECTING APPLICATIONS OF NIRS TO GRAINS AND SEEDS

Everything that affects a sample of grain (or anything else) is embodied in the spectrum. The spectrum of a sample can be affected by the instrument and by the operator. Assuming a consistent instrument performance, and consistent presentation by the operator, the spectrum of the sample then becomes the result of the genetic constitution of the sample, its composition and physical characteristics. These three factors have exerted their influence on samples before they reach the operator, and the resulting spectra are the ultimate things left for the operator to work with, for calibration model development or day-to-day analysis.

This section introduces some factors that can cause complications in calibration model development for grains and seeds. The factors (over 50) that have been listed (1) can exert direct or indirect influence on NIR spectra (and therefore on calibration and subsequent analysis). Most of the examples in this chapter are taken from experience with wheat. Other grains and seeds show great diversity in physical characteristics, but wheat is arguably the most diverse, especially in functionality. Pulses, such as the chickpea (*Cicer arietinum*), lentil (*Lens esculenta* and *L. culinaris*), and faba bean (*Vicia faba*) show even greater diversity in seed size, shape and color, but are rather less variable than wheat in kernel texture and functionality.

### Sizes, Shapes, and Colors

The crops included in Table 7.1.3 vary widely in size, shape, and color, as well as in composition. Seed size varies from 2–3 mg for canola seed to over 2 g per seed in the

Position 2 Sample number 21

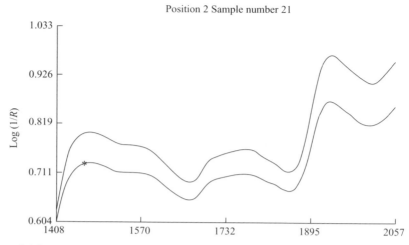

**Figure 7.1.3.** Influence of kernel size of wheat on spectra: different samples, 1400–2060 nm.

case of some faba bean varieties. Colours vary from green (peas, lentil, mung bean, some faba beans, and desi-type chickpea) to beige, red, yellow, and various shades of brown and black. Seed shape and size both influence sample presentation (mainly because of differences in packing density) and the diffuse reflectance or transmittance itself. Seed size in wheat usually varies by more than 300% within a given delivery. Small seeds can cause problems with cleaning, because they get removed by screening and must be recovered.

***Kernel or Seed Size.*** Figure 7.1.3 illustrates differences in spectra of two samples of wheat of the same class (Canada Western Soft White Spring, or CWSWS wheat). The upper spectrum is of large seeds (kernel weight 45.2 mg), and the lower of small seeds (kernel weight 28.3 mg). Both were of about the same protein and moisture contents, so that changes in spectral form could not be attributed to differences in composition. Figure 7.1.4 shows the spectra of the same two samples as Figure 7.1.3, but over the entire spectral range. Note that the spectra converge at about 1350 nm and then they become interchanged. Below about 1350 nm the spectrum of the larger kernels is uppermost, but that of the larger kernels is uppermost thereafter.

Figure 7.1.5 shows the effect of kernel size within a sample. The three samples were obtained by sieving the initial sample (in this case, Canada Western Red Spring, CWRS or hard red spring wheat). The sieves used were of 2.8, 2.5, and 2.2 mm. The upper spectrum is that of the large kernels (100-kernel weight, or TKW, 43 g), the overs of the 2.8-mm sieve. The lower spectrum is that of the smaller kernels (TKW 20 g), those kernels that passed through the 2.2-mm sieve, and the middle spectrum that of the intermediate kernels (TKW 33 g). The middle spectrum is actually two spectra superimposed—the spectrum of the intermediate kernels coincided (in this case) exactly with that of the original sample.

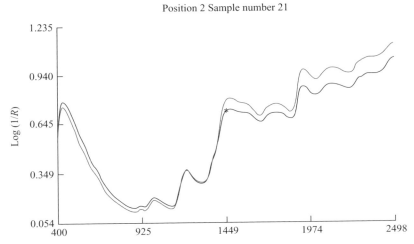

**Figure 7.1.4.** Influence of kernel size on wheat spectra, 400–2498 nm.

Figure 7.1.6 gives spectra of soybean samples of different sizes. Again, the spectrum of the larger seeds is uppermost, but seed size appeared to have less influence on the spectra of soybeans. Figure 7.1.7 shows spectra of large (uppermost), mainly green (macrosperma), and small red (microsperma) lentil seeds. The pronounced absorption band at 678 nm in the macrosperma lentil (the "bold" spectrum) is that of chlorophyll. Most of the "red" small-seeded lentils of commerce have red-orange endosperm, but the seed coats are a kind of rusty brown. This type of lentil is the most

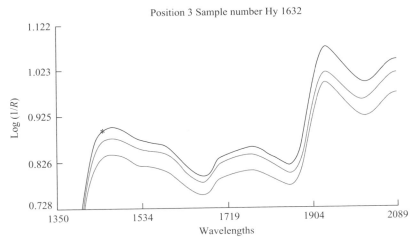

**Figure 7.1.5.** Influence of kernel size of wheat on spectra: effect of sizing within a sample.

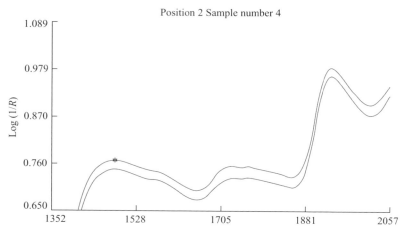

**Figure 7.1.6.** Influence of seed size of soybeans on spectra.

widely sought on the world market. The seeds are decorticated and split before final processing, for use in soups and a wide range of oriental foods.

***Seed Coat Color.*** Colour can also affect the shape of spectra. Figure 7.1.8 shows the spectra of two samples of wheat. Kernel size (weight per 1000 kernels) was practically identical for the two wheats, and the compositions of the two samples with respect to protein and moisture contents were similar (about 12.5% protein and moisture). The lower spectrum is that of white (CWSWS) wheat and the uppermost of red (CWRS) wheat. The absorbance of the spectrum of the white wheat was higher

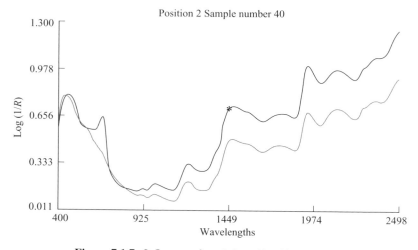

**Figure 7.1.7.** Influence of seed size of lentil on spectra.

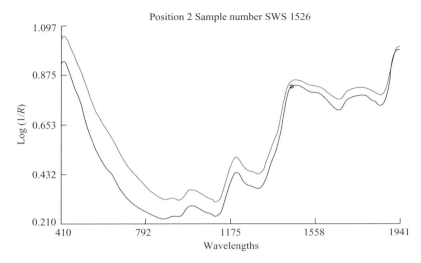

**Figure 7.1.8.** Influence of seed coat (bran) color on wheat spectra.

than that of the red wheat in the Vis-NIR range (400–1100 nm), and, unlike the spectra that showed the influence of kernel size, this persisted through the full range from 400 to 2498 nm.

To complicate matters, some of the new Canadian hard white spring (Canada Western Hard White Wheat, or CWHW) wheat lines have been found to turn red when grown under some conditions. Figure 7.1.9A shows spectra of samples of the

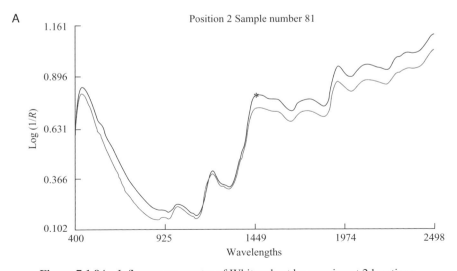

**Figure 7.1.9A.** Influence on spectra of White wheat by growing at 2 locations.

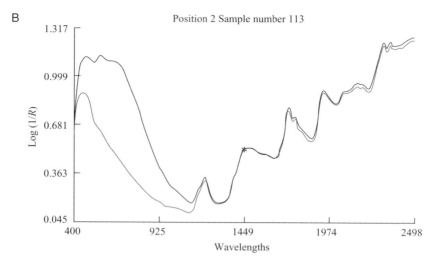

**Figure 7.1.9B.** Spectra of black and yellow canola seed.

same line grown in two locations, one of which suffered from moisture stress. In the vis-NIR range (400–1100 nm) the sample grown under moisture stress was of lower optical density than the sample grown under more available moisture, but the spectra converged at about 1156 nm (a water band) and thereafter the spectra were reversed in position.

Canola seed varies in color from yellow to black. Figure 7.1.9B shows spectra of black and yellow varieties of canola seed. The uppermost spectrum is of black-and the lower of yellow-seeded varieties. The differences are understandably particularly obvious in the vis-NIR region. The spectra converge at about 1340 nm. In this particular case the optical density of the black variety persisted at slightly higher than that of the yellow variety, but in commercial canola seed overlaps do occur, and some varieties have reddish-brown seeds, which are intermediate in the vis-NIR region.

**Seed Density.** Seed density is a quality factor that is often overlooked by cereal chemists, but not by wheat processors. Density is measured in kilograms per hectoliter and reported as "test weight," or "hectoliter weight." Flour millers recognize that the density of the wheat can affect flour yield, and wheat of low test weight is usually degraded and marketed at a lower price. Grain density affects the way the grain packs in a cell. The influence of density on NIR spectra is illustrated in Figure 7.1.10A, B, and C, which show spectra of hard red spring (CWRS), soft white (CWSWS), and durum (CWAD) wheat classes' respectively, of high and low test weight. Spectra of the samples with highest test weight were of higher optical density over the whole spectral range.

Seed density includes the way grains and seeds pack in a sample cell or move through a flow-through cell. The packing factor is affected by seed shape as well

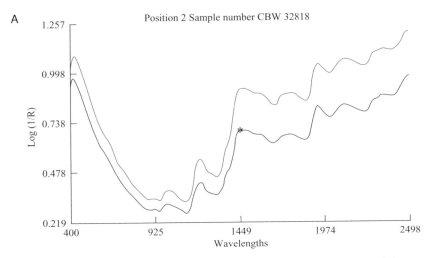

**Figure 7.1.10A.** Spectra of CWRS wheat of high (upper) and low test weight.

as density. Round seeds such as soybeans and canola pack in a different way than seeds such as wheat and barley, which are more or less oval. Oats pack differently because of the fibrous hull. Desi-type chickpeas can be almost triangular in shape. Seed size, density, and shape all interact to affect the packing factor, which can also be influenced by the operators and the way they load cells.

***Kernel/Seed Texture.*** Texture is a measure of the degree of hardness or softness of the kernel or seed. The hardest seeds are those of the faba bean (*Vicia faba*) and palm

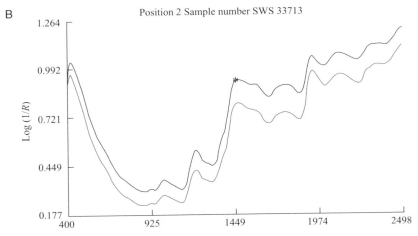

**Figure 7.1.10B.** Spectra of CWSWS wheat of high (upper) and low test weight.

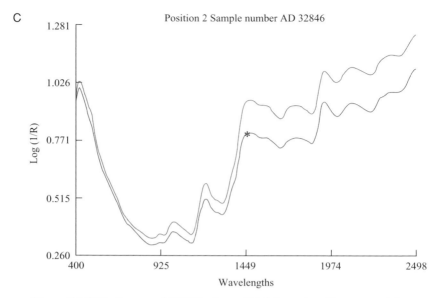

**Figure 7.1.10C.** Spectra of CWAD wheat of high (upper) and low test weight.

kernel (*Elaeis sp.*). Oil seeds, such as canola, rapeseed, or flax, are very soft. They can be crushed with the finger, and texture is not an important factor affecting spectra. Wheat and barley are intermediate in texture. Kernel texture in barley affects malt quality, and the softest kernels are associated with higher hot water extract (HWE) (107). High HWE gives higher alcohol content during fermentation, and consequently more beer or distilled liquors per ton of barley.

Wheat shows the biggest extremes in kernel texture. Kernel texture has a fundamental influence on end-use potential because it affects gas production during fermentation for bread baking and water absorption for preparation of practically all of the multitude of products prepared from wheat. Durum wheat (*Triticum durum* and other tetraploid species) is the hardest (108), and North American soft white winter and Australian Soft are the softest of the wheat classes. The most reliable reference method of measuring wheat kernel texture is the particle size index (PSI) method, based on a standardized grinding/sieving procedure (109). Extensive use is being made of the Perten Single Kernel Classification System (SKCS) for a rapid assessment of kernel texture in wheat and barley.

Figure 7.1.11 and B show the influence of kernel texture on the spectra of wheat within two classes of Canadian wheat, CWRS (hard red spring) and CWSWS (soft white spring). As a general rule spectra of harder wheats are of higher optical densities than those of softer wheats within a class. This can be complicated by factors such as test weight, kernel size, and moisture content. Figure 7.1.11C and D illustrate how such factors can result in significant exceptions to this rule. This can cause complications with NIRS calibrations for hardness that use the SKCS hardness index

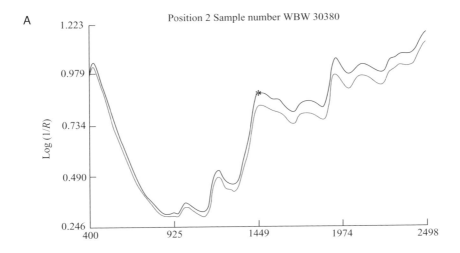

**Figure 7.1.11A.** Spectra of CWRS wheat of high and low (upper) PSI.

as reference data, because the SKCS calibration for kernel hardness index is based on spectral data alone, using a set of reference samples rather than a grinding/sieving test.

The PSI of both samples in Figure 7.1.11C was 49.0 (very hard). In Figure 7.1.11D the PSI of the sample with the "bold" spectrum was 73 (very soft) and that of the other sample was 47 (very hard).

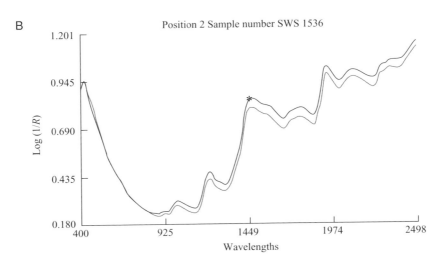

**Figure 7.1.11B.** Spectra of CWSWS wheat of high and low (upper) PSI.

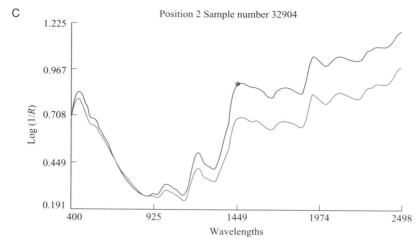

**Figure 7.1.11C.**  Spectra of wheat of same class, showing big difference in spectra but same PSI PSI of both samples was 49.0 (very hard).

***Growing Location.***  Figure 7.1.12 shows the influence of growing location on the spectra of CWRS wheat. The spectra represent the average of three growing locations (Central Prairie, Western Prairie, and Northern Prairie regions, covering about 1500 × 1200 km total area). Each spectrum represents over 100 samples grown over a period of six seasons. Individual seasons varied in the degree of influence caused by growing

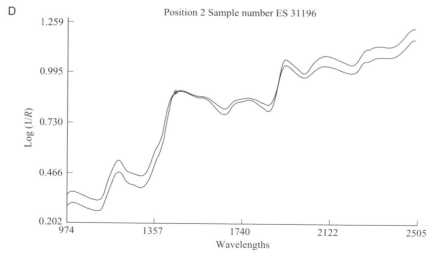

**Figure 7.1.11D.**  Spectra of wheat of two classes showing similar spectra but big difference in PSI. PSI of the sample with the "bold" spectrum was 73 (very soft) and that of the other sample was 47 (very hard).

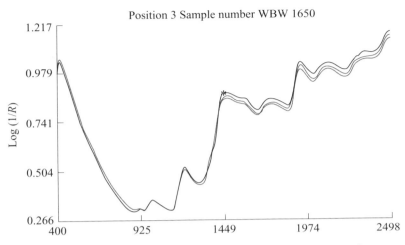

**Figure 7.1.12.** Influence of growing location on CWRS wheat spectra: 6 years.

location. Figure 7.1.13A and B demonstrates differences in the influence of growing location between the years 2000 and 1998. Clearly the impact of growing location on the spectra of the whole grains was strongly influenced by the growing season. This is mainly a function of interactions between temperature and precipitation. The trials were grown on the same fields in all seasons, so the differences in soil characteristics from season to season were much lower than differences among growing locations.

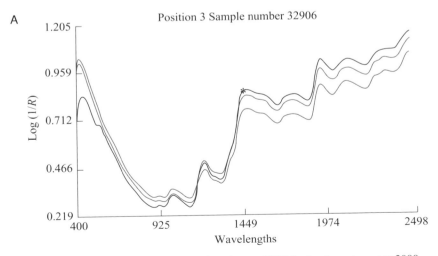

**Figure 7.1.13A.** Influence of growing location on CWRS wheat spectra: year 2000.

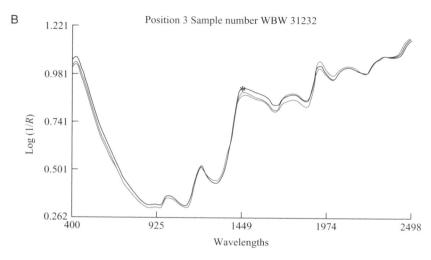

**Figure 7.1.13B.** Influence of growing location on CWRS wheat spectra: year 1998.

***Growing Season.*** Growing season has been found to have a significant influence on the spectral characteristic of wheat. Figure 7.1.14A–C shows the influence of growing season over 6 years (1996–2001) on three different wheat classes.

Each spectrum represents the average of about 140 samples of CWRS plant breeders' lines grown respectively on the eastern Prairie region over six growing seasons

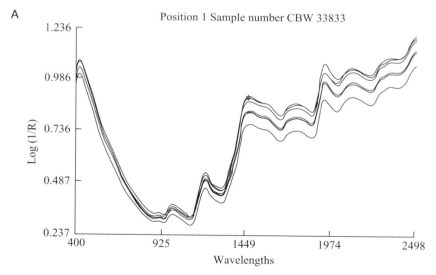

**Figure 7.1.14A.** Influence of growing season on spectra of CWRS wheat: Canadian Eastern Prairie Plant Breeders' lines.

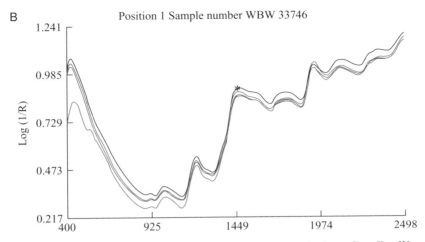

**Figure 7.1.14B.** Influence of growing season on spectra of CWRS wheat: Canadian Western Prairie Plant Breeders' lines.

(from 1996 through 2001). The spectra of Figure 7.1.14B each represent the average of about the same number of samples of CWRS wheat grown on the western prairies over the same growing seasons.

Figure 7.1.14C shows the influence of growing season on the western Canadian soft white wheat class. Figure 7.1.14A–C show that the influence of growing season depends on the type or class of wheat. The influence of growing season is further illustrated by Figure 7.1.15A and B, which give three-dimensional images of the

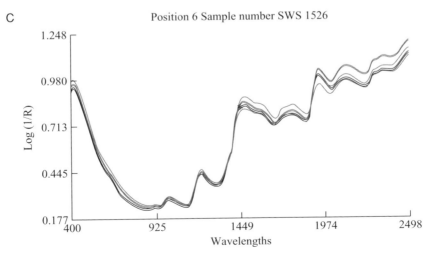

**Figure 7.1.14C.** Influence of growing season on spectra of CWSWS wheat.

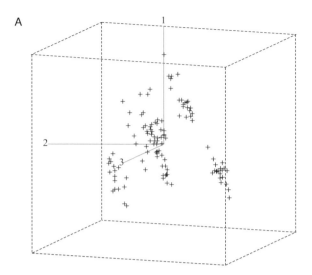

**Figure 7.1.15A.** Influence of season on spectra of CWRS wheat grown over 6 seasons: Eastern Prairie.

impact of growing season on the spectra of Eastern Prairie and Western Prairie CWRS wheat, respectively. The "strata" reflect different seasons.

The data points are computed from the PLS factors used in model development, using the log $1/R$. Despite the obvious differences in seasonal effect (as demonstrated by the strata), excellent calibration models could be developed for prediction of protein

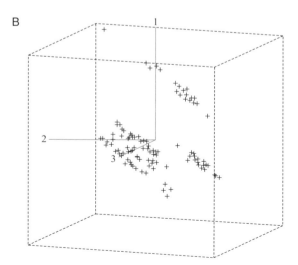

**Figure 7.1.15B.** Influence of season on spectra of CWRS wheat grown over 6 seasons: Western Prairie.

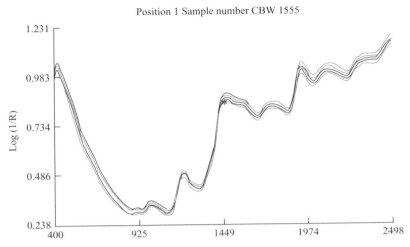

Position 1 Sample number CBW 1555

**Figure 7.1.16.** Spectra of seven western Canadian hard red wheat classes.

content. To emphasize the effect of the growing season/ location interaction further, the Eastern Prairie model would not predict the Western Prairie or Northern Prairie validation sets with accuracy, but a model computed from a combination of all three locations gave excellent predictions of protein content from each of the individual validation sample sets.

**Wheat Class.** Wheat is marketed in commerce by class, and the different exporting countries have devised their own systems for classifying their wheat on the basis of end-product potential. Australian states subdivide their wheat on the basis of kernel hardness, and the hard wheat classes are further subdivided on the basis of protein content. Western Canadian hard red wheat is subdivided on the basis of gluten characteristics, and the premium class (Canada Western Red Spring) is further segregated on the basis of protein content. Figure 7.1.16 shows the average spectra of the seven Canadian hard red wheat classes. Each spectrum represents at least 100 samples of wheat. Because of differences in kernel texture, size, and shape some of the overall differences among the classes are evened out.

## Composition

The literature contains references to over 30 constituents for which NIR calibration models have been developed (7, 8). The major constituents in grains and seeds are moisture, protein, oil, starch, and cellulose. The most prominent absorption bands for these are summarized in Table 7.1.6.

Moisture is the constituent with the single most influence on spectra. Figures 7.1.17 and 7.1.18 show the effect of big differences in moisture content on the spectra of whole wheat and maize.

**TABLE 7.1.6. Important Absorption Bands of Main Constituents in Grains and Seeds***

| Cellulose | Oil | Protein | Starch | Water |
|---|---|---|---|---|
| 678 | 758 | 770 | 754 | 718 |
| 758 | 816 | 868 | 914 | 758 |
| 816 | 928 | 982 | 980 | 810 |
| 914 | 1042 | 908 | 1062 | 964 |
| 982 | 1162M | 1016 | 1202M | 1044 |
| 1004 | 1724L | 1140S | 1360S | 1154S |
| 1156S | 1764L | 1186M | 1432L | 1410M |
| 1272S | 2308L | 1276S | 2282L | 1460M |
| 1364M | 2346L | 1426M | 2322M | 1778S |
| 1428L | | 1494M | 2376M | 1904L |
| 1488S | | 1692L | 2446 | |
| 2076M | | 1734L | | |
| 2104M | | 1978L | | |
| 2270L | | 2054L | | |
| 2332M | | 2168L | | |
| 2482L | | 2274L | | |
| | | 2466L | | |

* Spectra recorded on Foss/NIRSystems Model 6500 instrument. Bands were identified using second derivative 2 4 4 1; L = strong band, M = moderate intensity band; S = small band All of the bonds below 1100 nm are small.

In Figure 7.1.17 the moisture content of the sample represented by the upper spectrum was 21.7%. That of the sample represented by the lower spectrum was 11.9%.

The moisture content of freshly harvested maize (corn) can reach to over 50%. The samples represented by the two spectra of maize in Figure 7.1.18 differed by over 40%. The upper spectrum is that of a sample with 54% moisture, while the moisture

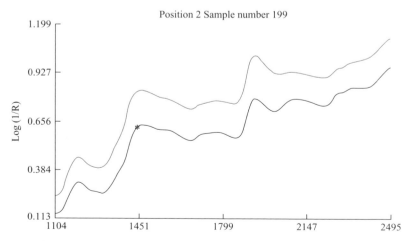

**Figure 7.1.17.** Influence of moisture content on whole wheat spectra.

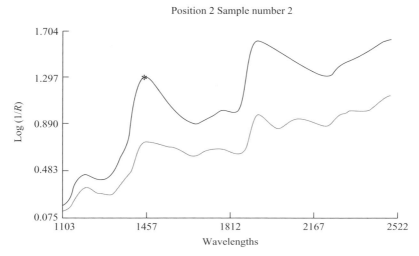

**Figure 7.1.18.** Influence of moisture content on whole maize spectra.

content of the sample represented by the lower spectrum was 12.0%. The second derivative of the spectra of Figure 7.1.18 (Fig. 7.1.19) shows that the considerable increase in moisture content was accompanied by a shift in the wavelengths at 1162, 1428, and 1916 nm toward the wavelengths of the main absorbers in pure water (1154, 1410, and 1906 nm respectively) by up to nearly 20 nm.

Figure 7.1.20 shows the influence of moisture content on the spectra of industrial hempseed. Note as well that in the cases of both maize and hempseed the influence of high moisture content has been to distort the area of the spectrum immediately above the water band in the 1910-nm region.

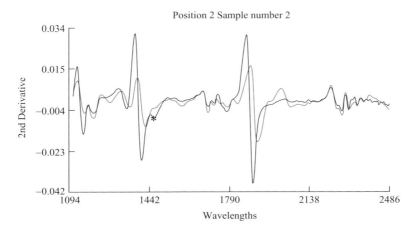

**Figure 7.1.19.** Influence of moisture content on whole maize spectra (D2OD).

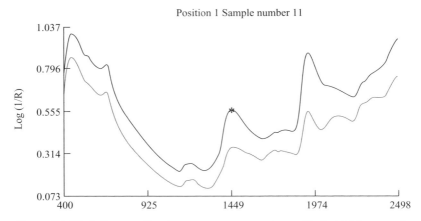

**Figure 7.1.20.** Influence of moisture content on spectrum of industrial hempseed.

Another intricacy in the complicated application of NIR technology to grains and seeds is that, because of inclusion of several major constituents in their makeup, wavelengths selected during model development do not necessarily conform to the absorbers of the constituents for which the models are developed. Table 7.1.7 shows the wavelengths selected by multiple linear regression (MLR) for prediction of moisture in maize, oil in canola seed, and protein content in wheat and lentil.

Although PLS regression includes all of the wavelengths in an array, most of the variance used in computation of a PLS model will be in the vicinity of the wavelengths selected by MLR (107). In the case of the predictions of water in maize and oil in canola, "protein" bands were important in the respective models, and the "protein" bands selected for the wheat and lentil models were all small bands. There is a strong negative correlation between oil and protein contents in canola ($r = 0.85 - 0.90$), which could account for the presence of a protein band in the canola oil model.

**TABLE 7.1.7. Wavelengths Selected for Prediction of Some Constituents in Wheat, Canola, and Lentil using log 1/R and MLR**

| Grain | λ/F | | | Wavelength selected (nm) | | | | | | $r^2$ | RPD |
|---|---|---|---|---|---|---|---|---|---|---|---|
| Moisture | λ | 1844 | 1524 | 2260 | 2356 | 2412 | 1692p* | | | 0.99 | 13.39 |
| Maize | F | 220 | 89 | 14 | 28 | 39 | 475 | | | | |
| Oil | λ | 1580 | 1108 | 1364 | 2052p | | | | | 0.95 | 3.74 |
| Canola | F | 136 | 342 | 209 | 253 | | | | | | |
| Protein | λ | 1660 | 1260 | 1180p | 1228 | 1700p | 1268 | 1236 | 1708 | 0.95 | 4.62 |
| Wheat | F | 60 | 697 | 139 | 74 | 51 | 489 | 126 | 34 | | |
| Protein | λ | 1508 | 1260 | 1276p | 1204s | 1188p | 1412w | 1404 | 1396 | 0.95 | 4.18 |
| Lentil | F | 102 | 220 | 211 | 73 | 275 | 381 | 276 | 161 | | |

* Small letters accompanying wavelengths indicate bands that correspond to bands in "pure" constituents, e.g., p = protein, s = starch and w = water; RPD = (SD of validation reference data)/SEP.

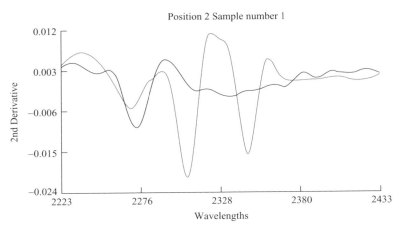

**Figure 7.1.21.** Soybean, whole and ground (prominent bands). D2OD, 2220–2450 nm.

Note that none of the "strong" absorbers identified in Table 7.1.6 was selected for any of the four calibrations. This calls for some speculation as to the reason. Earlier instruments, such as the DICKEY-john Models I and GAC III used six filters, all based on "classic" wavelengths originally identified by Karl Norris. The wavelengths were 1680, 1940, 2100, 2180, 2230, and 2310 nm, and the discrete filters corresponding to the peak absorbances. Backward stepwise MLR rarely used more than four of these in calibration model development. Logically, the bands selected in model development for a constituent should be those with the highest intensity for the constituent. One reason for the apparent collapse of that theory could be mutual interference among the major absorbers, so that the absorbers of lower intensity become more closely correlated to the reference data than those of higher intensity.

Composition may affect the appearance of spectra in other ways. Figures 7.1.21–7.1.25 show spectra of soybeans and three types of oilseed. In the case of whole canola seed the "oil" bands at 2306 and 2346 nm are prominent in the log $1/R$ spectrum (Figure 7.1.24 and 7.1.25), but these bands are not noticeable in whole soybean (Figure 7.1.21), sunflower (Figs. 7.1.22 and 7.1.23), or industrial hempseed (Fig. 7.1.24), despite the fact that these three commodities contain an average of 21%, 42%, and 36% oil, respectively. When these commodities are ground, on scanning the bands at 2306 and 2346 become prominent. The reason for the apparent reticence of these -CH- bands to appear in the whole grain spectra is believed to be caused by the presence in their seed coats of a nonabsorbing constituent. When the seeds are ground, the influence of this constituent is quashed by the intensity of the oil band absorptions at 2306 and 2346 nm.

Another example of this phenomenon is given in Figure 7.1.26, which gives the spectra of two different commercial animal feeds. The upper spectrum appears to be "flattened." This is caused by the presence in the feed of a small amount of carbon black, included to impart color to the feed. Carbon black is nonabsorbing and caused the apparent "flattening" of the spectrum.

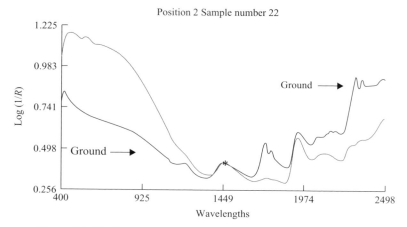

**Figure 7.1.22.** Sunflower whole and ground, Log $1/R$, 400–2498 nm.

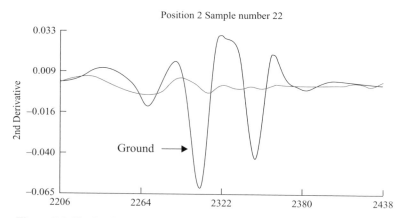

**Figure 7.1.23.** Sunflower whole and ground (Detail). D2OD. 2220–2450 nm.

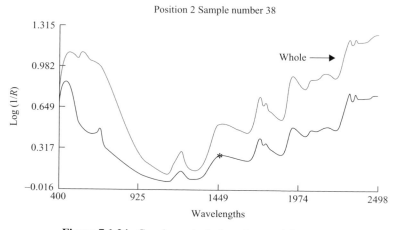

**Figure 7.1.24.** Canola seed whole and ground. Log $1/R$.

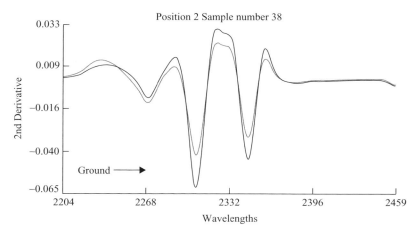

**Figure 7.1.25.** Canola seed whole and ground (Detail) D2OD.

Figure 7.1.27A and B shows the log $1/R$ spectra of the upper feed in log $1/R$ and D2OD formats, to demonstrate that all of the bands are present, despite the "flat" appearance of the spectrum.

This section has been included to illustrate the factors that can coalesce to defy application of NIR spectroscopy to grains and seeds—but the more rebellious believers in the early days of NIRS application managed to overcome the more hazardous hazards, and in doing so were able to uncover the plethora of factors that can affect the technology. For example, the importance of grain texture has been mentioned above, and over 20 factors have been shown to affect wheat kernel texture (110).

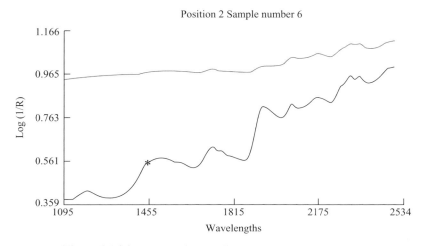

**Figure 7.1.26.** Spectra of two different pelleted animal feeds.

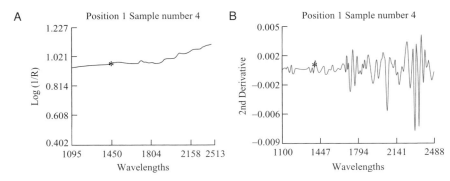

**Figure 7.1.27A.** Pelleted animal feed A, Log $1/R$   **B.** Pelleted animal feed, A D2OD.

## The Importance of Optimization

In the modern era of PLS regression, ANN calibration models, and chemometrics in general, optimization of mathematical pretreatment of spectral data and of wavelength selection is rarely discussed. Wavelength range is still an important aspect of instrument performance, and mathematical pretreatment within a wavelength range is of relatively less importance than the wavelength range itself, over which calibration models are developed (and instruments are designed). Table 7.1.8 gives the results for the prediction of protein content of whole-kernel wheat and oil content of whole canola seed using the log $1/R$ spectral data at different wavelengths, with no optimization of mathematical pretreatment.

Samples were scanned in a Foss Model 6500 scanning spectrophotometer, using the rectangular sample cell. Software was WINISI version 1.5. The wavelength ranges

**TABLE 7.1.8. Influence of Wavelength Range for Prediction of Wheat Protein and Canola Oil Content at Log 1/R with No Mathematical Pretreament of Spectral Log 1/R Data**

| | Wheat Protein Content | | | Canola Oil Content | | |
|---|---|---|---|---|---|---|
| Wavelengths | $r^2$ | SEP* | RPD | $r^2$ | SEP | RPD |
| 408–1092 | 0.912 | 0.237 | 3.37 | 0.889 | 0.981 | 3.00 |
| 408–2492 | 0.927 | 0.217 | 3.68 | 0.959 | 0.597 | 4.93 |
| 696–1792 | 0.962 | 0.156 | 5.12 | 0.967 | 0.535 | 5.50 |
| 708–2492 | 0.940 | 0.196 | 4.07 | 0.958 | 0.608 | 4.84 |
| 908–1692 | 0.968 | 0.144 | 5.54 | 0.964 | 0.560 | 5.25 |
| 1108–2492 | 0.947 | 0.184 | 4.34 | 0.955 | 0.622 | 4.73 |
| 1108–1392 | 0.957 | 0.166 | 4.81 | 0.960 | 0.596 | 4.94 |
| 1224–2224 | 0.924 | 0.220 | 3.63 | 0.956 | 0.619 | 4.75 |

* SEP = standard error of prediction; RPD = ratio of SEP to standard deviation of reference data for test-set samples.

**TABLE 7.1.9. Optimization of Mathematical Pretreatment and Wavelength Range for Prediction of Wheat Protein and Canola Oil Contents**

| Wavelengths (nm) | Wheat Protein Content | | | | Canola Oil Content | | | |
|---|---|---|---|---|---|---|---|---|
| | *Pretreatment* | $r^2$ | *SEP** | *RPD* | *Pretreatment* | $r^2$ | *SEP* | *RPD* |
| 408–1092 | 2 8 4 1** | 0.973 | 0.131 | 6.09 | 2 8 8 1 | 0.933 | 0.778 | 3.78 |
| 408–2492 | 1 4 4 1 | 0.972 | 0.135 | 5.91 | 2 4 8 1 | 0.963 | 0.571 | 5.15 |
| 696–1792 | 2 8 8 1 | 0.975 | 0.125 | 6.38 | 0 0 1 1 | 0.965 | 0.552 | 5.50 |
| 908–1692 | 2 8 8 1 | 0.972 | 0.133 | 6.00 | 0 0 2 1 | 0.965 | 0.553 | 5.32 |
| 1108–1392 | 2 8 8 1 | 0.959 | 0.162 | 4.93 | 0 0 8 1 | 0.965 | 0.565 | 5.21 |
| 1108–2492 | 1 4 8 1 | 0.964 | 0.151 | 5.28 | 1 2 2 1 | 0.965 | 0.553 | 5.32 |
| 1224–2224 | 2 4 8 1 | 0.954 | 0.172 | 4.64 | 2 8 8 1 | 0.961 | 0.582 | 5.05 |
| 708–2492 | 2 4 8 1 | 0.974 | 0.128 | 6.23 | 0 0 4 1 | 0.967 | 0.532 | 5.53 |

* SEP = standard error of prediction; $RPD^2$ = ratio of SEP to standard deviation of reference data for test-set samples.
** 0 0 4 1 = log $1/R$ with segment of 4 wavelength points; 1 4 4 1 = first derivative of log $1/R$ with gap of 4 and segment of 4 wavelength points; 2 8 8 1 = second derivative of log $1/R$ with gap and segment each of 8 wavelength points.

were selected to represent ranges typical of commercial NIR instruments operating in reflectance mode. Two other ranges are included. These are the 696–1792 range, which has been found to be useful, and the 708–2492 range. This range excludes the spectral data up to 700 nm, which often contain abnormal data.

The mathematical pretreatments of log $1/R$, first and second derivatives were optimized for each wavelength range. The size of the segments (degree of smoothing) and derivative sizes ("gaps") were optimized for each wavelength range. The data were based on about 400 samples of CWRS wheat assembled over six seasons and about 200 samples of canola seed representing two growing seasons. In all cases a test set was used for validation. Test sets were preferred to cross-validation for evaluation of calibration models because cross-validation does not provide information on features such as bias or slope irregularities.

The importance of optimizing the wavelength range is illustrated by the fact that the RPD values ranged from 3.37 to 5.54 in the case of wheat protein and from 3.00 to 5.50 for prediction of oil content in canola, depending on the wavelength range. Table 7.1.9 summarizes the results for prediction of protein content in whole wheat kernels and oil content in whole canola seeds after optimization of both wavelength range and optimization. In one case (prediction of oil content in canola seed using the wavelength range 696–1792 nm) the calibration based on "raw" spectral data proved to give the best statistics for prediction.

Wavelength selection was one of the most difficult conundrums to solve during the days when MLR was the only method for calibration model development. PLS regression was heralded as the answer to this question. But PLS uses variance in the same wavelength areas as MLR (107), and selection of the wavelength range most suitable for the task in hand is still essential to reliable model development. The results were all derived from PLS regression. Forward stepwise MLR was tested at the same time, but in most cases the results were inferior to those obtained by PLS regression. An interesting feature of the MLR exercise reported in Reference 106 was that for all wavelength ranges except 408–1092 nm log $1/R$ (with a segment of 4 wavelength points) was the most effective pretreatment.

Software systems such as the Bruker OPUS and Buchi software include options that optimize both mathematical pretreatment and wavelength range. These give very good guidelines as to the respective questions, and further fine-tuning can provide even better results.

## WHAT DOES INDUSTRY NEED FROM NIRS, AND WHAT CAN BE DONE?

### The Needs

To what extent is NIRS being used in the grain and seed industries? Delwiche and Dyer between them cite over 300 scientific publications on the application of NIRS to analysis of grains and seeds (7, 8). But most of the actual applications of NIRS in industry are those of testing grains for protein, oil, and moisture contents. Many operations still rely on dielectric meters for moisture testing, even though NIRS is more accurate over a bigger range than dielectric meters. A lot of flour mills use NIRS for analysis of incoming wheat, and for on-line prediction of flour moisture, protein, and ash contents during milling, and some feed manufacturers are using NIRS for ingredient analysis and feed formulation. Determination of ash by NIRS is an approved method of the American Association of Cereal Chemists (111), and is particularly useful to the miller. The standard ash test used to monitor the consistency of milling and for compliance with specifications takes a minimum of two hours, during which time a big mill will have processed 50 or more tons of flour. The NIRS test, which is actually more precise than the standard method, can provide this essential information in less than a minute, or even continuously on-line, and thereby enable the miller to avoid loss of valuable time and material.

What the industry needs from NIRS are methods for the prediction of parameters that are too time-consuming for effective determination at the time of grain delivery or shipping. Most of the grains of commerce are traded on the basis of physical, rather than composition, parameters. Test weight, seed size and uniformity, and grain class and grade are the most important factors that affect price. Other considerations include presence of foreign material and delivery times.

Regrettably, none of these lend themselves to resolution by NIRS, although Edney et al. (62) were able to achieve a reasonable prediction of barley plump kernel percentage. Reliable calibrations have not been reported for the measurement of parameters such as test weight or percentage of foreign material (FM) in grain deliveries (though preliminary studies on the prediction of test weight in Canadian CWRS wheat have been encouraging). Parameters such as FM involve accurate weighing. For example, FM must be determined accurately to tolerances of ±0.1%. At a value of $150.00

per ton 0.1% of a million tons translates into $150,000.00, and large grain terminals can receive and ship volumes of grain of this size and more every year. The error of a NIRS measurement at an SEP of even as low as 0.2% FM would not be practicable for industrial application at grain receiving and shipping points. The SEP is a standard deviation and only reflects the deviations from "true" values for about 67% of the total population. The 95% confidence limits are ±1.96 times the SD (in this case the SEP), so that a specification of 1% FM could be in error by over 0.4%. This is over 50% of the maximum amount of FM allowed in the top grade of CWRS wheat.

In terms of composition, the increased attention that is being paid to food safety has brought mycotoxins, such as deoxynivalenol (DON) increasingly into focus in recent years, while $\alpha$-amylase has long been a thorn in the side of the wheat industry in Australia, Canada, and Europe. Calibrations for the prediction of both DON (112, 113) and $\alpha$-amylase (as determined by the falling number (FN) test) ( 51, 52, 114) have been reported. "Wet chemistry" tests for DON (by ELISA) and $\alpha$-amylase by FN both take about 12–15 minutes (including sample preparation), which is too slow for use at the time of grain delivery, and a NIRS test would be very welcome. The general conclusion was that the calibrations were not sufficiently reliable for industrial application.

Calibrations of the accuracy reported could probably be put to use in screening incoming grain for $\alpha$-amylase (FN). Application in direct determination, or by discriminant analysis, could identify samples that were clearly too low in FN for acceptance, or clearly safe. The remaining 33% would require testing by the FN test, but two-thirds of the testing could be avoided by NIRS screening, and a NIRS method for FN would be useful at grain receiving and shipping points. The SEP for prediction of DON is too high for practical use in flour milling, because tolerances for acceptance for human food are very low, at less than 2.0 parts per million (ppm). The NIRS technique could be employed in the feed industry, where 5.0 ppm is the maximum allowable. Feed manufacturers and farmers could benefit by a rapid test to determine whether the wheat or barley is acceptable for use in feeds, either directly, or by blending down with grain of lower DON content. To comply with a specification of zero tolerance for DON would be impossible with a SEP for DON of even 0.5 ppm, because a NIRS predicted value of ppm could be ± more than 1 ppm.

An area that has probably not received the attention it merits in the grain industry is that of planning NIRS applications. This involves identifying where NIRS can be of economic benefit to the operation, what is needed in terms of physical and chemical analysis, where the most suitable areas of the processing plant are in which to locate the NIR instruments, sampling procedures, monitoring, and communication systems, and the errors of both reference and NIRS testing. The planning exercise should include the benefits of obtaining results important to maintaining efficient operation immediately, rather than after a few hours or days.

Sample presentation is a crucial aspect of a NIRS installation. In a bench-top instrument the rotating sample cell offers a more representative sample to the instrument than a stationary sample cell. Compared to the small "spinning sample cup" used in some instruments the rotating cell allows the instrument to scan more than 15 times the weight of grain (in this case wheat) than the spinning sample cup. At the same time the rotating cell operates with precision nearly three times that of the spinning sample cup (115). A system that could scan grain continuously on a moving conveyor belt or moving through a duct, recording spectral data for 1 second every 20 seconds or so would effectively scan about 1200 g of grain, evenly distributed throughout the bulk of grain. This is about 150 times the weight of wheat scanned in the rectangular sample cell used in the Foss Model 5000 scanning spectrophotometer. Sensing heads operating via fiber optics could be multiplexed to scan several conveyor belts, or different areas in a processing plant.

Applications of NIRS are limited by the combined errors of reference and NIRS testing, which includes the errors incurred by the instrument itself and its software, by the samples, and by the operator. The profusion of factors discussed in the section on factors affecting application of NIRS that can affect the samples adds to the tribulations of calibration model development. PLS regression, described by Martens and Naes (116) has been a boon to calibration model development, and most industrial applications of NIRS use models based on PLS regression. Artificial neural networks (ANN) (117) can effect still further improvements in model efficiency, and it is anticipated that extensive use will be made of ANN in future application. The most difficult obstacle to overcome in the adoption of ANN calibration is to obtain a sufficiently large database. Neural network calibrations are most effective with very large databases (at least 2000 samples recommended). Gaining access to that number of samples with accurate reference data is time-consuming and at least potentially expensive, particularly samples with reference data for parameters such as metabolizable energy or fiber components.

## What Can be Done for NIR Applications to Grains and Seeds?

*Calibration.* Refinements to calibration practice are constantly being introduced. The use of PLS regression and ANN has been of great benefit to NIRS users, and also to the instrument manufacturers, because of the improvements in calibration efficiency. A consequence of the establishment of ANN calibrations is the concept of marketing instruments with factory-developed calibrations, or with no calibration, with the calibrations developed, and operated by the instrument company. This is already in operation. One difficulty with this approach is the need to assemble large enough databases to permit its extension to prediction of the more "difficult" (and expensive to analyze for) constituents and functionality parameters, such as dietary fiber, or malting barley quality parameters. Some of the features in the process of being researched among software "gurus" will undoubtedly become widespread components of future NIRS software assemblies.

Some companies, such as Cognis, have introduced a system in which the operator has no need to calibrate the instrument. The sample is scanned and the signal goes directly to the company, is processed and the result(s) is relayed directly back to the operator. One problem that may emerge when this system becomes widely used is the time interval between scanning of the sample and receiving the result. If heavy use results in overloading the circuits so that lengthy delays (even of 4–5 min) are caused this could undermine the basic concept of a "rapid" analysis.

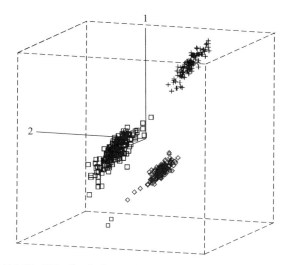

**Figure 7.1.28.** NIR discriminant classification of three white wheat types.

***Near-Infrared Discriminant Analysis.*** Near-infrared discriminant analysis (118) or classification was introduced earlier in this chapter. It is likely to become more widely used and may replace "classical" calibration for prediction of functionality in grains and seeds, and possibly even for screening for composition. Its use in classifying wheat has already been cited (83), and the technique would be a valuable addendum to NIRS instruments used at grain receiving stations to ensure that grain delivered is of the correct class, as well as composition. Figure 7.1.28 shows the clarity of identification of three classes of western Canadian white wheat that differ very widely in end-use potential. The two classes that differ to the greatest degree in functionality (and value) are similar in appearance and can cause some confusion at times of delivery. In plant breeding institutes discriminant analysis could be used in screening for quality parameters, such as kernel texture and water absorption, by developing files with high, medium, and low levels of the respective parameters.

***Use of Spectral Data Without Reference Data.*** Spectral data can be used advantageously with no need for reference data. Applications include monitoring of blending and mixing in grain processing and in precision control.

*Blending and Mixing Applications.* The efficiency of mixing in feed or flour mills is usually established by sending samples for analysis of a selected (more often than not a minor) constituent before and after the mixing or blending process. Apart from being prone to the errors of the sampling and testing, the results typically become available only after a delay of one or more days, far too late for effective remedy of inefficient mixing for a plant that is producing 100 tons or more per day. Blending efficiency can be monitored quickly and accurately by scanning samples by NIRS

**TABLE 7.1.10. Influence of Mix-Time on Spectral Data of Cattle Feed Samples**

| Spectral data CV % | Wavelength (nm) | | | | | | | | Average |
|---|---|---|---|---|---|---|---|---|---|
| | 1154 | 1210 | 1410 | 1692 | 1722 | 1978 | 2054 | 2310 | |
| Feed mix early* | 10.32 | 9.96 | 7.63 | 8.69 | 8.03 | 6.71 | 6.73 | 4.73 | 7.85 |
| Feed mix full mix | 1.83 | 1.73 | 1.29 | 1.31 | 1.26 | 1.13 | 1.05 | 0.79 | 1.30 |
| Ratio early/full mix | 5.64 | 5.76 | 5.91 | 6.63 | 6.37 | 5.94 | 6.41 | 5.99 | 6.08 |

* For the "Early" mix samples were taken after only 3 seconds of mixing. "Full" mix was for 3 minutes.

before and after mixing. The principle is that the spectral signals of a set of samples (at least 12) taken before mixing will be considerably more variable than those of a similar number of samples when mixing is complete, provided the mixer is working as it should be.

The data of Table 7.1.10 and Figure 7.1.29 illustrate the concept. Expressed as the coefficient of variability (CV%) the standard deviations (SDs) of the spectral data of the individual samples ($N = 35$) at the selected wavelengths early during the mixing were on the average more than six times higher than those of the spectral data after full mixing. The average CV of the spectral data of check samples of this type of material is 1.3–1.6%. The advantages of this system for the testing of mixing efficiency are that the data are not subject to the errors of wet chemistry testing and results are available within minutes. The same principle can likewise be applied in a flour mill to assure the efficiency of blending of mill streams. The ratio of the SDs of flour streams before and after blending is about sevenfold, and the CV of the spectral data of the check samples ranges between 0.6 and 1.2%.

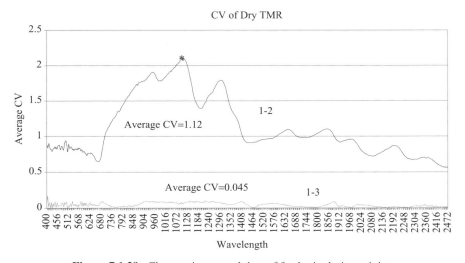

**Figure 7.1.29.** Changes in spectral data of feed mix during mixing.

**TABLE 7.1.11. Precision of Spectral Data from Check Samples of Some Commodities**

| | Wavelength (nm) | | | | | | | | | |
| | 1210 | | 1410 | | 1722 | | 1978 | | 2310 | | Average |
| Grain | Mean* | CV% | Mean | CV% | Mean | CV% | Mean | CV% | Mean | CV% | CV |
|---|---|---|---|---|---|---|---|---|---|---|---|
| Wheat | 0.4745 | 1.16 | 0.6895 | 1.26 | 0.8961 | 1.10 | 0.7623 | 1.20 | 0.9774 | 0.95 | 1.12 |
| Frost** | 0.3899 | 1.12 | 0.6180 | 1.10 | 0.6957 | 1.18 | 0.8111 | 1.19 | 0.9020 | 1.05 | 1.13 |
| FN*** | 0.3501 | 1.58 | 0.5935 | 1.50 | 0.6746 | 1.51 | 0.8147 | 1.33 | 0.9059 | 1.26 | 1.44 |
| Barley | 0.2427 | 1.78 | 0.3934 | 1.58 | 0.4356 | 1.83 | 0.5784 | 1.66 | 0.6897 | 1.52 | 1.67 |
| Maize | 0.3296 | 3.80 | 0.5873 | 3.32 | 0.6645 | 2.84 | 0.8831 | 2.59 | 0.9909 | 2.37 | 2.98 |
| Peas | 0.3187 | 1.87 | 0.5322 | 1.23 | 0.5991 | 1.07 | 0.8684 | 0.61 | 0.9928 | 0.50 | 1.06 |
| Canola | 0.2790 | 0.32 | 0.3873 | 0.18 | 0.7077 | 0.28 | 0.8193 | 0.29 | 1.1327 | 0.29 | 0.27 |
| Flour | 0.0273 | — | 0.1644 | 1.31 | 0.1945 | 1.22 | 0.4241 | 1.36 | 0.4905 | 0.58 | 1.12 |

\* Mean of 10 scans of check samples \*\* Frost-damaged wheat check; \*\*\* Sprouted wheat check

*Monitoring Instrument Precision.* The spectral data can also be very useful in monitoring instrument performance. The precision of the spectral data will vary depending on the commodity but will be independent of variance incurred as a result of the calibration model.

Table 7.1.11 summarizes the precision of spectral data at five wavelengths taken from analysis of check samples during calibration model development. With the exception of the wheat flour, all of the data are based on whole grains and seeds.

The precision of the spectral data is usually more reliable than that of the results obtained by prediction of constituents in the check samples because the results are susceptible to the error of the calibration models. Table 7.1.12 compares the precision of results for prediction of constituents in check samples of some commodities with that of spectral data (average taken at 5 wavelength points).

**TABLE 7.1.12. Comparison of SEP for Check Samples with SD of Mean Spectral Data at Five Wavelength Points for Different Commodities**

| Commodity | Constituent | CV % Constituent* | CV % Spectral Data |
|---|---|---|---|
| Sound wheat | Protein | 1.13 | 1.16 |
| Frosted wheat | Protein | 5.39 | 1.13 |
| Sprouted wheat | Protein | 2.12 | 1.44 |
| Barley | Protein | 1.77 | 1.67 |
| Maize | Moisture | 2.01 | 2.94 |
| Peas | Moisture | 0.90 | 1.06 |
| Canola seed | Oil | 0.50 | 0.50 |
| Wheat flour | Protein | 1.09 | 1.12 |
| Overall | — | 1.86 | 1.32 |

\* Based on precision (SEP) of results of prediction of the constituent in check samples, with the calibration model used for large-scale analysis.

In some cases the CVs of the predictions of constituents were actually better than the precision of the spectral data. The high CV for maize spectral data is attributable to the size and irregularity of the whole kernels of the check sample. For all commodities, the untreated spectral data are a reliable medium for the evaluation of the precision of the instrument for analysis of individual commodities.

*Outliers.*  Outliers have been discussed in other texts, and elegant methods have been developed to identify and remove them. But the outliers are caused by changes in the interaction between the instrument and the sample. During the early years it was largely the outliers that indicated the reality of new variables. Outliers can be a valuable source of new information, and new outliers must be researched to determine their origin and the reasons for their unreasonable behaviour! The new sources of variance that they present can then be added to strengthen existing calibration databases. For practical purposes in day-to-day commercial analysis there can be no outliers—every sample must be tested. Samples the spectral data of which do not appear to conform to those of the samples used in the development of the calibration, or to those of preceding samples, should be "flagged" and reanalyzed. In case of conflicts these samples can be sent for reference analysis. Companies that rely on NIRS analysis should identify a reliable laboratory to which samples can be sent for reference testing when such questions arise.

## REFERENCE ANALYSIS

A short note on reference analysis: The very first step in application of NIRS to analysis of a commodity should be that of determination of the error of the reference methods used in development of calibration models. This is the starting point of any NIRS analysis. Calibration model development and ongoing monitoring of accuracy and precision of analysis will always have to rely on reference analysis for verification of results. One of the "serendipitous" results of NIRS application worldwide has been the discovery in many operations of the need for improvement in their own hitherto sacrosanct wet chemistry laboratory practices.

## SOME ENIGMAS OF NEAR-INFRARED SPECTROSCOPY

This chapter ends with some food for thought for future research. One enigma is the reason underlying why wavelengths are often selected during calibration development that are apparently not closely related to any of the "classic" absorbers for any particular bond vibration, such as C-H stretching, etc. What are these wavelengths contributing, and why are the "classic" wavelengths ignored?

Another enigma that has persisted throughout the more than 30 years of NIRS application is the occurrence of situations in which two spectra of a commodity appear to be identical but give unacceptably big differences in the prediction of results between the two samples, despite retesting by both NIRS and reference methods. A possible reason is that although the wavelengths selected during calibration model development represent the best selections for the sample set, the spectrum of one of the two samples may conform closely to the spectra of the overall population, but that of the nonconformist differs slightly but enough in absorbance (recorded as the log 1/R) at some wavelengths to cause the difference in prediction.

The "Local" option of WINISI, which uses spectral data of a substantial data-base to predict a single sample, goes a long way to address this phenomenon, but refinement of this system is an area of research that should lead to a general enhancement in the application of NIRS to grains and seeds.

## REFERENCES

1. P. Williams. Variables affecting near-infrared spectroscopic analysis. In: *Near-Infrared Technology in the Agriculture and Food Industries*, 2nd Edition. P. Williams, K. Norris, co-eds. The American Association of Cereal Chemists, St. Paul, MN, 2001, p. 171–185.

2. P. C. Williams. Implementation of NIR Technology. In: *Near-Infrared Technology in the Agriculture and Food Industries*, 2nd ed. P. Williams, K. Norris, co-eds. The American Association of Cereal Chemists, St. Paul, MN, 2001, p. 165.

3. C. Starr, D. B. Morgan, D. B. Smith. An evaluation of near-infrared reflectance analysis in some plant breeding programmes. *J Agric Sci* **97**: 107–118, 1981.

4. *Collins Dictionary of the English Language*, 2nd ed. 1986, p. 1383.

5. *Collins Dictionary of the English Language*. 2nd ed. 1986, p. 661.

6. J. R. Hart, K. H. Norris, C. Golumbic. Determination of the moisture content of seeds by near-infrared spectrophotometry of their methanol extracts. *Cereal Chem* **39**: 94–99, 1962.

7. S. R. Delwiche. Analysis of Small Grain Crops. Ch. 11 In: *Near-Infrared spectroscopy in Agriculture*, Monograph series No. 44. C. A. Roberts, J. Jr. Workman, J. B. Reeves, III, co-eds. The American Society of Agronomy/The Crop Science Society of America/The Soil Science Society of America, 2004, p. 269–320.

8. D. J. Dyer. Analysis of Oilseeds and Coarse Grains. Ch. 12 In: *Near-Infrared spectroscopy in Agriculture*, Monograph series No. 44. C. A. Roberts, J. Jr. Workman, J. B. Reeves, III, co-eds. The American Society of Agronomy/The Crop Science Society of America/The Soil Science Society of America, 2004, p. 321–344.

9. R. A Stermer, Y. Pomeranz, R. McGinty. Infrared reflectance spectroscopy for estimation of moisture of whole grain. *Cereal Chem* **54**: 345–351, 1977.

10. S. G Stevenson, P. C. Williams. The determination of protein, moisture and oil in diverse cereals, cereal products and oilseeds by NIRS. *Cereal Foods World* **23**: 462, (abstr), 1978.

11. T. Hymowitz, J. W. Dudley, F. I. Collins, C. M. Brown. Estimations of protein and oil concentration in corn, soybean and oat seed by near-infrared light reflectance. *Crop Sci* **14**: 713–715, 1974.

12. G. R. Rippke, C. L. Hardy, C. R. Hurburgh Jr. Calibration and field standardization of Tecator infratec analyzers for corn and soybeans, In: *Near-Infrared Spectroscopy: The Future Waves: Proceedings of the Seventh International Conference on Near-Infrared Spectroscopy*, A. M. C. Davies and Phil Williams co-ed. NIR Publications, Chichester, UK, 1996, p. 122–131.

13. T. Tajuddin, S. Watanab, K. Harada, S. Kawano. Application of near infrared transmittance spectroscopy to the estimation of protein and lipid contents in single seeds of soybean recombinant inbred lines for quantitative trait analysis. *J Near Infrared Spectrosc* **10**: 315–325, 2002.

14. T. L. Hong, S.-J. Tsai, S. C. S. Tsou. Development of a sample set for soya bean calibration of near infrared reflectance spectroscopy. *J Near Infrared Spectrosc*. **2**: 223–227, 1994.

15. C. R. Hurburgh Jr. Identification and segregation of high-value soybeans at a country elevator. *J Am Oil Chem Soc* **71**: 1073–1078, 1994.

16. H. Abe. *et al.* Non-destructive determination of protein content in a single kernel wheat and soybean by near infrared spectroscopy In: *Near-Infrared Spectroscopy: The Future Waves: Proceedings of the Seventh International Conference on Near-Infrared Spectroscopy*, A.M.C. Davies, Phil Williams co-eds. NIR Publications, Chichester, UK, 1996, p. 457–461.

17. A. Salgo, Z. Fabian. Determination of trypsin inhibitor activity by NIR technique. In: *Proceedings of the Third International Conference on Near-Infrared Spectroscopy*, Brussels, Belgium. R. Biston, N. Bartiaux-Thill, co-eds. The Agricultural Research Centre Publishing, Gembloux, Belgium, 1990, p. 631–636.

18. G. Convertini, D. Ferri, F. Lanza, A. M. Cilardi. Determination by N. I. R. S. of P. K. Ca, and Mg contents in the various plant organs of sugar beet, wheat, sunflower, soybean and sorghum in continuous cropping and two-year rotations. *Proceedings of the Third International Conference on Near-Infrared Spectroscopy*, Brussels, Belgium. R. Biston, N. Bartiaux-Thill, co-eds. The Agricultural Research Centre Publishing, Gembloux, Belgium, 1990, p. 522–540.

19. R. K. Cho, M. Iwamoto, K. Saio. Determination of 7S and 11S globulins in whole ground soybeans by near infrared reflectance analysis. *Nippon Shokuhin Kogyo Gakkaishi* **34**(10): 666–672, 1987.

20. R. K. Cho, J. H. Lee, Y. Ozaki, M. Iwamoto. FT-NIR monitoring system for the study of denaturation of proteins at elevated pressures and temperatures. In: *Leaping ahead with Near Infrared Spectroscopy, Proceedings of the Sixth International Conference on Near-Infrared Spectroscopy*. G. D. Batten, P. C. Flinn, L. A. Welsh, A. B. Blakeney, co-eds. The NIR Spectroscopy Group, Royal Australian Chemical Institute, 1/21 Vale Street, North Melbourne, Australia, 1995, p. 483–486.

21. D. L. Pazdernik, S. D. Plehn, J. L. Halgerson, J. H. Orf. Effect of temperature and genotype on the crude glycinin fraction of (11S) soybean and its analysis by near-infrared reflectance spectroscopy. *J Agric Food Chem* **44**: 2278–2281, 1996.

22. T. Sato, H. Abe, S. Kawano, G. Ueno, K. Suzuki, M. Iwamoto. Near-infrared spectroscopic analysis of deterioration indices of soybeans for process control in oil milling plant. *J Am Oil Chem Soc* **71**: 1049–1055, 1994.

23. D. J. Dyer, P. Feng. Near infared applications in the development of genetically altered grains In: *Near-Infrared Spectroscopy: The Future Waves: Proceedings of the Seventh International Conference on Near-Infrared Spectroscopy*, A. M. C. Davies, Phil Williams co-eds. NIR Publications, Chichester, UK, 1996, p. 490–493.

24. A. M. C. Davies, D. J. Wright. Determination of protein in pea flour by near-infrared analysis. *J Sci Food Agric* **35**: 1034–1039, 1984.

25. R. Tkachuk, F. D. Kuzina. Analysis of protein in ground and whole field peas by near-infrared reflectance. *Cereal Chem* **64**: 418–422, 1987.

26. P. C. Williams, S. G. Stevenson, P. M. Starkey, G. C. Hawtin. The application of near infrared spectroscopy to protein testing in pulse breeding programmes. *J Sci Food Agric* **29**: 285–292, 1978.

27. O. Hermansson. Prediction of new constituents in peas using near infrared transmittance In: *Making Light Work, Proceedings of the 4th. International conference on Near Infrared Spectroscopy*, I. Murray, I. Cowe, co-eds. Ian Michael Publications, Chichester, U.K, 1991, p. 236–238.

28. A. M. C. Davies, D. T. Coxon, G. M. Gavrel, D. J. Wright. Determination of starch and lipid in pea flour by near infrared reflectance analysis. The effect of pea genotype on starch and lipid content. *J Sci Food Agric* **36**: 49–54, 1985.

29. R. Frankenhuizen, M. A. H. Tusveld, E. P. H. M. Schivens, R. G. van de Vaurst de Vries. Determination of the ripeness of peas by NIR-spectroscopy, In: *Proceedings of the Third International Conference on Near-Infrared Spectroscopy*, Brussels, Belgium. R. Biston, N. Bartiaux-Thill, co-eds. The Agricultural Research Centre Publishing, Gembloux, Belgium, 1990, p. 265–272.

30. T. Isaksson, L. Kjolstad. Prediction of sensory quality of peas by NIR and NIT, In: *Proceedings of the Second International Conference on Near-Infrared Spectroscopy*, Tsukuba, Japan. M. Iwamoto, S. Kawano, co-eds. 1990, p. 259–264.

31. A. Martens, H. Martens. Near-infrared reflectance determination of sensory quality of peas. *Appl Spectrosc* **40**: 303–310, 1986.

32. B. Barabas. Comparison of measurability of two produces of high protein content, In: *Near Infrared diffuse reflectance/transmittance spectroscopy, Proceedings of the International NIR/NIT conference*, Budapest, Hungary, May 12–16, 1986. J. Hollo, *et al* co-eds. Akademiai Kiado, Budapest. 1987, p. 169–176.

33. K. Kaffka. Determining amino acids in lupine by near infrared reflectance spectroscopy, *ibid*, 1987, p. 177–187.

34. A. B. Goodchild, F. Jaby El-Haramein, A. Abd El Moneim, H. P. S. Makkar, P. C. Williams. Prediction of phenolics and tannins in forage legumes by near infrared reflectance. *J Near Infrared Spectrosc* **6**: 175–181, 1998.

35. J. A. Panford, P. C. Williams, J. M. Deman. Analysis of oilseeds for protein, oil, fibre and moisture by near-infrared reflectance spectroscopy. *J Am Oil Chem Soc* **65**: 1627–1634, 1988.

36. J. B. Misra, R. S. Mathur, D. M. Bhatt. Near-infrared transmittance spectroscopy: a potential tool for non-destructive determination of oil content in groundnuts. *J Sci Food Agric* **80**: 237–240, 2000.

37. F. Jaby El-Haramein, A. Abd El Moneim, H. Nakkoul, P. C. Williams. Prediction of the neuro-toxin beta-N-oxalyl amonio-L-alanine in *Lathyrus* species, using near infrared reflectance spectroscopy. *J Near Infrared Spectrosc* **6**: A93–A96, 1998.

38. W. Saurer. Use of near-infrared reflectance measurement for the determination of protein and moisture content and kernel hardness in wheat (In German). *Getreide Mehl Brot* **32**: 272–276, 1978.

39. R. Cubadda, S. Douglas. Investigations on the near infrared reflectance techniques for measuring the properties of wheat and its products. In: *Proceedings of the 7th. World Cereal Bread Congress*, Prague, Czechoslovakia, 1982.

40. A. Höhne. Use of near infrared spectroscopy at cereal grain reception (wheat and barley) (In German). *Getreide Mehl Brot* **37**: 20–21, 1983.

41. P. C. Williams. Screening wheat for protein and hardness by near infrared reflectance spectroscopy. *Cereal Chem* **56**: 169–172, 1979.

42. P. C. Williams, D. C. Sobering. Attempts at standardization ofhardness testing of wheat. II. The near-infrared reflectance method. *Cereal Foods World* **31**: 417–420, 1986.

43. P. C. Williams. Prediction of wheat kernel texture in whole grains by near-infrared transmittance. *Cereal Chem* **68**: 112–114, 1991.

44. S. R. Delwiche. Measurement of single-kernel wheat hardness using near-infrared transmittance. *ASAE* **36**: 1431–1437, 1993.

45. M. Manley. Whole wheat grain hardness measurement by near-infrared reflectance spectroscopy. In: *Near-Infrared Spectroscopy: The Future Waves: Proceedings of the Seventh International Conference on Near-Infrared Spectroscopy*, A. M. C. Davies, Phil Williams, co-eds. NIR Publications, Chichester, UK, 1996, p. 466–470.

46. M. Manley, L. Van Zyl, B. G. Osborne. Using Fourier transform near infrared spectroscopy in determining kernel hardness, protein and moisture content of whole wheat flour. *J Near Infrared Spectrosc* **10**: 71–76, 2002.

47. H. Zwingelberg. Infrared spectroscopy – significance and measurement of grain hardness (In German). *Getreide Mehl Brot* **37**: 25–28, 1983.

48. P. C. Williams, F. Jaby El-Haramein, G. Ortiz-Ferreira, J. P. Srivastava. Preliminary observations on the determination of strength in wheat by near-infrared reflectance spectroscopy. *Cereal Chem* **65**: 109–114, 1988.

49. G. Rubenthaler, Y. Pomeranz. Near-infrared reflectance spectra of hard red winter wheats varying widely in protein content and bread-making potential. *Cereal Chem* **64**: 407–411, 1987.

50. T. Pawlinsky, P. C. Williams. Prediction of wheat bread-baking functionality in whole kernels, using near infrared reflectance spectroscopy. *J Near Infrared Spectrosc* **6**: 121–127, 1998.

51. H. Bolling, H. Zwingelberg. Infrared spectroscopy – applications in grain intake and end product control in milled products (In German). *Getreide Mehl Brot* **36**: 197–201, 1982.

52. B. G. Osborne. Investigations into the use of near infrared reflectance spectroscopy for the quality assessment of wheat with respect to its potential for bread-baking. *J Sci Food Agric* **35**: 106–110, 1984.

53. S. R. Delwiche, R. A. Graybosch, C. J. Peterson. Predicting protein composition, biochemical properties, and dough-handling properties of hard red winter wheat flour by near-infrared reflectance. *Cereal Chem* **75**: 412–416, 1998.

54. S. R. Delwiche, G. Weaver. Bread quality of wheat flour by near-infrared spectrophotometry: Feasibility of modeling. *J Food Sci* **59**: 410–415, 1994.

55. K. Sjöholm, J. Tenhunan, S. Home, K. Pietela. Wort analysis by near-infrared spectroscopy. *Proceedings of the EBC Congress*, Oslo, Norway, Oxford University Press, Oxford, 1993, p. 517–524.

56. K. Sjöholm, J. tenhunan, J. Tammisola, K. Pietela, S. Home. Determination of the fermentability and extract content of industrial worts by Near-infrared spectroscopy. *J Am Soc Brew Chem* **54**: 135–140.

57. J. Tenhunan, K. Sjöholm, K. Pietela, S. Home. Determination of fermentable sugars and nitrogenous compounds in wort by near- and mid-infrared spectroscopy. *J Inst Brew* **100**: 11–15, 1994.

58. S. Roumeliotis, L. D. Macleod, S. L. Graham, M. A. Dowling. Advances in the prediction of malting quality in barley, using Near infrared spectroscopy. In: *Near-Infrared spectroscopy Leaping ahead with*, Royal Australian Chemistry Institute, G. D. Batten, P. C. Flinn, L. A. Welsh, A. B. Blakeney, co-eds. Melbourne, Australia, 1995, p. 168–173.

59. E. V. Valdes, S. Leeson. Near-infrared reflectance analysis as a method to measure metabolizable energy in complete poultry foods. *Poultry Sci* **71**: 1179–1187, 1992.

60. E. V. Valdes, S. Leeson. The use of Near-infrared spectroscopy to measure metabolizable energy in poultry feed ingredients. *Poultry sci* **71**: 1559–1563, 1992.

61. R. J. van Barneveld, J. D. Nuttall, P. C. Flinn, B. G. Osborne. Near infrared measurements of the digestible energy content of cereals for growing pigs. *J Near Infrared Spectrosc* **7**: 1–7, 1999.

62. M. J. Edney, J. E. Morgan, P. C. Williams, L. D. Campbell. Analysis of Feed Barley by Near-infrared Reflectance Technology. *J Near Infrared Spectrosc* **2**: 33–41, 1994.

63. J. K. Daun, K. M. Clear, P. C. Williams. Comparison of three whole seed near-infrared analyzers for measuring quality components of canola seed. *J Am Oil Chem Soc* **71**: 1063–1068, 1994.

64. T. N. Pallot, A. S. Leong, J. A. Allen, T. M. Golder, C. F. Greenwood, T. Golebi. Precision of fatty acid analysis using near infrared spectroscopy of whole seed *Brassicas*. In: *Proc 10th Int Rapeseed Congr* Canberra, Australia, Regional Inst. Gosford, Australia, 1999.

65. T. C. Reinhardt, G. Robbelen. Quantitative analysis of fatty acids in intact rapeseed by near infrared reflectance spectroscopy. In: *Proc 8th Int Rapeseed Congr*. Saskatoon, SK, Canada. Regional Inst. Gosford, Australia. 1991, p. 1380–1384.

66. L. Velasco, H. C. Becker. Estimating the fatty acid composition of the oil in intact seed rapeseed (*Brassica napus L.*) by near-infrared reflectance spectroscopy. *Euphytica*, **101**: 221–230, 1998.

67. L. Velasco, A. Schierholt, H. C. Becker. Performance of near-infrared reflectance spectroscopy (NIRS) in routine analysis of C18 unsaturated fatty acids in intact rapeseed. *Fett/Lipid* **100**: 44–48, 1998.

68. J. K. Daun, P. C. Williams. Near-infrared analysis of oilseeds: current status and future directions. In: *New techniques and application in lipid analysis*, AOCS, Champaign, Illinois, 1997, p. 266–282.

69. B. L. Perez-Vich, L. Velasco, J. M. Fernandez-Martinez. Determination of seed oil content and fatty acid composition in sunflower through the analysis of intact seeds, husked seeds, meal and oil by near-infrared reflectance spectroscopy. *J Am Oil Chem Soc* **75**: 547–555, 1998.

70. L. Velasco, B. Perez-Vich, J. M. Fernandez-Martinez. Nondestructive screening for oleic and linoleic acid in single sunflower achenes by near-infrared reflectance spectroscopy. *Crop Sci* **39**: 219–222, 1999.

71. D. J. Dyer, P. Feng. Near infrared applications in the development of genetically-altered grains. In: *Near-Infrared Spectroscopy: The Future Waves: Proceedings of the Seventh International Conference on Near-Infrared Spectroscopy*, A. M. C. Davies, Phil Williams co-eds. NIR Publications, Chichester, UK, 1996, p. 490–493.

72. D. J. Dyer, P. Feng. NIR destined to be major analytical influence. *Feedstuffs* **69**: 20, 1997.

73. D. L. Pazdernik, A. S Killam, J. H. Orf. Analysis of amino acid and fatty acid composition in soybean seed, using near infrared reflectance spectroscopy. *Agron J* **89**: 679–685, 1997.

74. G. F. Tremblay, G. A. Broderick, S. M. Abrams, D. Pageau. Determination of ruminal protein degradability in roasted soybeans using near infrared reflectance spectroscopy.in Agronomy Abstracts. *ASA, CSSA, SSSA, Madison, WI* 1994, p. 194.

75. G. F. Tremblay, G. A. Broderick, S. M. Abrams. Estimating ruminal protein degradability of roasted soybeans using near infrared reflectance spectroscopy. *J. Dairy Sci.* **79**: 276–282, 1996.

76. T. Sato, H. Abe, S. Kawano, G. Ueno, K. Suzuki, M. Iwamoto. Near infrared spectroscopic analysis of deterioration indices of soybeans for process control in milling plant. *J Am Oil Chem Soc* **71**: 1049–1055, 1994.

77. L. Velasco, J. M. Fernandez-Martinez, A. DeHaro. The applicability of NIRS for estimating multiple seed quality components in Ethiopian mustard. In: *Proc 9th Int Rapeseed Congr.* Cambridge, UK. Regional Inst. Gosford, Australia, 1995, p. 867–869.

78. L. Velasco, J. M. Fernandez-Martinez, A. DeHaro. Determination of the fatty acid composition of the oil in intact-seed mustard by near-infrared reflectance spectroscopy. *J Am Oil Chem Soc* **74**: 1595–1602, 1997.

79. L. Velasco, J. M. Fernandez-Martinez, A. DeHaro. Application of near-infrared reflectance spectroscopy to estimate the bulk density of Ethiopian mustard seeds. *J Sci Food Agric* **77**: 312–318, 1998.

80. Y. S. Chen, A. O. Chen. Quality analysis and purity examination of edible vegetable oils by near infrared transmittance spectroscopy. In: *Near Infrared Spectroscopy: Leaping ahead with Proceedings of the Sixth International Conference on Near-Infrared Spectroscopy.* G. D. Batten, P. C. Flinn, L. A. Welsh, A. B. Blakeney, co-eds. The NIR Spectroscopy Group, Royal Australian Chemical Institute, 1/21 Vale Street, North Melbourne, Australia, 1995, p. 316–323.

81. J. Ha, J. Koo, H. Ok. Determination of the constituents of sesame oil by near infrared spectroscopy. *J Near Infrared Spectrosc* **6**: A371–A373, 1998.

82. J. J. Rose, T. Prusik, J. Markedian. Near infrared multicomponent analysis of parenteral products using the InfraAlyzer 400. *J Parenteral Sci* **36**: 71–78, 1982.

83. D. Bertrand, P. Robert, F. Loisel. Identification of some wheat varieties by near infrared reflectance spectroscopy. *J Sci Food Agric* **36**: 1120–1124, 1985.

84. H. Mark. Qualitative Near-infrared Analysis. In: *Near-Infrared Technology in the Agriculture and Food Industries*, 2nd ed. Phil Williams, Karl Norris, co-eds. The American Association of Cereal Chemists, St. Paul, MN, 2001, p. 233–238.

85. R. Kramer, J. K. Workman Jr., J. B. Reeves III (authors). Qualitative Analysis. In: *Near-Infrared Spectroscopy in Agriculture*, Monograph series No. 44. C. A. Roberts, J. K. Workman Jr., J. B. Reeves III, H. Mark, co-eds. The American Society of Agronomy/The Crop Science Society of America/The Soil Science Society of America, 2004, p. 175–206.

86. M. O. Westerhaus, J. K. Workman Jr., J. B. Reeves III, H. Mark. Quantitative Analysis. In: *Near-Infrared spectroscopy in Agriculture*, Monograph series No. 44. C. A. Roberts, J. K. Workman Jr., J. B. Reeves III, H. Mark, co-eds. The American Society of Agronomy/The Crop Science Society of America/The Soil Science Society of America, 2004, p. 147–148.

87. D. Bertrand, M. F. Devaux, P. Robert. Application of pattern recognition techniques in NIR spectroscopy. In: *Near Infrared Diffuse Reflectance/Transmittance Spectroscopy, Proceedings of the International NIR/NIT conference*, Budapest, Hungary, May 12–16, 1986. J. Hollo, K. J. Kaffka, J. L. Gonczy, co-eds. Akademiai Kiado, Budapest. 1987, p. 31–41.

88. P. Dardenne, R. Biston. Attempt to recognize wheat species by discriminant analysis. In: *Near Infrared Diffuse Reflectance/Transmittance Spectroscopy, Proceedings of the International NIR/NIT conference*, Budapest, Hungary, May 12–16, 1986. J. Hollo, *et al* co-ed. Akademiai Kiado, Budapest, 1987, p. 51–60.

89. P. C. Williams, D. C. Sobering. Objective prediction of functionality by near infrared technology. In: *Making Light Work, Proceedings of the 4th International conference on Near Infrared Spectroscopy*. I. Murray, I. Cowe, co-eds. Ian Michael Publications, Chichester, U.K, 1992, p. 217–222.

90. B. G. Osborne, B. Mertens, M. Thompson, T. Fearn. Authentication of Basmati rice by near infrared transmittance data of individual grains. In: *Near Infrared Spectroscopy: Leaping ahead with Proceedings of the Sixth International Conference on Near-Infrared Spectroscopy*, G. D. Batten, P. C. Flinn, L. A. Welsh, A. B. Blakeney, co-eds. The NIR Spectroscopy Group, Royal Australian Chemical Institute, 1/21 Vale Street, North Melbourne, Australia, 1995, p. 161–163.

91. M. Manley, A. E. J. McGill, B. G. Osborne. Wheat hardness by NIR: new insights. In: *Near Infrared Spectroscopy: Leaping ahead with Proceedings of the Sixth International Conference on Near-Infrared Spectroscopy*, G. D. Batten, P. C. Flinn, L. A. Welsh, A. B. Blakeney, co-eds. The NIR Spectroscopy Group, Royal Australian Chemical Institute, 1/21 Vale Street, North Melbourne, Australia, 1995, p. 178–180.

92. S. J. Tsai, T. L. Hong, S. C. S. Tsou. Discrimination in the processing quality of soybeans harvested at different seasons by near infrared reflectance spectroscopy. In: *Near Infrared Spectroscopy: Leaping ahead with Proceedings of the Sixth International Conference on Near-Infrared Spectroscopy*, G. D. Batten, P. C. Flinn, L. A. Welsh, A. B. Blakeney, co-eds. The NIR Spectroscopy Group, Royal Australian Chemical Institute, 1/21 Vale Street, North Melbourne, Australia, 1995, p. 218–221.

93. K. W. Bewig, A. D. Clarke, C. Roberts, N. Unklesbay. Discriminant analysis of vegetable oils using near infrared spectroscopy. In: *Near Infrared Spectroscopy: Leaping ahead with Proceedings of the Sixth International Conference on Near-Infrared Spectroscopy*, G.D. Batten, P. C. Flinn, L. A. Welsh, A. B. Blakeney, co-eds. The NIR Spectroscopy Group, Royal Australian Chemical Institute, 1/21 Vale Street, North Melbourne, Australia, p. 324–328.

94. D. Bertrand, B. Novales, M. -F. Devaux, P. Robert. Discrimination of durum wheat products for quality control. In: *Near-Infrared Spectroscopy: The Future Waves: Proceedings of the Seventh International Conference on Near-Infrared Spectroscopy*, A. M. C. Davies, Phil Williams, co-eds. NIR Publications, Chichester, UK, 1986, p. 430–443.

95. D. Bertrand, P. Courcoux, J. -C. Autran, R. Méritan. Stepwise canonical discriminant analysis of continuous digitalized signals application to chromatograms of wheat proteins. *J Chemometrics* **4**: 413–428, 1990.

96. S. Turza, A. I. Toth, M. Varadi. Multivariate classification of different soyabean varieties. *J Near Infrared Spectrosc* **6**: 183–187, 1998.

97. Y. Ootake, S. Kokot. Discrimination between glutinous and non-glutinous rice by vibrational spectroscopy. II Effects of spectral pre-treatment on the classification of the two types of rice. *J Near Infrared Spectrosc* **6**: 251–258, 1998.

98. T. L. Hong, C. S. T. Samson, S. -J. Tsai. Evaluation of soya bean quality for tofu processing by near infrared spectroscopy. *J Near Infrared Spectrosc* **6**: A325–A328, 1998.

99. Y. -K. Kwon, R. -K. Cho. Identification of the geographical origin of sesame seeds by near infrared spectroscopy. In: *Near Infrared Spectroscopy: Proceedings of the 9th*

*International Conference on Near Infrared Spectroscopy*, Verona, Italy. A. M. C. Davies, R. Giangiacomo, co-eds. NIR Publications, Chichester, UK, 2000, p. 551–555.

100. Z. Seregely. Investigation of near infrared characteristics of spring barley varieties. In: *Near Infrared Spectroscopy: Proceedings of the 9th International Conference on Near Infrared Spectroscopy, Verona, Italy*. A. M. C. Davies, R. Giangiacomo, co-eds. NIR Publications, Chichester, UK, p. 543–546, 2000.

101. K. J. Kaffka, L. S. Gyarmati. Qualitative (Comparative) Analysis by near infrared spectroscopy. In: *Proceedings of the Third International Conference on Near-Infrared Spectroscopy, Brussels, Belgium*. R. Biston, N. Bartiaux-Thill, co-eds. The Agricultural Research Centre Publishing, Gembloux, Belgium, 1991, p. 135–144.

102. J. S. Shenk, J. K. Workman Jr., M. O. Westerhaus. Application of NIR Spectroscopy to Agricultural Materials. In: *Handbook of Near-Infrared Analysis*. 2nd ed. A. Donald, Burns, Emil W. Ciurczak, co-eds. Marcel Dekker, Inc. New York, 2001, p. 431–433.

103. B. G. Osborne, T. Fearn, P. H. Hindle. Practical NIR Spectroscopy with applications in food and beverage analysis, 2nd ed. Longman Scientific & Technical, Harlow, Essex, England. 2001, p. 29–32.

104. C. E. Miller. Chemical Principles of Near-infrared Technology. In: *Near-Infrared Technology in the Agriculture and Food Industries*, 2nd ed. Phil Williams, Karl Norris, co-eds. The American Association of Cereal Chemists, St. Paul, MN. 2001, p. 19–37.

105. I. Murray, P. C. Williams. Chemical Principles of Near-infrared Technology. In: *Near-Infrared Technology in the Agriculture and Food Industries*, 1st ed. Phil Williams, Karl Norris, co-eds. The American Association of Cereal Chemists, St. Paul, MN. 1987, p. 17–34.

106. Phil Williams. Comparison of calibrations based on partial least squares and multiple linear regression for near infrared prediction of composition and functionality in grains. In: *Near Infrared Spectroscopy: Proceedings of the 9th International Conference on Near Infrared Spectroscopy*, Verona, Italy. A. M .C. Davies, R. Giangiacomo, co-eds. NIR Publications, Chichester, UK, 2000, p. 287–293.

107. R. J. Henry, I. A. Cowe. Factors influencing the hardness (milling energy) and malting quality of *J barley Inst Brew* **96**: 135–136, 1990.

108. P. C. Williams. The influence of chromosome number and species on wheat hardness. *Cereal Chem* **63**: 56–57, 1986.

109. *American Association of Cereal Chemists*. Approved Methods of the Association. 10th ed. The American Association of Cereal Chemists, St. Paul, MN, Method 2000, p. 55–30.

110. P. C. Williams. Variables affecting Near-infrared Reflectance Spectroscopic Analysis. In: *Near-Infrared Technology in the Agriculture and Food Industries*, 1st ed. Phil Williams, Karl Norris, co-eds. The American Association of Cereal Chemists, St. Paul, MN, 1987, p. 157.

111. *American Association of Cereal Chemists*. Approved Methods of the Association. 10th ed. The American Association of Cereal Chemists, St. Paul, MN, Method 2000, p. 50–30.

112. F. E. Dowell, M. S. Ram, L. M. Seitz. Predicting scab, vomitoxin and ergosterol in single wheat kernels using near-infrared spectroscopy. *Cereal Chem* **76**: 573–576, 1999.

113. Phil Williams. The status of research on application of NIRS to prediction of Fusarium Head Blight (Scab) in terms of DON. In: *Near Infrared Spectroscopy. Proceedings of the 11th International Conference on Near Infrared Spectroscopy*, Cordoba, Spain. A. M. C. Davies, A. Garrido-Varo co-eds. NIR Publications, Chichester, UK, 2004, p. 1051–1055.

114. S. Stopford, G. Hillier, A. Clucas. The Correct Citation is: "Attempts at prediction of alpha-amylase in wheat samples" NIR 84. Proceedings of an International Symposium on Near-infrared Reflectance Spectroscopy, Melbourne, Australia, D. Miskelly, D. P. Law, A. Clucas, co-eds. The Cereal Chemistry Division, Royal Australian Chemical Institute, 1985, p. 173–174.

115. Phil Williams. Seeing agricultural products through the NIR window. In: *Near Infrared Spectroscopy. Proceedings of the 11th International Conference on Near Infrared Spectroscopy*, Cordoba, Spain. A. M. C. Davies, A. Garrido-Varo co-eds. NIR Publications, Chichester, UK, 2004, p. 313–318.

116. H. Martens, T. Naes. Multivariate Calibration by Data Compression. In: *Near-Infrared Technology in the Agriculture and Food Industries*, Ist ed. Phil Williams, Karl Norris, co-eds. The American Association of Cereal Chemists, St. Paul, MN, 1987, p. 57–87.

117. C. Borggaard. Neural Networks in Near-Infrared Spectroscopy. In: *Near-Infrared Technology in the Agriculture and Food Industries*, 2nd ed. Phil Williams, Karl Norris, co-eds. The American Association of Cereal Chemists, St. Paul, MN, 2001, p. 101–107.

118. H. L. Mark. Qualitative Discriminant Analysis. In: *Handbook of Near-Infrared Analysis*. 2nd ed, D. A. Burns, E. W. Ciurczak, co eds. Marcel Dekker, Inc., New York. 2001, p. 363–400.

# Fruits and Vegetables

SIRINNAPA SARANWONG AND SUMIO KAWANO

## INTRODUCTION

The application of NIR technology to the fruit and vegetable area attracts considerable attention today. Once thought to be of little significance, especially in the presence of water in fresh fruits and vegetables, NIR spectra have been found to contain considerable information related to quality. Applications are too numerous to consider in this chapter. Instead, we will focus on general guidelines needed for the successful application of NIRS to fruits and vegetables.

## HISTORY

From the 1920s to the 1970s, measurement of internal or eating quality of fruits and vegetables nondestructively became a very active area of research. Among many studies well described by Slaughter and Abbott (1), a well-known study was by Yeatman and Norris (2). By using the absorbance values measured in transmittance mode at 695 and 740 nm, the authors developed the "automatic internal quality (IQ) sorter" that could predict the internal quality of an apple by measuring chlorophyll. At this early stage, the study did not attempt to directly determine chemicals related to taste and palatability, such as sugar or acid. Instead, the chlorophyll measurement was an indirect or secondary correlation with product quality. In 1985, Birth et al. developed the "Biospec" (short for "biological spectrometer") that was later used to measure the dry matter of onions (3). Still later, they showed that Brix (a measure of soluble solids) in cantaloupes could be determined with the same instrument (4). During 1990s, numerous applications of NIR technology to fruits and vegetables were published, dealing mainly with the determinations of Brix and dry matter. Table 7.2.1 summarizes this work.

*Near-Infrared Spectroscopy in Food Science and Technology*, Edited by Yukihiro Ozaki,
W. Fred McClure, and Alfred A. Christy.
Copyright © 2007 John Wiley & Sons, Inc.

**TABLE 7.2.1.  Reported NIR Applications to Intact Fruits and Vegetables**

| Product | Author (Year) | NIR technique | Constituents | Special Features |
|---|---|---|---|---|
| Apple | Lovasz et al. (1994) (5) | Transmittance Infratec 1225 | Firmness, AIS, DM, refractive index, pH, TA | Effect of storage condition, cultivars, and seasons |
| | Lammertyn et al. (1998) (6) | Interactance OSA 6602 | Brix, pH, texture | |
| | Ventura et al. (1998) (7) | Reflectance Ocean Optics SD-1000 | Brix | |
| | Lammertyn et al. (2000) (8) | Interactance, Reflectance OSA 6602 | Brix | Analysis of penetration depth |
| | Lu et al. (2000) (9) | Reflectance Oriel MS257 | Brix, firmness | Effect of variety and peel development of an universal calibration |
| | Peirs et al. (2001) (10) | Reflectance Rees OSA6602 | Streif index (maturity), Brix, TA, firmness | |
| | Peirs et al. (2003) (11) | NIR Imaging NIRCamera: SU320-1.7RT-V Spectrograph: ImSpector | Starch pattern compared with iodine stain | |
| | Temma et al. (2002a) (12) | Interactance Pacific Scientific 6250 | Brix | |
| | Temma et al. (2002b) (13) | Interactance Own designed portable instrument | Brix | |
| | Sanchez et al. (2003) (14) | Interactance Zeiss MMS1 | Brix | Effect of sample temperature, instrument temperature, and ambient light |
| | Walsh et al. (2004) (15) | Reflectance Zeiss MMS1 | Brix | |
| | McGlone and Martinsen (2004) (16) | Transmittance MK-Photonics | DM | Comparison of two high-speed measurements |
| Apricot | Carlini et al. (2000) (17) | Interactance Pacific Scientific 6250 | Brix | |

| Fruit | Reference | Instrument | Parameter | Notes |
|---|---|---|---|---|
| Banana | Tarkosova and Copikova (2000) (18) | Reflectance NIRSystems 6500 | Sucrose, glucose, fructose, total sugar | |
| | Walsh et al. (2004) (15) | Reflectance Zeiss MMS1 | Brix, DM | Peeled banana was used. Peeled banana was used for DM calibration. |
| Cantaloupe, melon, and watermelon | Dull et al. (1989a) (4) | Interactance Biospect | Brix | |
| | Guthrie et al. (1998) (19) | Interactance NIRSystems 6500 | Brix | Effect of variety |
| | Walsh et al. (2000) (20) | Interactance Zeiss MMS1 Ocean OpticsS2000 Oriel FICS | Brix | Instrument comparison |
| | Greensill et al. (2001) (21) | Interactance Zeiss MMS1 | Brix | Calibration transfer |
| | Tsuta et al. (2002) (22) | Interactance NIRSystems 6500 NIR Imaging Own designed instrument | Brix | |
| | Walsh et al. (2004) (15) | Reflectance Zeiss MMS1 | Brix | |
| Cherry | Carlini et al. (2000) (17) | Interactance Pacific Scientific 6250 | Brix | |
| | Lu et al. (2001) (23) | Reflectance Oriel MS257 | Brix, firmness | |
| Cucumber | Miller et al.(1995) (24) | Transmittance Trebor 101 | Internal bruise | |
| Date | Dull et al. (1991) (25) | Interactance Biospect | Moisture | |
| | Schmilovitch et al. (1999) (26) | Interactance LTI Quantum 1200 | Brix, moisture | Development of a sorting machine |

(Continued)

**TABLE 7.2.1.** (*Continued*)

| Product | Author (Year) | NIR technique | Constituents | Special Features |
|---|---|---|---|---|
| Kiwi fruit | McGlone et al. (1997) (27) | Reflectance Sanyo SDL–4032 | Firmness (stiffness and rupture force) | Laser NIR at 864 nm |
| | Mowat and Poole (1997) (28) | Interactance Ocean Optics PS1000 | Ripening quality | |
| | Martinsen and Schaare (1998) (29) | NIR Imaging Own designed instrument | Brix | Cut fruit was used. |
| | McGlone and Kawano (1998) (30) | Interactance NIRSystems 6500 | Brix, DM, firmness | Effect of orchard origin and fruit size |
| | Martinsen et al. (1999) (31) | NIR Imaging Own designed instrument | Moisture-related image | Low-cost instrument |
| | Osborne et al. (1999) (32) | Interactance Zeiss MMS1 | Brix, DM | |
| | Schaare and Fraser (2000) (33) | Reflectance, transmittance, interactance Zeiss MMS1 | Brix, density, flesh color | Comparison of mode of measurement |
| | Walsh et al. (2004) (15) | Reflectance Zeiss MMS1 | Brix, DM | |
| Mango | Guthrie and Walsh (1997) (34) | Interactance NIRSystems 6500 | DM | Comparison of calibration method |
| | Schmilovitch et al. (2000) (35) | Interactance LTI Quantum 1200 | Brix, TA, firmness, storage period | Comparison of calibration method |
| | Saranwong et al. (2001) (36) | Interactance NIRSystems 6500 | Brix, DM | Comparison of wavelength region |
| | Saranwong et al. (2003a) (37) | Interactance NIRSystems 6500 FANTEC FT20 | Brix | Comparison of instrument |
| | Saranwong et al. (2003b) (38) | Interactance FANTEC FT20 | DM, starch | On-tree NIR measurement |
| | Saranwong et al. (2004) (39) | Interactance NIRSystems 6500 | DM, starch, ripening Brix | Ripening Brix was predicted from green-stage spectra. |
| | Walsh et al. (2004) (15) | Reflectance Zeiss MMS1 | Brix, DM | |

| | | | | |
|---|---|---|---|---|
| Onion | Birth et al. (1985) (3) | Interactance Biospect | Brix, DM | |
| | Walsh et al. (2004) (15) | Reflectance Zeiss MMS1 | Brix | |
| Orange, mandarin, nectarine | Kawano et al. (1993) (40) | Transmittance Pacific Scientific 6250 | Brix | Method to remove sample size effect |
| | Miyamoto and Kitano (1995) (41) | Transmittance Pacific Scientific 6250 | Brix | Development of an universal calibration |
| | Slaughter (1995) (42) | Interactance NIRSystems 6500 | Brix, fructose, glucose, sucrose, sorbitol | |
| | Miyamoto et al. (1998) (43) | Transmittance Pacific Scientific 6250 | TA | |
| | McGlone et al. (2003) (44) | Transmittance, interactance, reflectance Zeiss MMS1 | Brix, TA | Comparison of mode of measurement |
| Papaya | Slaughter et al. (1999) (45) | Interactance NIRSystems 6500 | Brix | |
| | Greensill and Newman (1999) (46) | Transmittance Own designed instrument | maturity | |
| | Walsh et al. (2004) (15) | Reflectance Zeiss MMS1 | Brix | |
| Peach | Kawano et al. (1992) (47) | Interactance Pacific Scientific 6250 | Brix | |
| | Bellon et al. (1993) (48) | Interactance Jobin Yvon CP200 | Brix | Instrumental development |
| | Kawano et al. (1995) (49) | Interactance Pacific Scientific 6250 | Brix | Sample temperature compensation |
| | Slaughter (1995) (42) | Interactance NIRSystems 6500 | Brix, fructose, glucose, sucrose, sorbitol | |
| | Peiris et al. (1998) (50) | Interactance Acousto-optical instrument | Brix | Development of an universal calibration |
| | Walsh et al. (2004) (15) | Reflectance Zeiss MMS1 | Brix | |

*(Continued)*

**TABLE 7.2.1.** (*Continued*)

| Product | Author (Year) | NIR technique | Constituents | Special Features |
|---|---|---|---|---|
| Pineapple | Guthrie and Walsh (1997) (33) | Interactance NIRSytems 6500 | Brix | Comparison of calibration method |
| | Guthrie et al. (1998) (18) | Interactance NIRSystems 6500 | Brix | Effect of growing season |
| | Walsh et al. (2004) (15) | Reflectance Zeiss MMS1 | Brix | |
| Potato and sweet potato | Katayama et al. (1996) (51) | Transmittance Pacific Scientific 6250 | Starch, moisture, Brix | Sliced sample was used. |
| | Mehrubeoglu and Cote (1997) (52) | Transmittance SPEX 270M | Total reducing sugar | Sliced sample was used. |
| | Walsh et al. (2004) (15) | Reflectance Zeiss MMS1 | DM | |
| Prune | Slaughter et al. (2003) (53) | Interactance Ocean Optics S1000-TR | Total solids content (TSC), Brix | |
| Tomato | Slaughter et al. (1996) (54) | Interactance NIRSystems 6500 | Brix | Effect of sample presentation |
| | Walsh et al. (2004) (15) | Reflectance Zeiss MMS1 | Brix, DM | |
| | Khuriyati et al. (2004) (55) | Transmittance Zeiss MMS1 | Brix | |
| | Khuriyati et al. (2005) (56) | Interactance Zeiss MMS1 | DM | Effect of sample presentation |

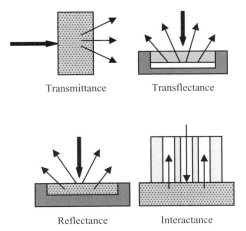

**Figure 7.2.1.** Modes of NIR measurements.

## QUALITY FACTORS

### Mode of Measurement

The most important guideline for success in the development of calibrations for quality factors is to make certain that the sampling area and the NIR *illuminating area* match. This guideline is related to the measurement mode which largely determines the depth of penetration of the NIR energy into the sample. Depending on the type of sample and the chemical being analyzed, there are four modes of measurement available: reflectance, transmittance, interactance, and transflectance modes (Fig. 7.2.1). Each mode provides different benefits and drawbacks. In this chapter we will focus on the first three modes, as the transflectance mode was originally designed for measuring spectra of thin and/or clear samples having characteristics quite different from the characteristics of fruits and vegetables.

Intact fruits and vegetables can have a thin peel (apples, mangoes, peaches), a medium-thick peel (oranges, bananas), or a very thick peel (cantaloupes, watermelons). Thin-peel samples might fit the interactance mode, whereas samples with thicker peels may require the transmittance mode in order to obtain suitable spectra. A discussion of each type of sample will be presented in a later paragraph. If the chemical of interest is located just under the peel (such as anthocyanin in blueberries), the less penetrating "reflectance mode" provides numerous advantages. Unfortunately, the chemical that is related directly to taste and palatability is located deep within the flesh. Hence, a measuring mode different from the reflectance mode is needed to ensure significant penetration of the NIR energy into the sample.

The interactance mode is the most popular mode for use on fruits and vegetables. This mode is suitable for detecting diffusely reflected energy from deep within a sample while excluding surface-reflected energy. The first instrument utilizing an

Cylindrical case

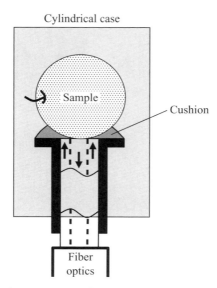

**Figure 7.2.2.** Spectral measurement of intact peaches using "interactance mode" (47).

interactance geometry was the "Biospect" for DM determination in onions (3). In the Biospec, the illuminator and detectors are separately mounted with the detectors placed in contact with the surface of the onion. Birth and his coauthors reported the prediction of dry matter in onions with an SEP of 0.79%w/w. Later, utilizing a configuration similar to the Biospect, Kawano et al. (47) reported a similar performance for peaches. Utilizing a commercially available fiber-optics *interactance probe* provided by Foss NIRSystems and an in-house designed fruit holder (Fig. 7.2.2), the authors were able to satisfactorily determine Brix of peaches from short-wavelength spectra (700–1100 nm) with $R = 0.97$, SEC = 0.48° Brix, SEP = 0.50° Brix, bias = 0.01° Brix). A cushion made of urethane foam was glued to the end of the probe to hold the sample in order to exclude energy not interacting with internal sample structure. Reference spectra were recorded from an 8-cm Teflon sphere having similar characteristic light scattering of the fruit. Kawano and his associates have shown that the interactance mode is suitable for determining the internal quality of several kinds of thin-peel fruits, including apples, peaches, mangoes, and papayas.

The transmittance mode is recommended for thick-peel fruit. In case of oranges, in which the peel is quite thick with a white mesocarp layer, NIR radiation striking the surface is strongly reflected. Thus spectra obtained in the interactance mode will not provide spectral information pertaining to the flesh of the orange (44). Kawano et al. (40) demonstrated that, by placing a silicon detector under the sample, useful spectral information transmitted by satsuma oranges could be obtained (Fig. 7.2.3). However, unlike the interactance and reflectance modes, the spectra collected in the transmittance mode were severely affected by the sample thickness (Fig. 7.2.4). Kawano further showed that the sample thickness effect could be minimized by

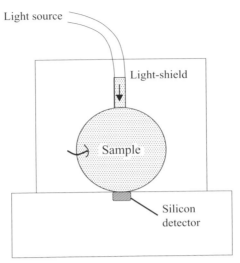

**Figure 7.2.3.** Spectral measurement of intact oranges using "transmittance mode" (40).

utilizing second-derivative spectra normalized to a specific wavelength. Kawano et al. (40) chose the water band at 844 nm as the denominator because water content varied little from one sample to another and the 844-nm band has a strong correlation to sample diameter. With the normalized spectra and a multiple linear regression (MLR) calibration equation containing four wavelengths, impressive results for Brix determinations ($R = 0.99$, SEC $= 0.28°$ Brix, SEP $= 0.32°$ Brix, bias $= -0.02°$ Brix) were obtained.

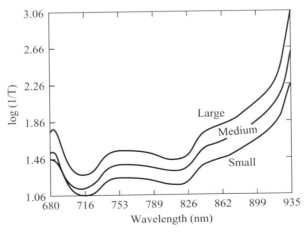

**Figure 7.2.4.** NIR spectra of intact oranges with different sizes but with same Brix value (40).

## Wavelength Region

As the NIR region is normally divided into two subregions, the short wavelength region from 780 nm to 1100 nm (sometimes called the "Herschel region") and the long wavelength region from 1100 nm to 2500 nm (57), a frequently asked question is, "Which wavelength region is better?" There is no one single answer. The answer must come from an analysis of the fundamental guideline (or rule) requiring the measured energy and the chemical analysis to be determined from the same sample volume. If the measured energy comes from one volume of the sample and the chemical analysis from another volume of the sample, results will always be poor.

The basis for the above studies came from a recognition of the contributions of Planck that showed that energy in the electromagnetic spectrum increases with decrease in wavelength. Hence, the shorter the wavelength the higher the energy. Therefore, if the chemical is located deeply in the flesh, the use of the short-wavelength region must provide more reliable information. On the other hand, the long-wavelength region will be useful if the chemical is located in the peel. Saranwong et al. (36) proved that the short-wavelength region had enough penetration depth into mango flesh, giving a superior Brix determination result compared to the long-wavelength region (Fig. 7.2.5, Table 7.2.1).

## Instrument

Proper instrumentation is always a problem. Unfortunately, limited budgets force researchers to use any instrumentation and accessories available. A word of caution

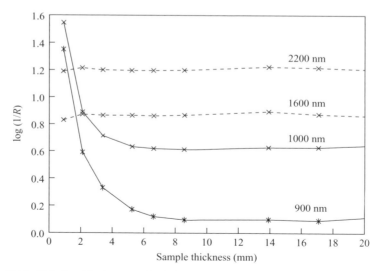

**Figure 7.2.5.** Relationship between $\log(1/R)$ value and sample thickness as a function of wavelength (36).

is due here. Before utilizing any available instrumentation and spending the time to develop calibrations, one should first consider whether or not the available instrument provides adequate spectral information. Three questions must be considered for any NIR instrument: 1) Does the instrument allow for the three basic measuring modes? 2) Does the instrument cover the needed wavelength region? 3) Do the irradiation point and detector point correspond?

Research instruments are very expensive, sometimes costing in excess of $US 100,000. In the past few years, portable and/or hand-held NIR instruments have become an alternative solution for NIR workers. Many of the portable instruments, costing far less than research instruments, utilize the Herschel region to measure the spectra of fruits and vegetables in the interactance mode. In Japan, Kubota and FANTEC companies have developed a series of portable instruments. Kubota manufactures a *dream-come-true* instrument for the housewife to measure Brix of thin-peel fruit in a supermarket (Model K-FS200). For researchers and growers, Kubota produces a in-field-type instrument called "the fruit selector" (the Model K-BA100, Fig. 7.2.6A). The instrument predict the sweetness of many kinds of fruits, such as apple, orange, peach, melon, and watermelon (58, 59). Fantec has developed an in-field instrument (called the Fruit Tester 20, FT20 (Fig. 7.2.6B) and the smaller sized FQA-NIR Gun (Fig. 7.2.6C). Saranwong (37, 38) and her fellow researchers utilized the FT20 for the determination of many mango quality parameters. The authors proved

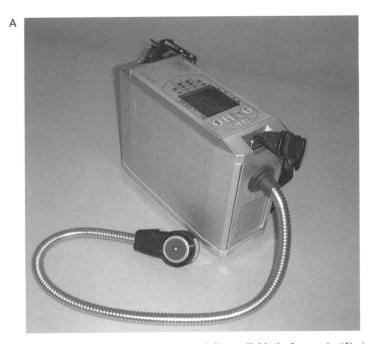

**Figure 7.2.6.** Portable NIR instruments commercially available in Japan. A: "Fruit selector (K-BA100)" developed by Kubota.

**Figure 7.2.6.** (*Continued*) B: "Fruit Tester (FT20)." C: "FQA-NIR Gun" developed by FANTEC.

that the FT20 had efficiency similar to the research-grade 6500 (FOSS NIRsystems) for Brix determination of mangoes (Table 7.2.2). Also, the FT20 has demonstated its usefulness for determining on-tree maturity.

The NIR-Gun is smaller and weighs much less and costs less than the the FT20. The "Gun" is new and appears to have a very promising future for determining quality parameters of fruits and vegetables in the laboratory and in the field. Presently, the

**TABLE 7.2.2. Calibration and Validation Results for Brix Value of Mangoes Using Spectral Data Measured with FT20 and NIRS6500 Instruments**

| Instrument | F | R | SEC[1] | SEP[1] | Bias[1] | RPD[2] |
|---|---|---|---|---|---|---|
| FT20 | 5 | 0.98 | 0.39 | 0.40 | 0.02 | 3.98 |
| NIRS6500 | 5 | 0.97 | 0.43 | 0.40 | 0.02 | 3.98 |

The wavelength region used for PLS regression is 850–1000 nm [37].

[1] Unit: °Brix.

[2] RPD is the ratio of standard deviation of reference data, say Brix value, in validation set to SEP.

NIR-Gun has a function to connect with a cellular phone for sending the spectra or NIR-predicted results to remote areas.

## Measuring Conditions

*Sample Temperature.* Even though NIR spectroscopy is a robust technology, compared to others such as mid-infrared (MIR) spectroscopy, there are certain measuring conditions that must be considered in order to obtain satisfactory results. The most prominent factor for fruit and vegetable analysis is sample temperature. Kawano et al. (49) demonstrated that the calibration equation developed from spectra of samples with a temperature of 26°C would show significant biases when applied to spectra of samples with temperatures of 21°C and 31°C (Table 7.2.3). To remove these biases, samples with a wide range of temperature, say from 21°C to 31°C, must be used to develop a calibration equation, referred to as a "temperature-compensated equation." The authors elucidated the effect of sample temperature on calibration efficiency related to the role of water and its hydrogen bonds. The difference spectra between second-derivative spectra, $D^2[\log(1/R)]$, of peaches with temperatures between 21°C and 26°C and between 21°C and 31°C showed characteristics similar to the difference spectra of water of different temperatures (Fig. 7.2.7). Peaks at 841 nm and 966 nm could be notd in both the peach and water spectra, indicating that water absorption was the key factor causing the biases. The relationship between water absorption and its temperature was proposed by Iwamoto et al. (60). and Abe et al. (61). The

**TABLE 7.2.3. Effect of Sample Temperature on Validation Result When Using the Calibration Equation Without and With Temperature Compensation**

| Calibration | Without temperature compensation | | With temperature compensation | |
|---|---|---|---|---|
| Sample temp. | SEP[1] | Bias[1] | SEP[1] | Bias[1] |
| 21°C | 0.49 | −0.33 | 0.46 | −0.03 |
| 26°C | 0.44 | 0.05 | 0.39 | −0.01 |
| 31°C | 0.50 | 0.20 | 0.41 | −0.02 |

The calibration equation without temperature compensation was developed from a set of peaches with sample temperature of 26°C, whereas that with temperature compensation was developed from the same set of peaches with different sample temperatures of 21°C, 26°C, and 31°C (49).

[1] Unit: °Brix.

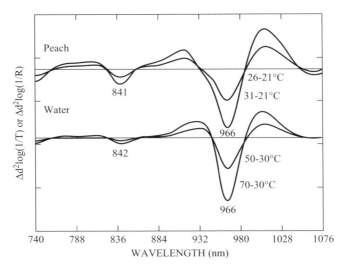

**Figure 7.2.7.** Difference spectra for peaches and water calculated from the second-derivative spectra (49). The term "26–21°C" means the difference spectra between spectra measured at 26°C and at 21°C, for example.

authors decomposed the water band into five components, which were assigned to five different species of water differing in the number of hydrogen bonds in which they involved, that is, free water molecules (S0) and molecules with one (S1) to four (S4) hydrogen bonds (Fig. 7.2.8). On increase of temperature, the proportion of water molecules with a smaller number of hydrogen bonds would increase, causing shifts of

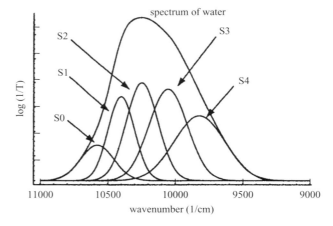

**Figure 7.2.8.** Spectrum decomposition of water band in the wavenumber region from 11,000 cm$^{-1}$ to 9000 cm$^{-1}$. The terms $S_0$, $S_1$, $S_2$, $S_3$, and $S_4$ are different species of water differing in the number of hydrogen bonds (61).

water spectra—both the absorption magnitude and wavelength position. Similar effects appear in the spectra of fruits and vegetables because 80–90% of these products is water. This sample temperature effect cannot be ignored in situations where sample temperature cannot be controlled—for example, in NIR sweetness machines located in a packing houses and when using portable or hand-held NIR instruments in an orchard.

***Sunlight.*** Unlike neon or fluorescent lamps that produce very little NIR radiation, energy from the sun can perturb the output of NIR instruments operated out of doors. Instruments operated in the laboratory or in a packing house have very few problems with external NIR energy. However, when NIR instruments are operated in open spaces, the effect of sunlight should always be evaluated. Saranwong et al. (38) evaluated the effect of sunlight when measuring NIR spectra of mangoes. Figure 7.2.9 indicates that, if a proper light shield was not provided, either in the open-field or shade conditions, the diffuse reflectance spectra ($R$) of mango would become elevated and noisy, deleterious effects caused by the sun acting as an auxillary source of NIR radiation. This spectral inconsistency was determined to have a deterioriating effect on calibrations. Table 7.2.4 shows that when the calibration developed from spectra measured indoors is applied to spectra measured outdoors without a light

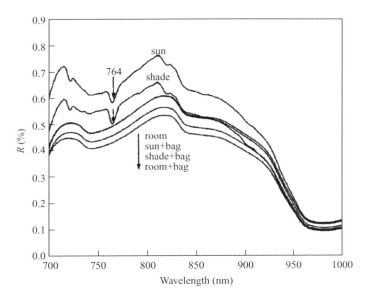

**Figure 7.2.9.** Diffuse reflectance spectra of a typical mango fruit measured at various sunlight conditions; open space under strong sunshine without a light-tight bag (sun), open space under strong sunshine with a light-tight bag (sun+bag), shade without a light-tight bag (shade), shade with a light-tight bag (shade+bag), laboratory without light-tight bag (room), and laboratory with a light-tight bag (room+bag) (38).

**TABLE 7.2.4. Prediction Results for Brix Value of Mangoes from NIR Spectra Acquired under Various Sunlight Conditions**

| | Prediction result | |
|---|---|---|
| Sunlight condition[1] | SEP[2] | Bias[2] |
| Sun | 2.00 | −3.33 |
| Sun+bag | 0.41 | 0.43 |
| Shade | 3.46 | −4.24 |
| Shade+bag | 0.51 | 0.48 |
| Room | 0.51 | 0.70 |
| Room+bag | 0.53 | 0.72 |

The calibration equation used was developed from the spectra measured in the "room" condition. The prediction results were calculated from a set of 10 mangoes (38).
[1] Sunlight condition; open space under strong sunshine without a light-tight bag (sun), open space under strong sunshine with a light-tight bag (sun+bag), shade without a light-tight bag (shade), shade with a light-tight bag (shade+bag), laboratory without a light-tight bag (room), and laboratory with a light-tight bag (room+bag).
[2] Unit: °Brix

shield, satisfactory calibration results cannot be obtained. There are two solutions to the sun problem: 1) A light shield such as a light-tight bag should be used to shield the fruit and the measuring head during an outdoor spectra acquisition (Fig. 7.2.10). Figure 7.2.9 and Table 7.2.4 show that spectra obtained with proper shielding from the sun, sun+bag and shade+bag, are free from the sunlight effect. 2) The second possibility to remove this effect is through hardware. If possible, the researcher/user should select an instrument that has a function to detect and remove the background light.

**Calibration Structure.** After the above anomalies have been accounted for and satisfactory spectra are obtained, it is important to develop the best (correct and robust) calibrations possible. In high-moisture products such as fruits and vegetables, there is a high negative correlation between Brix value and moisture content. Moreover, the absorption band for in the 960- to 970-nm region is stronger than that at the sugar band in the 910- to 920-nm region (Fig. 7.2.11). Therefore, optical data (say, log $(1/R)$ or $D^2[\log(1/R)]$) at a moisture band is often selected by mistake as the first variable term of a calibration equation for Brix determination. The calibration equation developed by using this secondary correlation is not stable and produces a large error in predictions. Figure 7.2.12 shows the simple correlation coefficients, so called "correlation plots," between $D^2[\log(1/R)]$ and the Brix value of intact peaches when the first optical term is selected. Positive and negative high correlation coefficients can be observed at specific wavelengths. For example, a positive high correlation coefficient of 0.83 is obtained at 950 nm, whereas the negative high correlation coefficient can be obtained at 906 nm. By comparing these correlation plots with the spectra of standard chemicals (shown in Fig. 7.2.11), it is readily apparent that the 950-nm peak is under the effect of water. Hence, the absorption at the designated wavelength is stronger for water but weak for amorphous sucrose, the chemical that

**Figure 7.2.10.** Spectral acquisition of mango fruit in the orchard with the FT20 portable instrument (38).

contributes to Brix determinations. Therefore, the stronger correlation at 950 nm must occur by the reverse correlation between the Brix value and moisture content. On the other hand, Figure 7.2.11 indicates that in the vicinity of 918 nm, the absorption peak of amorphous sucrose, the absorption due to water is not strong; therefore the high correlation at 906 nm must be due to the absorption of sucrose. Table 7.2.5 shows the calibration results of MLR equations having $d^2\log(1/R)$ at 906 or 950 nm as the first optical term. The calibration equation based on a sugar band is more stable than that based on internal or secondary correlation.

It is important here to consider how to develop a direct relationship (calibration equation) using partial least-squares regression (PLSR). Even with PLSR's numerous advantages, the selection of improper wavelength region(s), either too wide or too narrow, can lead to erroneous or weak calibrations. To make sure that the PLSR equation utilizes the absorption bands related to the chemical of interest, it is important to take a look at the *regression coefficient plots*, regression coefficients plotted against

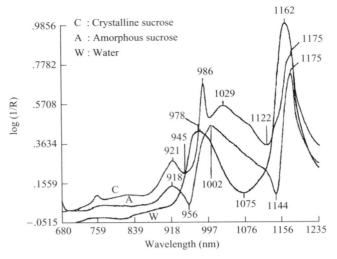

**Figure 7.2.11.** NIR spectra of crystalline (C), amorphous sucrose (A), and water (W) (47).

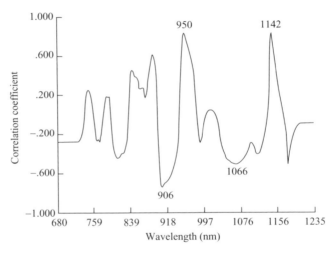

**Figure 7.2.12.** Correlation plots for Brix calibration of peaches (47). The plots are the correlation coefficients between $d^2\log(1/R)$ and Brix value plotted against wavelength when the first optical term is selected for MLR calibration.

**TABLE 7.2.5. Validation Results for Brix Value of Peaches Calculated Using MLR Calibration Equations Having $d^2\log(1/R)$ at 906 or 950nm as the First Optical Term**

| | Selected wavelength (nm) | | | | |
|---|---|---|---|---|---|
| λ1 | λ2 | λ3 | λ4 | SEP[1] | Bias[1] |
| 906 | 878 | 870 | 889 | 0.50 | 0.01 |
| 950 | 877 | 1146 | 884 | 0.53 | −3.45 |

[1]Unit: °Brix.

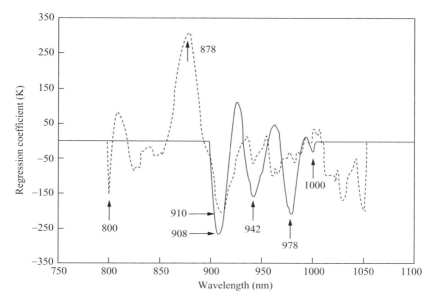

**Figure 7.2.13.** Regression coefficient plots for PLS calibrations of Brix value (solid line) and DM (dashed line) of mangoes (36).

wavelength. Therefore, among several peaks indicated in the plots, there ought to a peak or peaks related to the chemical of interest. Figure 7.2.13 shows that in the calibration equations for Brix value and DM of mangoes, the equations utilize the absorption band of sucrose in the area of 908–910 nm.

**Constituents.** Under the conditions described above, there are some constituents that NIRS can precisely predict, including Brix, DM, and starch content. On the other hand, there are some constituents that NIRS can predict with less accuracy—say, titratable acidity (TA) and firmness. Therefore, before the design of a new NIRS experiment, it would be advisable to consider whether the research objective is to tackle the challenge of determining the hard-to-detect constituents or just to screen fruits for the major components.

*Brix Value.* Brix is one of the most common quality factors of fruits. Beside water, soluble sugars, determined as Brix, is the major component in most fruits—say 10–20% by fresh weight. Therefore, in a spectrum of typical fruit, there is an adequate amount of information related to the sugar absorption. The reference measurement of Brix is also simple. The value can be precisely determined by duplicate analyses of fruit juice with a temperature-compensated digital refractometer. The critical factor is to measure Brix value as fast as possible after juice extraction. The evaporation of water from the juice will increase the Brix reading. The use of gauze to squeeze the juice from the flesh may seem to be a convenience and is a common way to obtain

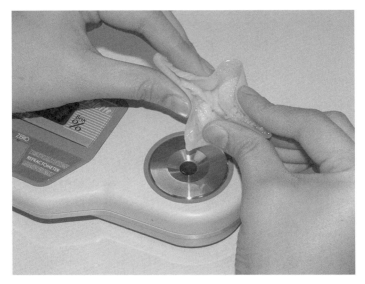

**Figure 7.2.14.** Measurement of Brix value from an intact fruit.

juice. However, for reference values used in NIRS, it might be better to squeeze the juice from the flesh directly by hand (Fig. 7.2.14) and immediately pour the juice onto the refractometer. Gauze can absorb water from the juice, increasing the Brix reading, thus leading to a mismatch between Brix readings determined with a refractometer and Brix obtained by NIRS.

*Dry Matter (DM).* For many fruits and vegetables, DM determinations include sugars, starch, and fiber. The content will vary depending on the species, variety, and maturity. It is very important to construct a DM calibration equation from a calibration set representing the samples to be measured. The DM equation developed for unripe apples should consider the absorption band of starch, whereas that developed for ripe apples should consider the absorption band of sugar. Another factor affecting DM determination is the structure of fruits or vegetables. If the fruit is homogeneous, such as apples or mangoes, it is quite simple to measure a spectrum and obtain a precise DM calibration result by a single NIR measurement per fruit. However, if the fruit does not have a homogeneous structure, for example, tomatoes that have a starlike structure, determination of DM becomes complicated. Khuriyati (56) showed that the use of tomato spectra measured from a single position along the equator could result in an inconsistency of SEPs from DM equations. However, if the spectra were derived from multiposition NIR measurements, stable determinations of DM were possible.

*Titratable Acidity.*   In orange marketing, the measurement of acid content, represented by the term of titratable acidity (TA), is as important as Brix determinations. Unfortunately, NIR measurement of acid is much more difficult. One reason has to do with the acid functional group -COOH. The covalent bond between big atoms, carbon and oxygen, has a smaller dipole moment and very low absorptivity when compared to the bond between big and small atoms such as C-H or O-H. Another reason for the difficulty in measuring TA is the low concentration of acid, typically less than 2%. Given these two causes, coupled with the low absorptivities in the short-wavelength region, it is difficult to obtain reliable/repeatable information related to TA in orange spectra. In the past decade, when it seemed impossible to obtain precise calibrations, some researchers showed that the NIRS could be used as an acid screener. Using peeled oranges and the Pacific Scientific 6250 in transmittance mode, Miyamoto et al. (43) showed that satisfactory results could be obtained with PLS regression ($R = 0.93$, $SEP = 0.146\%$, Bias $= -0.013\%$, RPD $= 2.27$). The regression coefficient plots indicated that the model developed was utilizing some absorption peaks related to the absorption bands of citric acid. However, the accuracy dropped considerably when the transmitted spectra of intact oranges measured by a commercial on-line instrument were used ($R = 0.83$, $SEP = 0.147\%$, Bias $= 0.024\%$, RPD $= 1.82$). The structure of the calibration indicated that the model for intact orange also utilized wavelengths at or near the acid absorption bands. Hence, the calibration problem related to TA may be related to the high reflection of the orange peel. In another study by McGlone et al. (44), the calibration was quite acceptable (RMSEP $= 0.147\%$) but the structure of the calibration might have been improved if the authors had utilized the secondary absorption of chlorophyll. The attempt to use an alternative technique such as time to flight (TOF) has also emerged for TA determination of apples. Tsuchikawa and Hamada (62) demonstrated the potential of TOF measurements by obtaining $r = 0.92$ and SEC $= 0.023\%$. However, the authors did not validate their results, leaving the matter of calibration stability to be determined.

*Firmness.*   Several efforts have been made to predict flesh firmness from NIR spectra, but few have attempted to interpret the meaning of the calibration equation. From a physiological viewpoint, the loss of firmness is related to the loss of cell wall structures such as pectin, cellulose, and hemicellulose. Some absorptions related to those components should be involved in the calibration structure. Sohn and Cho (63) showed that with an appropriate reference analysis a fairly precise calibration equation for apple firmness from long-wavelength NIR spectra could be developed ($R = 0.84$, $SEC = 0.84N$, $SEP = 0.92N$). The authors described that an equation with better accuracy ($SEP = 0.86N$) could be developed from NIR spectra of alcohol-insoluble solid (AIS) cell wall components. The spectra of AIS in intact apples has major differences in the region 2000–2400 nm that may relate to the absorption of methoxyl content, a pectin composition having absorption around 2250 nm. For kiwi fruit, McGlone et al. (27, 30) tried to used both conventional interactance measurement in the short-wavelength spectra and a laser beam at 864 nm to measure fruit firmness.

The first was designed with the hope of measuring some cell wall-related chemicals, whereas the latter was for determining scattering properties. Unfortunately, neither technique could precisely predict firmness. Causes for this poor performance are probably related to the low concentration of pectin, too low for good NIR measurements; in addition, more detailed studies and optimization are needed for the NIR-laser study.

## FUTURE TRENDS

### NIR Sweetness Sorting Machine

In 1989, the Mitsui Mining and Smelting Co. Ltd. developed and introduced the first peach sweetness sorting machine with which fruits can be sorted nondestructively depending on Brix. Projecting the output of two focused tungsten halogen lamps onto the moving fruits, the scattered reflected radiation is measured by a sensor unit that consists of a grating and a diode array (see Fig. 7.2.15A). A lens in the sensor unit causes the reflected radiation to converge onto a grating to extract the required wavelength and intensity data. The intensity of the radiation at any one wavelength is measured by the diode array. The Brix of a peach was calculated from the measured reflection intensity of NIR radiation by using a previously developed calibration equation. Both the reflectance and transmittance modes were studied with the sorting machine. Presently, the FANTEC Research Institute is developing a series of sorting machines using the transmittance mode for apples and oranges. For apples, instead of placing a detector on the measuring apparatus, the fiber optics is located under a rubber tray conveyer (Fig. 7.2.15B). NIR energy transmitted through the fruit travels via the fiber optics to an NIR sensor unit. During the measurement, the fruit is illuminated by 12 100-W tungsten halogen lamps mounted inside the measuring apparatus. The NIR spectrum acquired in this fashion relates to whole fruit quality including internal browning. In the case of oranges, the geometry may vary differently. The halogen lamp(s) and the detector are located in the measuring chamber in the opposite direction (Fig. 7.2.15C). This arrangement accommodates the introduction of sweetness sorting into a conventional color grating machine while increasing the sorting speed to 6 fruits/min. In Japan, three companies provide a sorting machine with calibration and maintenance packages for many kinds of fruits and vegetables, including but not limited to apples, peaches, pears, oranges, potatoes, and watermelon. In case of watermelon, 32 100-W halogen lamps are used as illuminators.

### Field Measurements

Complementary to the development of portable and hand-held instruments is research to move NIR measurements from the laboratory into the field. This move is complicated by sunlight and sample temperature over which the user has little control. Future work for field applications must concern itself with ways to compensate for these two effects. Although it is not necessary to develop calibration equations

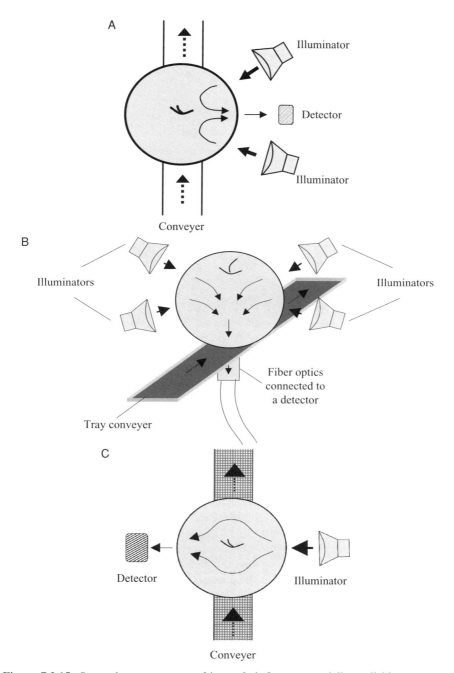

**Figure 7.2.15.** Spectral measurements of intact fruit for commercially available sweetness sorting machine; for example, peach sweetness sorting machine in reflectance mode (A; top view), apple sweetness sorting machine in transmittance mode with many lamps, say 12 lamps, and fiber optics (B; side view), and orange sweetness sorting machine in transmittance mode where opposite-side geometry between an illuminator and a detector (C; top view).

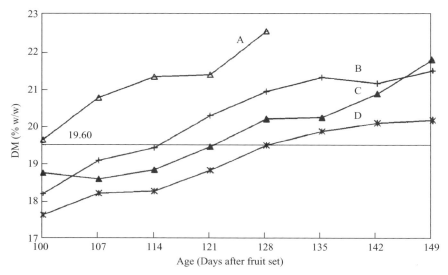

**Figure 7.2.16.** Monitoring of DM content of hard green mangoes on trees. Each line represents the data of one fruit during maturation. Dotted line is the boundary between immature (<19.60%w/w) and mature fruit (≥19.60%w/w) fruits (38).

from the spectra measured out of doors, it is necessary that the spectra measured out of doors have characteristics similar to those obtained in the laboratory and that the calibrations developed in the lab have samples that vary in temperature from 25°C to 35°C. Saranwong et al. (38) demonstrated that by using calibration equations developed from the indoor spectra with simple bias corrections, DM and starch content of on-tree mangoes could be precisely predicted. On-tree predictions of chemical compositions can be utilized both for maturity determination and for physiological monitoring during growth and development and for other experimental treatments, such as light or fertilizer treatments (Fig. 7.2.16).

Another factor that can fluctuate out of doors is relative humidity (RH). Fortunately, the NIR analysis is quite stable against this factor. In the long-wavelength region, Davies and Grant (64) showed that the changes of RH or atmospheric water vapor concentration caused by the operation of an air conditioner could alter the spectra. However, the phenomenon does not seem to be a problem in fruits and vegetables because the short-wavelength region of NIR is relatively insensitive to low levels of moisture in the atmosphere. In the long-wavelength region, the largest water vapor peaks have magnitudes around 0.0015 log $1/R$ units, but the magnitude of the water vapor bands may be 100 times smaller in the short-wavelength region. Actually, the main problem related to water in the outdoor measurements is rain. It is common knowledge that there should be no water layer or droplets on sample surface of the fruits during NIR measurement. Care should be taken to see that the fruit surfaces are dry during spectra collection.

## NIR Imaging

For fruits and vegetables, the imaging technique appears to be of little use for improving consumer satisfaction; however, the imaging technique has a great potential as a tool in physiological and agricultural analyses. As the movement of biochemical agents during growth and development becomes more complex, a biochemical map, both for type and for concentration, provided by NIR imaging would be very useful. A good example is the work of Tsuta et al. (22). Tsuta and his colleagues constructed a Brix map of a melon using an in-house NIR imaging instrument operating in the short-wavelength region. Another example is the uses of NIR imaging for the search of defects, bruises, or insects.

## Food Safety

After a decade of use NIRS for improving the quality of fruits and vegetables (ensuring consumer satisfaction), there now appears to be a logical extension of NIRS for monitoring food safety. The authors of this chapter are working on the rapid NIR measurement of health hazard materials contaminating on or in fruits and vegetables—including pesticides (65).

## REFERENCES

1. D. C. Slaughter, J. A. Abbott. In: *Near-Infrared Spectroscopy in Agriculture*, American Society of Agronomy, Crop Science Society of America, Soil Science Society of America, 2004.
2. J. N. Yeatman, K. H. Norris. *Food Technol* **19**: 123–125, 1965.
3. G. S. Birth, G. G. Dull, W. T. Renfroe, S. J. Kays. *J Am Soc Hort Sci* **110**: 297–303, 1985.
4. G. G. Dull, G. S. Birth, D. A. Smittle, R. G. Leffler. *J Food Sci* **54**: 393–395, 1989.
5. T. Lovasz, P. Meresz, A. Salgo. *J Near Infrared Spectrosc* **2**: 213–221, 1994.
6. J. Lammertyn, B. Nicolai, K. Ooms, V. D. Smedt, J. D. Baerdemaeker. *Trans ASAE* **41**: 1089–1094, 1998.
7. M. Ventura, A. D. Jager, H. D. Putter, F. P. M. M. Roelofs. *Postharvest Biol Technol* **14**: 21–27, 1998.
8. J. Lammertyn, A. Peirs, J. D. Baerdemaeker, B. Nicolai. *Postharvest Biol Technol* **18**: 121–132, 2000.
9. R. Lu, D. E. Guyer, R. M. Beaudry. *J Texture Studies* **31**: 615–630, 2000.
10. A. Peirs, J. Lammertyn, K. Ooms, B. M. Nicolai. *Postharvest Biol Technol* **21**: 189–199, 2000.
11. A. Peirs, N. Scheerlinck, J. D. Baerdemaeker, B. M. Nicolai. *J Near Infrared Spectrosc* **11**: 379–389, 2003.
12. T. Temma, K. Hanamatsu, F. Shinoki. *Optical Rev* **9**: 40–44, 2002a.
13. T. Temma, K. Hanamatsu, F. Shinoki, *J Near Infrared Spectrosc* **10**: 77–83, 2002b.
14. N. H. Sanchez, S. Lurol, J. M. Roger, V. Bellon-Maurel *J Near Infrared Spectrosc* **11**: 97–107, 2003.

15. K. B. Walsh, M. Golic, C. V. Greensill, *J Near Infrared Spectrosc* **12**: 141–148, 2004.

16. V. A. McGlone, P. J. Martinsen. *J Near Infrared Spectrosc* **12**: 37–43, 2004.

17. P. Carlini, R. Massantini, F. Mencarelli. *J Agric Food Chem* **48**: 5236–5242, 2000.

18. J. Tarkosova, J. Copikova. *J Near Infrared Spectrosc* **8**: 21–26, 2000.

19. J. Guthrie, B. Wedding, K. Walsh. *J Near Infrared Spectrosc* **6**: 259–265, 1998.

20. K. B. Walsh, J. A. Guthrie, J. W. Burney. *Aust J Plant Physiol* **27**: 1175–1186, 2000.

21. C. V. Greensill, P. J. Wolfs, C. H. Spiegelman, K. B. Walsh. *Appl Spectrosc* **55**: 647–653, 2001.

22. M. Tsuta, J. Sugiyama, Y. Sagara. *J Agric Food Chem* **50**: 48–52, 2002.

23. R. Lu. *Trans ASAE* **44**: 1265–1271, 2001.

24. A. R. Miller, T. J. Kelley, B. D. White. *J Am Soc Hort Sci* **120**: 1063–1068, 1995.

25. G. G. Dull, R. G. Leffler, G. S. Birth, Z. Zaltzman, Z. Schmilovitch. *HortSci* **26**: 1303–1305, 1991.

26. Z. Schmilovitch, A. Hoffman, H. Egozi, R. Ben-Zvi, Z. Bernstein, V. Alchanatis. *J Sci Food Agric* **79**: 86–90, 1999.

27. V. A. McGlone, H. Abe, S. Kawano. *J Near Infrared Spectrosc* **5**: 83–89, 1997.

28. A. D. Mowat, P. R. Poole. *J Near Infrared Spectrosc* **5**: 113–122, 1997.

29. P. Martinsen, P. Schaare. *Postharvest Biol Technol* **14**: 271–281, 1998.

30. V. A. McGlone, S. Kawano. *Postharvest Biol Technol* **13**: 131–141, 1998.

31. P. Martinsen, P. Schaare, M. Andrews. *J Near Infrared Spectrosc* **7**: 17–25, 1999.

32. S. D. Osborne, R. Kunnemeyer, R. B. Jordan. *J Near Infrared Spectrosc* **7**: 9–15, 1999.

33. P. N. Schaare, D. G. Fraser. *Postharvest Biol Technol* **20**: 175–184, 2000.

34. J. Guthrie, K. Walsh. *Aust J Exp Agric* **37**: 253–263, 1997.

35. Z. Schmilovitch, A. Mizrach, A. Hoffman, H. Egozi, Y. Fuchs. *Postharvest Biol Technol* **19**: 245–252, 2000.

36. S. Saranwong, J. Sornsrivichai, S. Kawano. *J Near Infrared Spectrosc* **9**: 287–295, 2001.

37. S. Saranwong, J. Sornsrivichai, S. Kawano. *J Near Infrared Spectrosc* **11**: 175–181, 2003a.

38. S. Saranwong, J. Sornsrivichai, S. Kawano. *J Near Infrared Spectrosc* **11**: 283–293, 2003b.

39. S. Saranwong, J. Sornsrivichai, S. Kawano. *Postharvest Biol Technol* **31**: 137–145, 2004.

40. S. Kawano, T. Fujiwara, M. Iwamoto. *J Japan Soc Hort Sci* **62**: 465–470, 1993.

41. K. Miyamoto, Y. Kitano. *J Near Infrared Spectrosc* **3**: 227–237, 1995.

42. D. C. Slaughter. *Trans ASAE* **38**: 617–623, 1995.

43. K. Miyamoto, M. Kawauchi, T. Fukuda. *J Near Infrared Spectrosc* **6**: 267–271, 1998.

44. V. A. McGlone, D. G. Fraser, R. B. Jordan, R. Kunnemeyer. *J Near Infrared Spectrosc* **11**: 323–332, 2003.

45. D. C. Slaughter, C. G. Cavaletto, R. E. Paull. *J Near Infrared Spectrosc* **7**: 223–228, 1999.

46. C. V. Greensill, D. S. Newman. *J Near Infrared Spectrosc* **7**: 109–116, 1999.

47. S. Kawano, H. Watanabe, M. Iwamoto. *J Japan Soc Hort Sci* **61**: 445–451, 1992.

48. V. Bellon, J. L. Vigneau, M. Leclercq. *Appl Spectrosc* **47**: 1079–1083, 1993.

49. S. Kawano, H. Abe, M. Iwamoto. *J Near Infrared Spectrosc* **3**: 211–218, 1995.

50. K. H. S. Peiris, G. G. Dull, R. G. Leffler, S. J. Kays. *J Am Soc Hort Sci* **123**: 898–905, 1998.

51. K. Katayama, K. Komaki, S. Tamiya. *HortSci* **31**: 1003–1006, 1996.

52. M. Mehrubeoglu, G. L. Cote. *Cereals Foods World* **42**: 409–413, 1997.

53. D. C. Slaughter, J. F. Thompson, E. S. Tan. *Postharvest Biol Technol* **28**: 437–444, 2003.

54. D. C. Slaughter, D. Barrett, M. Boersig. *J Food Sci* **61**: 695–697, 1996.

55. N. Khuriyati, T. Matsuoka. *J Env Ctrl Biol* **42**: 217–223, 2004.

56. N. Khuriyati, T. Matsuoka, S. Kawano. *J Near Infrared Spectrosc* 2005, in press.

57. N. Sheppard, H. A. Willis, J. C. Rigg. *International Union of Pure and Applied Chemistry Agreements. Pure Appl Chem* **57**: 105–120, 1985.

58. H. Ito, S. Morimoto, R. Yamauchi, K. Ippoushi, K. Azuma, H. Higashio. *Acta Hort* **588**: 353–356, 2002.

59. S. Morimoto, H. Ishibashi, T. Takada, Y. Suzuki, M. Kashu, R. Yamauchi. In: *Near Infrared Spectroscopy: Proceeding of the 10th International Conference*, NIR Publications, 2002.

60. M. Iwamoto, S. Kawano, H. Abe. *NIR News* **6**: 10–12, 1995.

61. H. Abe, T. Kusama, S. Kawano, M. Iwamoto. *Bunkou-kenkyu (J Spectrosc Soc Japan)* **44**: 247–253, 1995.

62. S. Tsuchikawa, T. Hamada. *NIR News* **14**: 8–11, 2003.

63. M. R. Sohn, R. K. Cho. In: *Near Infrared Spectroscopy: Proceeding of the 9th International Conference*, NIR Publications, 2000.

64. A. M. C. Davies, A. Grant. *Appl Spectrosc* **41**: 1248–1250, 1987.

65. S. Saranwong, S. Kawano, *J Near Infrared Spectrosc* **13**: 169–175, 2005.

# Meat and Fish Products

TOMAS ISAKSSON AND VEGARD H. SEGTNAN

## INTRODUCTION

The meat and fish processing industry has shown an increasing demand for fast and reliable methods to determine product quality characteristics during the last few decades. Traditional quality analyses based on wet chemistry have several drawbacks, the most significant of which are low speed (several hours or days), use of chemicals, destruction of the sample, and the physical distance between the process and the analytical instrument. Several fast and nondestructive instrumental methods have been reported, such as use of X-ray energy, ultraviolet energy, fluorescence, visual light, Raman scatter, infrared energy, radio waves, nuclear magnetic resonance, dielectricity, and ultrasonic waves. A few of these methods are in use in the processing industry, and some may have a potential use. NIR spectroscopy has proven to be an interesting and good analytical method, for off-, at-, on-, and in-line analyses of a variety of meat and fish products and quality parameters. Over the last two to three decades, a very large number of NIR applications on meat products have been reported. It is beyond the scope of this chapter to give a full overview of all these reported applications. Our attempt is to present some milestones and some central reported applications of NIR to determine some quality attributes in meat and fish products.

First we need to state some basic statistics, which is of crucial importance in the following sections. When determining a continuous variable such as fat or moisture content, etc. with a wet chemistry laboratory reference method, it is important to estimate the standard error of the laboratory reference method (SEL or $S_{ref}$). Assume that there are $N[1, 2, \ldots i, \ldots N]$ samples, each measured in $M[1, 2, \ldots j, \ldots M]$ replicates and that the average from the replicates is used in the multivariate NIR

*Near-Infrared Spectroscopy in Food Science and Technology*, Edited by Yukihiro Ozaki,
W. Fred McClure, and Alfred A. Christy.

regression. The standard error of the laboratory reference method ($S_{ref}$) is defined as:

$$S_{ref} = \left[ \frac{\sum\limits_{i=1}^{N} \sum\limits_{j=1}^{M} (y_{ij} - \bar{y}_i)^2}{M \left( \sum\limits_{i=1}^{N} M_i - N \right)} \right]^{1/2}$$

If the number of replicates is equal (i.e., $M$) for all samples, the equation above can be written as:

$$S_{ref} = \left[ \left( \sum s_i^2 \right) \middle/ \left( \sum s_i^2 \right) MN \, MN \right]^{1/2} \left[ \left( \sum_{i=1}^{N} s_i^2 \right) \middle/ MN \right]^{1/2}$$

where $s_i$ is the standard deviation between the replicates for sample $i$. The standard error of the laboratory reference method is an estimate of the precision of the reference method.

In a multivariate calibration, where a set of NIR spectra ($\mathbf{X}_{N \times K}$, $N$ samples and $K$ variables) is regressed onto a continuous variable ($\mathbf{y}_{N \times 1}$) such as the fat or moisture content, the statistical errors, the accuracy, are most often used as a quality measure of the calibration. The absolutely most common quality measure of a multivariate calibration is the prediction error, expressed either as root mean square error of prediction (RMSEP) or standard error of performance (SEP). Both are calculated and are the result of a validation process, such as test set or cross-validation. These prediction errors are defined as:

$$RMSEP = \left[ \frac{\sum\limits_{i=1}^{N} (y_i - \hat{y}_i)^2}{N} \right]^{1/2}$$

$$SEP = \left[ \frac{\sum\limits_{i=1}^{N} (y_i - \hat{y}_i - bias)^2}{N - 1} \right]^{1/2}$$

$$bias = \frac{\sum\limits_{i=1}^{N} (y_i - \hat{y}_i)}{N}$$

where $y_i$ is the reference value and $\hat{y}_i$ is the NIR predicted value. The bias is the systematic error. The SEP value is always lower or equal to the RMSEP and can be overoptimistic if the calculations are based on a low number of samples. Another common measure of the quality of a calibration is the validated correlation coefficient ($R$) between the predicted $\hat{y}_i$ and $y_i$, defined as:

$$R = \left[ \frac{S_{tot}^2 - RMSEP^2 \times \frac{N}{N-1}}{S_{tot}^2} \right]^{1/2}$$

where $S_{tot}$ is the standard deviation of $y_i$.

**Figure 7.3.1.** Plastic bags with about 13-mm-thick ground beef with different chemical composition (MATFORSK/Kjell J. Merok).

The prediction error is a sum of several sources of errors such as:

1. Standard error of the laboratory reference method ($S_{ref}$) as defined above. This error term is or should be easy to calculate.
2. Errors from the NIR instrument including electronic noise, spectral noise, and variation in sample presentation. It is often difficult to get an estimate of this error.
3. Sampling error, meaning that the sample measured on the NIR instrument is not identical to the sample measured with the reference method because of heterogeneity in the sample. Meat and fish products have typically a substantial heterogeneity due to the structural composition of meat (Fig. 7.3.1). It is difficult to calculate this error term.
4. Model error, meaning that the preprocessing and statistical model (e.g., linear models) is not correct or optimal for the actual data. This error term can be substantial, clearly indicated when different preprocessing methods or regression methods give different prediction error results.

It is important to keep in mind that the prediction error is a sum of all these error sources. Consequently, to improve the quality of a calibration and hence the prediction error, one must have control over and minimize the largest or each single error source.

A prediction error value (SEP) can be interpreted in the following manner: If a NIR prediction of a sample is $\hat{y}_i$, then one can expect with about 95% certainty that the reference method will give a result within the interval $[\hat{y}_i + 2*\text{SEP} \; \hat{y}_i - 2*\text{SEP}]$. For example, with a prediction error of 0.5% moisture; if the NIR prediction of a new sample is 20.0% moisture, the reference method is expected with about 95% confidence to give a result between 19.0% and 21.0% moisture.

When using discriminant analysis on NIR data ($\mathbf{X}$), each sample is allocated or classified into a categorical variable ($\mathbf{y}$), For example, beef or pork. The quality of the estimated model, the prediction error, is typically presented as either correct or noncorrect classification of a single sample. As for regression, all classification results should always be a result of a validation process. If the number of correctly classified samples is $C$ and the number of noncorrectly classified samples is $E$, then $C + E = N$ and the percentage of correct and noncorrect for a set of samples are calculated as $100C$ and $100E/I$, respectively.

## MEAT AND FISH PRODUCTS

In the broadest sense, meat can be defined as the edible postmortem components originating from animals. A large variety of different species can be used to produce meat and meat products. In the following we will only focus on meat from domesticated beef cattle, pork, and poultry and farmed salmon.

About 35% of live cattle and pork is carcass muscle. From 40% to 50% of live chicken and turkey is carcass muscle, and over 50% of live fish is carcass muscle. The meat from these species is composed 97–99% of water, fat, and protein. The remaining 1–3% constitutes carbohydrates, ash (i.e., mainly inorganic salts and oxides), vitamins, hormones, etc. The protein fraction, which is around 20% of the muscle, is composed of several different types of proteins: myofibrillar proteins, connective tissue proteins, etc. The most common method to determine the amount of protein in meat and fish products is to determine the total reducible nitrogen content and recalculate to protein by multiplication with a factor, often called the Kjeldahl method. The fat content, which varies from 1% to more than 80% of the muscle weight, is composed of several different lipids, fatty acids, acylglycerols, phospholipids, etc. The fat content can be determined in many ways, but the most common methods are based on extraction with organic solvents. The water molecules in meat and fish products are in different states, that is, free water and bound water with different degrees of hydrogen bonding due to temperature, ion strengths, etc. Because of the reference measurement method to determine water, drying in vacuum or at elevated temperature, the amount of water is referred to in the following as moisture.

The amounts of fat, moisture, and protein in raw meat and fish muscles are highly intercorrelated. As an example, is the data from the Isaksson et al. report (23), 100 beef samples gave high intercorrelations between fat, moisture, and protein (Table 7.3.1). If other additional ingredients, like starch, water, and spices, are mixed to give meat or fish products, such as sausages, meat spreads, and fish cakes, the intercorrelation between fat, moisture, and protein can be much lower.

Determining the main chemical constituents, namely, fat, moisture, and protein, in meat and fish muscle is often the most important point of analysis in a production process. Knowing the chemical composition of the raw materials to be used in a process makes it possible to optimize the composition of the final product. However, quality characteristics other than chemical composition may also be important for some products, such as sensory attributes, texture, added NaCl, and added starch.

**TABLE 7.3.1. Fitted Correlation Coefficients Between the Main Chemical Components in 100 Beef Samples (data from Ref. 23)**

|          | Fat      | Moisture | Protein |
| -------- | -------- | -------- | ------- |
| Fat      | 1        |          |         |
| Moisture | − 0.978  | 1        |         |
| Protein  | − 0.780  | 0.678    | 1       |

NIR can be used for production control, for commercial trade control, and for official authority control and screening of meat and fish products.

## APPLICATIONS: MEATS AND MEAT PRODUCTS

### Fat, Moisture, and Protein

Several applications for determination of moisture, fat, and protein using off- and at-line NIR spectroscopy have been reported over the last few decades. In the following we focus on two reported applications (23, 27). Lanza (27) presents a thorough study on the prediction of moisture, protein, fat, and calories in raw emulsified pork and beef by diffuse transmittance and reflectance NIR spectra. Beef samples were selected from 14 different cuts from 11 carcasses, USDA grades prime through standard. The 99 beef samples were divided into a calibration set (63 samples) and a test set (36 samples). Pork samples were selected from 7 retail cuts from 71 carcasses, USDA quality grades 1, 2 and 3. The 91 pork samples were divided into a calibration set (64 samples) and a test set (27 samples). All the samples were emulsified in a Robot-Coupe food processor for 12 seconds at 1500 rpm and for 18 seconds at 3000 rpm. The samples were analyzed for moisture and nitrogen (Kjeldahl) by AOAC procedures. Total fat was determined by the Folch procedure. Energy (calories) in the samples was determined by an Atwater method (50). The NIR spectra were measured at 20–23°C, on a standard Pacific Scientific instrument equipped with sample cups with quartz covers, in the 1100- to 2500-nm range. Figure 7.3.2 illustrates the similarity of a beef spectrum and a pork spectrum. The NIR transmittance was measured with 2.0 mm thick samples. A stepwise multiple linear regression (MLR) was used to select four wavelengths.

Table 7.3.2 gives an overview of the chemical reference analysis. Reflectance NIR measurements gave better prediction results compared to transmittance NIR measurements. Transmittance measurements will not be discussed further. The lowest prediction error for reflectance measurements for pork was 0.60% for moisture, 0.66% for protein, and 0.16% for fat (Table 7.3.3). The latter is probably somewhat overoptimistic, because the standard error of calibration was significantly higher, that is, SEC = 0.27%. The corresponding prediction errors for beef were comparable, namely, 0.52%, 0.61%, and 0.31% for moisture, protein, and fat, respectively. By

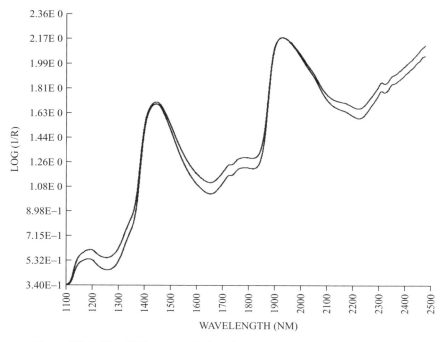

**Figure 7.3.2.** Two NIR spectra of minced meat, beef, and pork (from Ref. 27).

using a combined calibration of pork and beef, the prediction errors increased or were similar compared to local calibrations for each species.

Isaksson et al. (23) reported a study using NIR to determine the main three chemical components in beef. Here, 100 beef samples were mixed from 50 original beef samples. These 50 samples were commercial minced beef products. The test set

**TABLE 7.3.2. Composition of Samples Analyzed by Chemical Methods (27)**

| | | Calibration Set / Test Set | | |
|---|---|---|---|---|
| Meat | Components | Mean | SD | Range |
| Beef (63/36 samples) | Moisture | 71.2/69.0 | 1.9/2.9 | 65.7–74.5/65.8–75.0 |
| | Protein | 20.5/20.0 | 1.3/1.2 | 17.8–23.0/18.1–22.3 |
| | Fat | 7.7/9.3 | 2.6/3.4 | 3.1–14.0/4.4–14.7 |
| | Calories | 155.8/168.9 | 19.7/27.6 | 123.5–211.1/125.7–212.8 |
| Pork (64/27 samples) | Moisture | 71.3/72.7 | 2.2/2.0 | 66.9–75.9/65.2–75.5 |
| | Protein | 20.5/20.3 | 1.2/1.3 | 17.5–25.6/18.4–22.5 |
| | Fat | 7.9/7.3 | 2.9/2.9 | 3.9–15.5/3.9–11.3 |
| | Calories | 159.2/150.9 | 24.6/33.8 | 128.0–226.8/127.3–122.8 |

All numbers are in weight-%.

**TABLE 7.3.3. Prediction Results (27)**

| | | | Transmittance | | Reflectance | |
|---|---|---|---|---|---|---|
| Calibration | Test Set | Components | Prediction Error (SEP)/Bias [weight-%] | $R$ | Prediction Error (SEP) [weight-%] | $R$ |
| Beef | Beef | Moisture | | | 0.52/−0.01 | 0.982 |
| | | Protein | | | 0.61/−0.01 | 0.865 |
| | | Fat | | | 0.31/0.18 | 0.996 |
| | | Calories | | | 4.28/1.00 | 0.988 |
| Pork | Pork | Moisture | 0.59/0.01 | 0.979 | 0.60/−0.03 | 0.975 |
| | | Protein | 0.99/0.73 | 0.739 | 0.66/0.04 | 0.870 |
| | | Fat | 1.12/1.37 | 0.828 | 0.16*/0.06 | 0.999 |
| | | Calories | 5.33/2.48 | 0.973 | 4.92/−0.06 | 0.978 |
| Combined beef and pork | Beef | Moisture | | | 0.59/0.42 | 0.960 |
| | | Protein | | | 1.15/−0.06 | 0.630 |
| | | Fat | | | 0.27/−0.10 | 0.995 |
| | | Calories | | | 5.30/−1.02 | 0.970 |
| | Pork | Moisture | | | 0.66/−0.26 | 0.930 |
| | | Protein | | | 0.92/−0.19 | 0.530 |
| | | Fat | | | 0.28/0.15 | 0.990 |
| | | Calories | | | 3.80/1.73 | 0.970 |

validation of the regression models was restricted such that no original beef sample was included in both the calibration set (68 samples) and the test set (32 samples). The reflectance NIR spectra were measured on a Technicon 500 instrument in an open standard black rubber cup, from 1100 to 2500 nm, in 4-nm steps. The transmittance NIR spectra were measured in a 1250 Infratec Food and Feed Analyser (Tecator AB, Höganäs, Sweden), from 850 to 1050 nm in 2-nm steps, using an open rotating sample cup. In transmittance, the spectra were measured through 13 mm of minced beef. The NIR measurements were done both with and without adding an 85-$\mu$m layer of laminate, extruded from polyamide (PA), ethylenevinylalcohol (EVOH), and polyethylene (PE), on top of each sample. The reason for using NIR measurements of plastic-wrapped beef samples was to study the effect of the laminate on the predictive ability for the main chemical components. All NIR measurements were done at about 22°C.

The chemical reference measurements were done in triplicate: a semiautomatic Kjeldahl method for protein, ethylene tetrachloride extraction for fat, and air drying at 105°C for moisture. Principal component regression (PCR) was used for calibration. Table 7.3.4 gives an overview of the chemical reference analysis.

This study used beef samples spanning a relatively wide range of the components, for example, 1–23% fat (Table 7.3.4). The standard error of the laboratory reference method (SEL or $S_{ref}$) was 0.11% for moisture, 0.075% for protein, and 0.054% for fat.

**TABLE 7.3.4. Composition of Beef Samples analyzed by Chemical Methods (23)**

| Components | Calibration Set / Test Set | | |
|---|---|---|---|
| | Mean | SD | Range |
| Moisture | 70.42/70.43 | 3.53/3.26 | 58.98–75.94/62.58–75.33 |
| Protein | 20.32/20.35 | 1.12/1.09 | 16.62–22.99/18.25–22.94 |
| Fat | 7.91/7.75 | 4.69/4.22 | 0.94–23.17/1.30–17.47 |

All numbers are in weight-%.

Because of large physical and chemical differences between the samples, the spectra showed large absorbance variations due to light scattering effects. One method to reduce such spectral variation is the multiplicative scatter correction (MSC) method (13). The lowest prediction error without laminate was 0.37% for moisture with MSC transmittance measurements, 0.27% for protein with MSC transmittance measurements, and 0.28% for fat with reflectance measurements (Tables 7.3.5–7.3.8). The prediction results show that reflectance gave lower prediction errors compared to transmittance for some components, and vice versa for other components and preprocessing methods. However, prediction of fat gave much lower error results for reflectance compared to transmittance when no preprocessing method was used. No unanimous conclusion could be drawn in favor of reflectance or transmittance. Preprocessing the data with MSC gave an overall improvement for transmittance measurements but is nonconclusive for reflectance measurements. It can be concluded that the main chemical components in minced meat can be measured by NIR through the laminate with only a slight loss of predictive ability compared to measurements directly on the meat. This conclusion can have large importance both from a nondestructive and a hygienic point of view. Similar prediction error results for transmittance NIR measurements of 13-mm minced beef in plastic bags (Fig. 7.3.1), 0.30% protein, 0.53% fat, and 0.56% moisture, were found in a subsequent study (22).

Being aware of the intercorrelations between fat, moisture and protein, it is interesting to compare NIR models with models based only on the other constituents. Prediction of the fat content from the moisture content in the the Isaksson et al. (23) data, using full cross-validation, gave RMSEP = 0.96 and $R = 0.977$. Prediction of

**TABLE 7.3.5. Prediction Results for Minced Beef Without Laminate Using log(1/R) and log(1/T) Transformations of NIR Spectra (23)**

| Components | Transmittance | | Reflectance | |
|---|---|---|---|---|
| | Prediction error (RMSEP) | $R$ | Prediction error (RMSEP) | $R$ |
| Moisture | 0.48 | 0.989 | 0.48 | 0.989 |
| Protein | 0.33 | 0.984 | 0.42 | 0.921 |
| Fat | 0.74 | 0.984 | 0.28 | 0.998 |

All RMSEP values are in weight-%.

**TABLE 7.3.6.  Prediction Results for Minced Beef Without Laminate Using MSC log(1/R) and MSC log(1/T) Transformations of NIR Spectra (23)**

| | Transmittance | | Reflectance | |
|---|---|---|---|---|
| Components | Prediction error (RMSEP) | R | Prediction error (RMSEP) | R |
| Moisture | 0.37 | 0.994 | 0.45 | 0.990 |
| Protein | 0.27 | 0.967 | 0.40 | 0.929 |
| Fat | 0.38 | 0.993 | 0.37 | 0.996 |

All RMSEP values are in weight-%.

protein content from moisture content, using full cross-validation, gave RMSEP = 0.82 and $R = 0.660$. These are substantially higher prediction errors compared to NIR predictions.

The studies discussed above gave similar prediction error results for off-/at-line determinations of minced pork and beef samples, namely, 0.37–0.60% for moisture, 0.27–0.66% for protein, and 0.16–0.28% for fat. Other reports have presented similar results. These prediction errors should give fair expectations for NIR predictions of the main chemical contents in minced pork and beef meat.

Products made from meat and other additional ingredients such as salts, starch, and spices, giving products such as sausages, meat cold cuts, and spreads, are designed according to recipes. However, because of varying raw materials and process vs. product relations (water holding capacity, etc.) the final chemical composition may vary substantially. In the following two examples will be discussed. Cornish (5) reports NIR reflectance measurements on an Australian sausage with a typical composition of 60% water, 20% fat, 11% protein, and 9% starch. NIR spectra were measured with an InfraAlyser 450 (Technicon, Terrytown, NY) with 19 filters, 33–35 samples were used for calibration, and 44 samples were used as a test validation set, selected by a calibration sample selection algorithm. The prediction results (Table 7.3.9) were good but somewhat higher compared to raw meat.

Isaksson and Hildrum (21) reported NIR transmittance measurements on smoked sausages (Figure 7.3.3), meat cold cuts, and meat sausage spreads. All the samples

**TABLE 7.3.7.  Prediction Results of Minced Beef with Laminate Using log(1/R) and log(1/T) Transformations of NIR Spectra (23)**

| | Transmittance | | Reflectance | |
|---|---|---|---|---|
| Components | Prediction error (RMSEP) | R | Prediction error (RMSEP) | R |
| Moisture | 0.47 | 0.989 | 0.50 | 0.988 |
| Protein | 0.39 | 0.933 | 0.51 | 0.878 |
| Fat | 0.70 | 0.986 | 0.29 | 0.998 |

All RMSEP values are in weight-%.

**TABLE 7.3.8. Prediction Results of Minced Beef with Laminate Using MSC log(1/R) and MSC log(1/T) Transformations of NIR Spectra (23)**

| | Transmittance | | Reflectance | |
|---|---|---|---|---|
| Components | Prediction error (RMSEP) | R | Prediction error (RMSEP) | R |
| Moisture | 0.42 | 0.991 | 0.50 | 0.988 |
| Protein | 0.31 | 0.956 | 0.45 | 0.908 |
| Fat | 0.52 | 0.992 | 0.37 | 0.996 |

All RMSEP values are in weight-%.

**TABLE 7.3.9. Prediction Results for Sausages (5)**

| | Calibration | Test Set | |
|---|---|---|---|
| Constituent | Range | Range | SEP/Bias |
| Moisture | 44.1–69.0 | 51.0–66.5 | 1.18/0.26 |
| Protein | 8.8–16.7 | 9.3–15.9 | 0.60/0.01 |
| Fat | 5.5–34.3 | 14.5–29.7 | 0.64/0.07 |
| Defatted meat | 38.4–76.2 | 40.1–73.7 | 3.27/0.28 |
| Total meat | 64–92.2 | 55.8–94.6 | 3.07/1.03 |

All numbers are in weight-%.

**Figure 7.3.3.** A platter with smoked sausages (NLH/Håkon Sparre).

**TABLE 7.3.10. Prediction Results for Third Control Test Set of Smoked Sausages (21)**

| Constituent | Range | $S_{ref}$ | RMSEP | SEP/Bias |
|---|---|---|---|---|
| Moisture | 53.9–69.3 | 0.33 | 0.65 | 0.64/ − 0.1 |
| Protein | 9.2–12.4 | 0.10 | 0.41 | 0.38/0.1 |
| Fat | 10.2–24.9 | 0.37 | 0.92 | 0.88/0.3 |
| Carbohydrates | 4.0–7.4 | — | 0.77 | 0.66/ − 0.4 |

All numbers are in weight-%.

were collected from different manufacturers and production lines, and at different times. The reference standard chemical methods were duplicate measurements, Fosslet for fat, Kjeldahl for protein, air drying for 14 hours at 102–105°C for moisture, and carbohydrates as glucose after acid hydrolysis. Spectra were measured after homogenization with a Tecator Infratec Food and Feed Analyzer, from 850 to 1050 nm, in 2-nm steps. Diffuse transmittance spectra were measured through 17-mm-thick samples. This study used a very thorough validation method: each product group was divided into three data sets: one for calibration (70–79 samples), one test set (37–54 samples) for choosing the optimal number of partial least squares (PLS) factors, and a third control test set (24–25 samples, no control set for the meat sausage spreads set) for the final estimation of the prediction errors.

The range of the chemical components was wide (Tables 7.3.10–7.3.12), as could be expected because the samples originated from different product lines and producers. The overall NIR predictive ability was good, with reasonably low prediction errors.

To illustrate the use of these prediction errors, let us look at an example. The Norwegian authority regulation for smoked sausages allows a maximum of 22.0% fat. The SEP for fat is 0.88%, and hence the "error window" for fat analysis by NIR is ± 1.8%, using a 95% confidence interval. This means that if a sample was determined between 20.2% and 23.8% for fat by NIR, it would need to be analyzed by the classic technique to determine whether this sausage meat conforms to regulations. For samples with less than 20.2% fat one can be 97.5% (half-sided) confident that they pass fat requirements. Likewise, for samples greater than 23.8% one can be 97.5% confident that they fail the food regulations.

These prediction error results indicate similar or somewhat higher errors for moisture, higher errors for protein, and clearly higher errors for fat, compared to measurements on raw beef or pork. One explanation is the significantly higher standard

**TABLE 7.3.11. Prediction Results of Third Control Test Set of Meat Cold Cuts (21)**

| Constituent | Range | $S_{ref}$ | RMSEP | SEP |
|---|---|---|---|---|
| Moisture | 62.1–74.8 | 0.33 | 0.60 | 0.60 |
| Protein | 16.6–25.8 | 0.10 | 0.74 | 0.50 |
| Fat | 0.8–14.1 | 0.37 | 0.61 | 0.54 |

All numbers are in weight-%.

**TABLE 7.3.12. Prediction Results of Third Control Test Set of Meat Sausage Spreads (21)**

| Constituent | Range | $S_{ref}$ | RMSEP |
|---|---|---|---|
| Moisture | 59.7–69.6 | 0.33 | 0.40 |
| Protein | 9.6–17.4 | 0.10 | 0.35 |
| Fat | 5.0–19.8 | 0.37 | 0.91 |
| Carbohydrates | 3.8–7.3 | — | 0.53 |

All numbers are in weight-%.

error of the laboratory reference method ($S_{ref}$) for these processed meat products compared to the raw meat applications discussed above. Another reason could be complex variability of the samples due to large variation in raw materials, processes, and productions.

On- or in-line NIR determinations of the main chemical components in beef (25, 48) and mixtures of beef and pork (47) have been reported. These three studies report the use of a NIR gauge, a monopixel camera (MM55/M55E NCD Infrared Engineering LTD, Maldon, UK) (Fig. 7.3.4 and 7.3.5), which uses a rotating five-filter wheel to determine the five-wavelength NIR spectra. The NIR spectra, which were recorded directly and noninvasively at the outlet of a meat grinder, were modeled against the moisture, protein, and fat. The distance between the gauge and the meat stream was about 20 cm. The filter wheel, with five selected filters, rotated at 20 Hz, giving a single spectrum every 1/20 s. Average spectra from several readings were used

**Figure 7.3.4.** The NIR gauge placed about 20 cm from the outlet of a meat grinder and the authors Tomas Isaksson (left) and Vegard H. Segtnan (right) (NLH/Håkon Sparre).

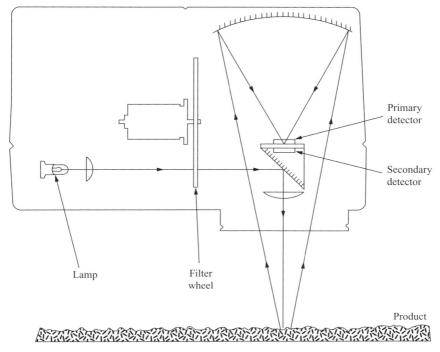

**Figure 7.3.5.** The principle of the NIR gauge.

in the calibrations. The first challenge was to select the five filters. The five filters or wavelengths were selected with stepwise multiple regression methods and prior knowledge about NIR spectroscopy. The selected diffraction filters had maximum throughput at 1441, 1510, 1655, 1728, and 1810 nm for the nonfrozen samples and 1630, 1728, 1810, 2100, and 2180 nm for the semifrozen samples. The different selection of filters is due to the shift of water absorption peaks during freezing. Figure 7.3.4 illustrates how the NIR gauge is placed at the outlet of the meat grinder, and Figure 7.3.5 illustrates the principle of the NIR gauge.

In the first study, a pilot plant grinder was used. Beef batches of 20 kg with ranges of 6.2–21.7% fat, 59.6–72.9% moisture, and 18.1–20.7% protein were analyzed. The grinder was equipped with several different plate hole diameters: 4, 8, 13, and 19 mm. Spectra were taken with all four grinder plates. The standard error of the laboratory reference method ($S_{ref}$) varied with the grinder plate hole diameter and was 0.12–0.24% for fat, 0.12–0.25% for moisture, and 0.07–0.10% for protein.

The prediction ability for this noninvasive, in-line NIR measurement gave low prediction errors (Table 7.3.13). The prediction errors increased with increasing hole diameter. The reason for this decrease in accuracy is the increasing heterogeneity of the distribution of fat, moisture, and protein. The instrument illuminated a 40-mm-diameter spot on the continuous meat stream. Using the data from measurements of

**TABLE 7.3.13. Prediction Errors Expressed as RMSEP (in weight-%) for Different Grinder Plate Hole Diameters (25)**

| Constituent | Diameter of grinder plate holes (mm) | | | |
|---|---|---|---|---|
| | 4 | 8 | 13 | 19 |
| Moisture | 0.75 | 0.81 | 1.05 | 1.25 |
| Protein | 0.23 | 0.27 | 0.32 | 0.27 |
| Fat | 0.73 | 0.88 | 1.14 | 1.39 |

4-, 8-, and 13-mm grinder plate hole diameters gave prediction errors of 0.91% for moisture, 0.29% for protein, and 1.03% for fat (Fig. 7.3.6).

Using the same instrumentation at the outlet of an industrial scale grinder also gave good prediction ability (47). A set of 112 batches of fresh beef (61 batches) and pork (51 batches) ranging from 200 to 800 kg were ground on a combined mixer/grinder (Maschinenfabrik Laska, Linx, Germany, Type LER WMW 200/2000 B5241). The grinder was equipped with a 25- and a 13-mm hole plate in consecutive order. Additionally, 21 beef batches and 21 pork batches were measured and used as an independent test set (Table 7.3.14). During the grinding in the process hall, small samples were continuously taken out from the grinder outlet, giving a 5- to

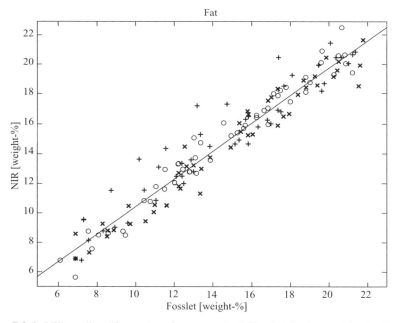

**Figure 7.3.6.** NIR predicted fat vs. the reference method (Fosslet) for the combined calibration on samples ground at 4 (o)-, 8 (+)- and 13 (x)-mm hole diameter grinder plates. The correlation coefficient was 0.967 (from Ref. 25).

**TABLE 7.3.14. Statistics for Reference Analysis of Test Set (47)**

| Constituent | Range | $S_{ref}$ |
|---|---|---|
| Moisture | 58.4–70.1 | 0.19 |
| Protein | 16.1–20.0 | 0.16 |
| Fat | 9.1–25.0 | 0.17 |

All numbers are in weight-%.

10-kg subsample for each batch that was further homogenized and sent for reference analysis. The subsamples were analysed in duplicate for fat (Fosslet, Foss Electric, Hillerød, Denmark), moisture (105°C for 18 h), and protein (Kjeltec Auto 1030, Tecator AB, Höganäs, Sweden).

Compared to the pilot plant study using 13-mm-diameter plate holes, these industrial-scale data gave very similar prediction error results (Tables 7.3.15 and 7.3.16). Pork and beef gave very similar prediction error results.

Many producers of ground meat feed the grinder with frozen meat, resulting in semifrozen ground meat at the outlet of the grinder. The temperature-dependent water band shifts result in difficulties in building robust calibrations when the ratio of frozen and nonfrozen meat is varying between the batches (Fig. 7.3.7). Choosing other wavelength filters, namely, 1630, 1728, 1810, 2100, and 2180 nm, gave good prediction error results (48) (Table 7.3.17). For the in-line calibration, 55 industrial-scale (400–800 kg) batches of frozen and semifrozen beef at about −7°C were ground with 4- and 13-mm-diameter grinder plates.

These three studies indicate that noninvasive in-line NIR measurements can be used successfully to determinate accurately the main chemical components in both ground beef and pork.

Valdes and Summers (49) reported NIR determination of fat and protein in poultry. Whole carcasses from 3-week-old broilers and from 7-week-old broilers were ground, and NIR spectra were determined (19-wavelength filter instrument). The prediction errors were reported to be 2.0–2.6% for fat and 1.0–2.0% for protein. Windham et al. (51) report a study with a fat range from 1% to 65%, giving SEP values from 0.38% to 0.84% fat.

**TABLE 7.3.15. Prediction Errors Expressed as SEP and Correlation Coefficients for Beef Test Set (47)**

| Constituent | No Outliers Removed | | Outliers (no. in parenthesis) Removed | |
|---|---|---|---|---|
| | SEP | R | SEP | R |
| Moisture | 1.21 | 0.95 | 0.94 (2) | 0.96 |
| Protein | 0.45 | 0.79 | 0.45 (0) | 0.79 |
| Fat | 1.30 | 0.96 | 1.16 (1) | 0.96 |

All SEP values are in weight-%.

**TABLE 7.3.16. Prediction Errors Expressed as SEP and Correlation Coefficients for Pork Test Set (47)**

| Constituent | No Outliers Removed | | Outliers (no. in parenthesis) Removed | |
|---|---|---|---|---|
| | SEP | R | SEP | R |
| Moisture | 1.18 | 0.92 | 1.03 (1) | 0.93 |
| Protein | 0.57 | 0.68 | 0.35 (1) | 0.87 |
| Fat | 1.35 | 0.94 | 0.82 (3) | 0.98 |

All SEP-values are in weight-%.

**TABLE 7.3.17. Full Cross-Validation Prediction Error (RMSEP in Weight-%) Results in Semifrozen Ground Beef (48)**

| Constituent | Diameter Hole in Grinder Plate (mm) | | | |
|---|---|---|---|---|
| | 4 | | 13 | |
| | RMSEP | R | RMSEP | R |
| Moisture | 0.59 | 0.98 | 0.97 | 0.95 |
| Protein | 0.47 | 0.83 | 0.45 | 0.80 |
| Fat | 0.54 | 0.99 | 1.11 | 0.96 |

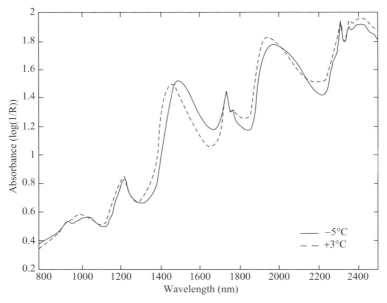

**Figure 7.3.7.** NIR spectra of frozen ($-5°C$) and nonfrozen ($3°C$) minced beef.

**TABLE 7.3.18. Test Set Prediction Errors for Fatty Acids in Back and Breast Fat from Pork, Measured in NIR Transmittance Mode (39)**

| Fatty Acid | Range (weight-%) | RMSEP (weight-%) | R |
|---|---|---|---|
| Saturated | 32.2–45.7 | 0.66 | 0.97 |
| Monounsaturated | 40.5–57.8 | 0.60 | 0.99 |
| Polyunsaturated | 8.0–19.6 | 0.50 | 0.99 |
| C16:0 | 20.8–28.3 | 0.49 | 0.93 |
| C18:0 | 9.0–17.9 | 0.63 | 0.94 |
| C18:1 | 37.8–54.7 | 0.58 | 0.99 |
| C18:2 | 7.2–18.4 | 0.51 | 0.90 |

## Determining Minor Components

Fatty acid profiles have been studied. Windham and Morrison (52) report prediction of saturated fatty acids in the range of 43.2–62.0% and unsaturated fatty acids in the range of 38.3–56.2% in fat from beef. Prediction error expressed as SEP was 1.10% ($R = 0.88$) for saturated acids and 1.13% (no chance) for unsaturated fats. Individual fatty acids like oleic and palmitic acids gave similar prediction errors, whereas other fatty acids gave lower accuracy.

NIR determination of monounsaturated, polyunsaturated, and saturated fatty acids and some specific acids gave good cross-validated prediction errors in chloroform and methanol (2:1)-extracted pork breast and back fat (39) (Table 7.3.18). The extracted pork fat was measured in a 5-mm transmission cell at 35°C and scanned from 900 to 2500 nm. Three wavelength ranges were used in the calibrations; 1362–1480, 1687–1855 and 2115–2172 nm, characteristic for $CH_3$ and $CH_2$ stretch, CH stretch, and CH=CH stretch, respectively. Forty samples were used in an independent test set for validation. The PLS models used nine factors for all components. Attenuated total reflection (ATR) measurement in the IR range, 6000–900 cm$^{-1}$, gave slightly lower or similar prediction error results. NIR reflectance measurements directly on meat slices gave much higher prediction errors compared to the extract transmission measurements.

Determination of added NaCl in the range of 1.4–2.2% in 57 sausages is discussed by Ellekjær et al. (11). Analysis of the sausages gave a range in the test set of 11.9–13.3% protein, 9.3–28.6% fat, 47.1–68.7% moisture, and 2.8–9.2% starch. A test set validated PLS regression gave good predictions of the NaCl content. NIR reflectance in the 1100- to 2500-nm range gave lower prediction errors compared to NIR transmittance of 15- and 17-mm samples in the 850- to 1050-nm range. NIR reflectance measurements of the sausage mix, that is, before thermal processing to 72°C, gave a SEP value of 0.06% NaCl with 14 PLS factors. Reflectance measurements of the final thermal processed sausages gave SEP = 0.04% NaCl with 15 PLS factors. NaCl itself is not an NIR active compound. The report concludes that the predictive ability of NaCl was found mainly in the 1364- to 1420-nm water band shifts because of the varying NaCl content. Also, other spectral regions (e.g., 1700- to 1760-, 1808-, 1876-, and 2300- to 2330-nm bands) contributed to the predictive ability of NaCl, probably because of swelling of proteins and carbohydrates.

NIR determination of collageneous proteins in meat has been reported (2). The results in this report have not been reproduced by other researchers and are not discussed further here.

## Determining Sensory Attributes

When consuming whole meat, like steak, roasted beef, and stews, sensory tenderness, toughness, and juiciness are often considered the most important quality attributes. Traditional sensory analysis is both time consuming and sample destructive. Other rheological methods to determine tenderness and toughness are also destructive analysis methods. A noninvasive, nondestructive measurement method is desired by the industry.

One of the first attempts to model NIR data and texture properties in beef was presented by Mitsumoto et al. (33). This report presented fitted correlations between four wavelengths and shear values from beef to 0.83 for NIR reflectance, 0.80 for NIR transmittance and 0.80 for NIR with a reflectance fibre-optic probe.

Hildrum et al. (15) presented a study determining sensory tenderness, hardness, and juiciness in M. longissimus dorsi from seven bulls. Slices (1 cm thick) from the muscles were measured with NIR reflectance (Technicon 500, Foss Electric, Hillerød, Denmark). Sensory and texture analyses were performed after 1, 8, and 14 days of aging at 2°C. The NIR reflectance spectra (1100–2500 nm) were collected in a specially made cup, constructed such that the light beam was parallel with the fibers in the muscle slices. Sensory analysis was done in duplicate servings, after heat treatment and cooling to 20°C, by nine trained assessors. Average values for all measurements on a 1–9 scale were used in the calibrations. Hardness was defined as the resistance of the first bite, whereas tenderness refers to the whole chewing process. Texture was measured in five replicates on a Warner-Bratzler shear force device in an Instron Materials Testing Machine (Model 4202, Instron, Engineering Corporation, High Wycombe, UK). The shear force was measured perpendicular to the muscle fibers on a rod with a cross section of $1 \times 1$ cm$^2$.

NIR measurements on frozen and thawed meat gave slightly higher correlations between NIR predictions and reference values compared to measurements on fresh meat (Table 7.3.19) (Fig. 7.3.8). The destructive shear force texture measurements gave a fitted correlation coefficient of $-0.87$ with the sensory tenderness. This study

**TABLE 7.3.19. Cross-Validated NIR Predictions of Sensory Tenderness, Hardness, and Juiciness in M. Longissimus Dorsi (15)**

| Variable to be Predicted | Range | Fresh Meat | | Frozen/Thawed Meat | |
|---|---|---|---|---|---|
| | | RMSEP | R | RMSEP | R |
| Hardness | 2.9–7.1 | 0.58 | 0.87 | 0.54 | 0.88 |
| Tenderness | 2.9–7.4 | 0.67 | 0.85 | 0.64 | 0.86 |
| Juiciness | 4.6–6.1 | 0.51 | ns[*] | 0.36 | 0.49 |

[*]Nonsignificant.

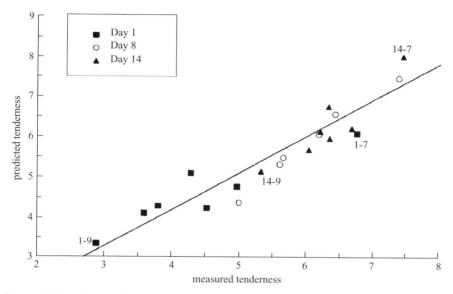

**Figure 7.3.8.** NIR predicted versus measured sensory tenderness for fresh beef (from Ref. 15).

was later extended with 120 beef samples (16). The latter reports slightly lower predicted correlation coefficients, 0.74, 0.61, and 0.61 for frozen and thawed samples for sensory hardness, tenderness, and juiciness, respectively. NIR measurements on fresh beef samples gave correlation coefficients of 0.61, 0.58, and 0.47 for sensory hardness, tenderness, and juiciness, respectively. The second principal component explained only 9% of the spectral variance but almost all of the explained sensory ($y$) variance. The second PC loading was almost identical to the average NIR spectrum, which indicated that light scattering properties of the samples held predictive information. More precisely, the NIR penetration depth is greater for tender meat than for tough meat. This optical effect can be seen with the naked eye, that is, prerigor and tough beef appears to be more transparent compared to postrigor and tender beef. The NIR spectra concealed little or no chemical information that could be related to the sensory attributes.

In a subsequent publication (35), discriminant analysis showed that the NIR spectra gave good allocation into three classes; Tender, intermediate, and tough samples. Up to 60% of the samples were correctly classified for all three classes. Note that one cannot expect very high correct classifications of continuous classification variables. The number of classifications in correct or neighboring subgroups for the two extreme subgroups was equal to 97%.

In a study published by Park et al. (36), 119 *M.* longissimus thoracis muscles were measured with a NIR reflectance instrument (1100–2500 nm) and a Warner-Bratzler shear force instrument for tenderness. The data were split into a calibration set (80 samples) and a test set (39 samples). The test set gave SEP = 1.3 kg and

$R = 0.79$. In a discriminant analysis the data was divided in two classes, 1) <6kg and 2) >6kg shear force. For the first class, 19 of 23 samples were correctly allocated. For the second class, 12 of 16 samples were correctly allocated, giving a 79% correct classification. The report concludes that these data are capable of predicting Warner-Bratzler shear force values of longissimus steaks. The report also states that "It is unclear whether this technology can be applied to carcasses at packaging plants within 1 to 5 days after slaughter and used to accurately predict how tender the longissimus steaks from those carcasses would be after ageing." Other studies have shown similar results for NIR prediction and classification of beef (3, 29, 40, 41.) Altogether, NIR reflectance measurements seem to give reasonably good predictions for both sensory and instrumental texture values for beef samples.

Prediction of 16 different sensory attributes in meat sausages (Fig. 7.3.3) was reported by Ellekjær et al. (12). In this study, 57 meat sausage batches were produced in accordance with an experimental design based on fresh beef, pork fat, potato starch, skim milk powder, NaCl, spices, ascorbate, and ice. The experimental design was set to vary from 8 to 28% fat, 1.3 to 1.9% NaCl, and 1.5 to 7.5% starch. The remaining ingredients were kept constant, with a protein content of 11%. The batters were placed in plastic casings and thermally processed to 72°C, smoked, and partially dried. Both NIR reflectance (1100–2500 nm) and transmittance (850–1050 nm, 17-mm-thick samples) spectra were measured on the batters and the sausages (without casing) at 20°C. All the sausages were evaluated by a nine-membered sensory panel using a 1–9 scale. The attributes were whiteness, color, color strength, odor intensity, meat odor, smoke odor, off-odor, flavor intensity, meat flavor, saltiness, smoke flavor, spiciness, off-flavor, firmness, juiciness, and greasiness. The sausages were presented to the panel held at a temperature of 72°C. Because of the relatively low precision in sensory analysis, the NIR predictive ability was presented as relative ability of prediction (RAP), which is defined as:

$$RAP = \frac{S_{tot}^2 - RMSEP^2}{S_{tot}^2 - S_{ref}^2}$$

RAP is similar to predicted correlation coefficients ($R$) but is weighted for the standard error of the laboratory reference method ($S_{ref}$) and $0 \leq RAP \leq 1$. As for $R$, a high RAP indicates good predictive ability. The prediction results were generally better for NIR transmittance than for NIR reflectance. This is probably due to better handling of the heterogeneity of the samples. The highest prediction abilities for NIR were the color attributes on both batter and sausages with RAP larger than 0.9, followed by texture attributes with RAP values around 0.80. For odor and flavor the highest predictive abilities for NIR transmittance were found for meat odor and smoke odor in sausages with RAP values of 0.70 and 0.83, respectively. Even off-flavor and off-odor in sausages gave high RAP values, 0.71 and 0.72, respectively. These high predictive abilities were probably due to the correlation between odor and fat content. Flavor attributes were predicted better with NIR reflectance compared to NIR transmittance. The main variation in sensory attributes was related to fat and starch contents of the sausages, which could explain much of the predictive ability.

Sensory characteristics and NIR spectroscopy of broiler breasts for various chill-storage regimes is reported by Lyon et al. (30).

## Other Quality Attributes

In this section NIR prediction of some other quality attributes are discussed, such as prediction of previous heat treatments, freeze determination, and meat speciation. State authorities recommend that meat patties (e.g., hamburgers) for food service institutions should be heated to a minimum of 68–72°C. Temperature recommendations differ for different countries and products. Because of thermal denaturation and destruction of meat cell structures, NIR spectroscopy has been suggested as a method to control whether samples reach the recommended temperatures. Isaksson et al. (20) reported a preliminary study of the ability of NIR to determine the maximum temperature of previously heated meat. In this study, three ground beef batches were tempered to 20, 50, 60, 75, 85, and 95°C. The cross-validated RMSEP was 3.9°C ($R = 0.987$) for direct NIR measurements and 2.5°C ($R = 0.992$) for NIR measurements of freeze-dried samples. In a subsequent study (10), ground (5-mm-hole plate) *M. longissimus dorsi* samples from 33 bulls were measured. Eight-millimeter-thick meat patties were cooked in two or three replicates in plastic bags to approximately 50, 55, 60, 65, 68, 72, 75, 80, and 85°C and cooled to 4°C. Both cooked and freeze-dried samples were measured with NIR spectroscopy. The samples were measured at 20°C, for both NIR reflectance (1100–2500 nm) and NIR transmittance (850–1050 nm, 15 mm thick for direct and 5 mm for freeze-dried samples) instruments. The data were divided into a calibration set with samples from 25 bulls and a validation test set with samples from the remaining bulls. Different preprocessing methods and regression methods were compared. The lowest prediction errors were found for NIR reflectance measurements of freeze-dried samples, giving an RMSEP of 1.4°C with 12 PLS factors. For direct measurements on "wet" meat samples, the lowest RMSEP was 2.0°C for both reflectance and transmittance. In a third report from this study, Thyholt et al. (46) discuss the use of different sample presentation method, namely, dry extract spectroscopy by infrared reflection (DESIR) (Fig. 7.3.9). After heating (53.5–85.5°C), cooling, centrifugation, and filtering, 0.5 ml of beef juice was pipetted on the center of a 55-mm glass microfiber filter. After the filters were dried in vacuum at 20°C, they were measured with a reflectance cell from 400 to 2500 nm (NIRSystems 6500, NIRSystems Inc., Silver Spring, MD, USA). The 400- to 2500-nm spectra from DESIR gave RMSEP = 0.74°C ($R = 0.965$) for the 65.6–75.6°C range with 8 PLS factors (Fig. 7.3.10). The reason for this substantially improved prediction performance compared to the earlier reports may be the DESIR method itself, but better precision and control of the reference method, that is, the temperature control, may also have played a role.

The freezing process may impose a few negative effects on the meat, such as freeze burns, decreased juiciness, increased drip loss, and cooking loss. The microbial quality might also be affected. After thawing, nutrients in the exudates on the meat surface promote growth of mesophilic bacteria, which may be pathogenic. For these reasons it may be important to control whether meat has been previously frozen. Slices from

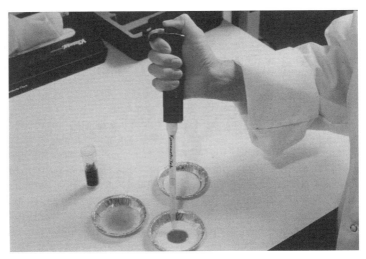

**Figure 7.3.9.** Illustration of a dry extract spectroscopy by infrared reflection (DESIR) application. Meat juice is pipetted on the center of a 55-mm glass microfiber filter. After drying the filters in vacuum, it can be measured in NIR reflectance (MATFORSK/Kjell J. Merok).

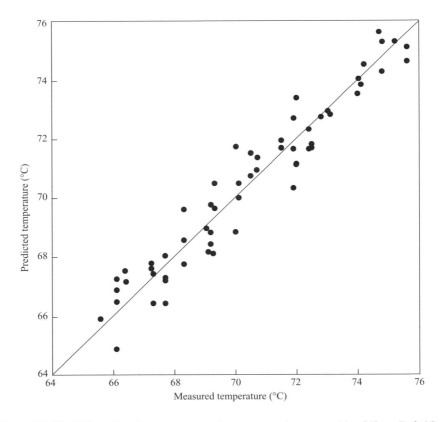

**Figure 7.3.10.** NIR predicted versus measured temperature heat-treated beef (from Ref. 46).

*M. longissimus dorsi* from 40 beef cattle were frozen at $-20°C$, and reference slices were stored at $4°C$ for 4–5 days (44). After thawing at $4°C$, 0.6 ml of meat juice from all samples was applied to glass fiber filters, and the DESIR measurements were taken from 400 to 2500 nm. Intact beef samples were also measured in reflectance with the light beam parallel to the muscle fiber, from 1100 to 2500 nm. The discriminant analysis method, *k*-nearest neighbor (kNN) allocated up to 100% of the samples correctly for fresh and previously frozen beef samples with the DESIR technique. Up to 98% of all samples were correctly allocated for measurements performed directly on the beef samples. A high number of principal components (15–25) were needed to get this high accuracy, which indicates that the chemical differences between frozen and fresh beef are complex. No hypothesis was suggested that sought to explain why it is possible to discriminate between frozen and fresh beef. Downey and Beuchêne (8, 9) performed a similar study using both DESIR and a fiber-optic surface interactance probe, scanning from 650 to 1100 nm. The DESIR samples were 100% correctly classified, whereas the fiber-optic probe resulted in slightly lower accuracy. The 16 previously frozen samples were all allocated as "previously frozen," and 13 of 16 fresh beef samples were allocated as "fresh" samples, giving an overall classification rate of 92%.

When meat batches are finely ground or homogenized it can be difficult for the naked eye to determine what species the meat originates from. For economic or other reasons it can happen that minced meat is adulterated by meat from other, low-cost species. Consequently, there is a need for a fast, nondestructive analytical method for meat authentication and adulteration. Thyholt et al. (45) collected 68 different meat samples over a 2-year period from 4 different animal species. These were from 17 beef carcasses, 17 pork carcasses, 17 mutton carcasses, and 17 batches of mechanically recovered turkey and chicken meat. Meat samples were centrifuged to give meat juice. Meat juices were mixed to 350 samples according to a design to simulate adulterated meat samples. After filtering, 0.5 ml of meat juice was pipetted onto glass fiber filters, vacuum dried, and measured by NIR reflectance (400–2500 nm, NIRSystems 6500, NIRSystems Inc., Silver Spring, MD). Several discriminant analysis methods were compared. Quadratic discriminant analysis (QDA) using PLS loadings gave the overall best classification results. Classification of single-species samples gave good results (Table 7.3.20). Only one sample, a poultry meat sample, was mis-classified

**TABLE 7.3.20. Classification of 68 Meat Juice Samples from Four Different Species; Beef, Pork, Mutton and Poultry (45)**

| Classes | No. Misclassified out of 68 Samples |
|---|---|
| Beef vs. not beef | 0 |
| Pork vs. not pork | 0 |
| Mutton vs. not mutton | 1 |
| Poultry vs. not poultry | 1 |
| Beef, pork, mutton, and poultry | 1 |

as a mutton sample; all the other samples were correctly classified. Predictions of the amount of pork, mutton, and poultry in the beef samples gave RMSEPs down to 6.8%. These prediction errors indicate that it is not possible to detect low percentages of pork, mutton, and poultry added to beef samples.

Other authors (37) have reported classification of homogenized chicken breast, turkey breast, and pork loin chops with up to 91.9% correct classification. McElhinney et al. (32) reported 97.4% correct classification of homogenized beef, pork, lamb, and poultry. Arnalds et al. (1) used a more complex hierarchical decision tree approach. They reported 100% correct classifications between red and white meat samples and between lamb and beef samples; 98.8% and 94.4% of pork/poultry samples and turkey/chicken samples, respectively, were correctly classified.

## APPLICATIONS: FISH AND FISH PRODUCTS

Prediction of chemical components in fish is important throughout the whole production chain. Fish farms may use nondestructive NIR analysis of live fish in breeding to optimize the quality of the offspring, for optimizing feeding regimes, and for sorting of slaughtered fish. In professional fishing, one may use NIR to sort the different species and qualities. The fish processing industry may use NIR analysis to optimize processes and to classify different samples for different markets. Fish distributors may want to use NIR technology to sort fish for different markets and into different price classes.

NIR studies have been published on different species, such as rainbow trout, Atlantic farmed salmon, halibut, mackerel, herring, and whiting, in addition to salmon roe. In the following we focus on applications related to European and American salmon farming.

### Chemical Components

As for meat and meat products, the most common reason for applying NIR spectroscopy on fish and fish products is to measure the fat, moisture, and protein contents. First, destructive NIR measurements on freeze-dried, cross-sectioned, and homogenized samples will be discussed. Second, nondestructive NIR measurements on fillets, and whole postmortem and live fish will be discussed. Finally, NIR prediction of minor chemical components will be briefly discussed.

Gjerde and Martens (14) reported NIR prediction of fat, moisture, and protein in homogenized freeze-dried rainbow trout (*Salmo gairdneri*). They used a 19-filter NIR instrument (InfraAlyser 400, Technicon, Terrytown, NY), in the wavelength range 1445–2350 nm. The prediction errors were 0.45% for fat, 0.35% for moisture, and 0.50% for protein. At about the same time, Mathias et al. (31) reported NIR measurements on homogenized freeze-dried freshwater fishes. Rasco et al. (38) analyzed 1- to 3-mm-thick cross sections of frozen and thawed rainbow trout (*Oncorhynchus mykiss*), using a scanning reflectance NIR instrument in the 900- to 1800-nm wavelength range. The reported prediction errors were expressed as weight-% in dry weight of fish flesh. For comparison, the reported values are recalculated to percentage of wet weight using 70% moisture in the fishes as a recalculating basis. The recalculated prediction errors were 0.1% for fat, 0.37% for moisture and 0.18% for protein. Sollid and Solberg (43) measured 23-mm-thick homogenized Atlantic farmed salmon (*Salmo salar*), using a NIR transmittance instrument (Tecator Infratec Food and Feed Analyzer, Höganäs, Sweden) in the 850- to 1050-nm range. They reported a prediction error of 0.7% for fat. Wold et al. (54) reported similar prediction error results measuring NIR transmittance spectra of cuts from Atlantic farmed salmon. Ground (2-mm hole diameter grinder) fillets from 50 farmed Atlantic salmon (*Salmo salar*) from 4 to 5 kg were measured in NIR reflectance in the region of 400–2500 nm (NIRSystems 6500, NIRSystems Inc.) (24). The main chemical constituents ranged from 9.1 to 20.5% for fat, 59.9–70.9% for moisture, and 18.6–20.9% for protein. The cross-validated prediction errors (using MSC and PCR) were down to 0.66% for fat, 0.38% for moisture, and 0.20% for protein in the 760- to 1100-nm wavelengths range. The short-wavelength range (760–1100 nm) gave lower prediction errors compared to the longer-wavelength region (1100–2500 nm). This is probably due to differences in penetration depths, giving different pathlengths.

Nondestructive NIR analysis of whole fillets from farmed Atlantic salmon (*Salmo salar*) was performed with a fiber-optic probe in which the detector is located in the probe head (NIRSystems no. NR-6539, NIRSystems 6500, NIRSystems Inc.) (24). This arrangement allows measurements from 400 to 2500 nm. For further details about the samples, see Isaksson et al. (24). NIR spectra were measured at seven different locations on the inside of the fillets. The reference analyses were performed on whole ground fillets. Prediction errors were similar for NIR measurements at the different locations on the fillets, ranging from 1.4% to 1.8% for fat and from 1.2% to 1.5% for moisture with the 760- to 1100-nm region. As expected, average NIR spectra from all measurements gave the lowest prediction errors, namely, 1.1% for fat and 0.85% for moisture. The study did suggest an optimal measurement location for fat or moisture in whole fillets.

Nondestructive NIR analysis of whole intact fish was first reported by Lee et al. (28). They measured 52 frozen and thawed rainbow trout (*Oncorhynchus mykiss*) in the 66.5- to 883-g range. A bifurcated fiber-optic probe, connected to a scanning (700–1050 nm) instrument, was used. The fiber-optic reflectance measurements were done through the skin of the intact fish at three different locations. The fat range (recalculated from dry weight to wet weight, assuming 70% moisture) was 2.0–13%. The prediction errors ranged from 0.7% to 2.2% for fat, giving correlation coefficients between 0.73 and 0.90, varying with the measurement locations on the fish body. Whole farmed Atlantic salmon (*Salmo salar*) were measured both postmortem (7, 42, 53) and live (42). Downey (7) collected salmon from retail stores over a 1-year period. The salmon were measured through the skin and scales by NIR at several sites, normally six sites along the dorsal surface, six sites on each side, and six sites along the ventral surface, giving 294 NIR measurements in total. NIR spectra were recorded with a NIRSystems 6500 spectrophotometer equipped with a surface interactance fiber-optic probe, in the wavelength range 400–1100 nm. The sites were excised for moisture and oil chemical reference analysis. The data were divided into a calibration and a validation set with a selection algorithm. Both dorsal and ventral calibrations were validated because of large spectral differences. The prediction error (SEP) results for the dorsal measurements were 1.45–1.75% for moisture ($R = 0.79$–$0.85$) and 1.89–2.16% for oil ($R = 0.79$–$0.84$), varying for different preprocessing and regression methods. The corresponding values for the ventral measurements were 1.90–2.79% for moisture ($R = 0.73$–$0.89$) and 2.41–3.74% for oil ($R = 0.62$–$0.86$). Wold and Isaksson (53) reported NIR measurements of 49 salmon, each salmon was measured at one site only, about 1 cm behind the dorsal fin, midway on the epaxial part. The same instrumental setup as that of Downey (7) was used, measuring through the skin and scales. The reference analysis was done on a homogenized 8-cm cross cut at the NIR measurement region. The cross-validated prediction error results were 1.12% ($R = 0.87$) for fat and 0.98% ($R = 0.86$) for moisture.

NIR measurements on 100 live salmon were compared to postrigor measurements (42). The live salmon were anesthetized by placing the salmon in 12°C seawater with 0.02% $m$-amino-benzoic-acid-ethyl-ester-methanesulfonate. The salmons were measured both with a fiber-optic probe as described above and with a noncontact, diffuse reflectance fixed grating diode array spectrophotometer (DA 7000 Flexi-mode, Perten Instruments, Huddinge, Sweden). The diode array instrument scanned from 400 to 1700 nm in 5-nm steps. The salmon were measured at one spot as described above (53). The range of the salmon was 0.73–10.4 kg carcass weight and 8.2–23.2% fat. The cross-validated prediction error results were very similar for both instruments, namely, 1.4% ($R = 0.90$) fat for live salmon and 1.3–1.5% ($R = 0.88$–$0.91$) fat for postrigor salmon (Fig. 7.3.11).

The use of NIR to determine other quality characteristics for different kinds of fish has also been reported. Atlantic halibut (*Hippoglossus hippoglossus*) fillets in the range of 1.0–12.3% fat, 16.5–27.4% protein and 23.4–33.5% dry matter gave cross-validated prediction errors of 0.27% for fat, 0.52% for protein, and 0.42% for dry matter (34). Studies have also been performed on European sea bass (*Dicentrarchus labrax L.*, Ref. 55). Free fatty acid determination of fish oil and mackerel quality has been reported (56). Minced raw fish from different species, such as mackerel (*Scomber scombrus*), herring (*Clupea harangus*), salmon (*Salmo salar*), blue whiting (*Micromesistius poutassau*), and other species gave prediction errors of 0.39% ($R = 0.99$) for moisture and 0.80% ($R = 0.98$) for oil. Determination of NaCl content in cured salmon roe was predicted with a correlation coefficient of up to 0.93 (17). NaCl content determination in cold smoked Atlantic salmon was predicted with a correlation coefficient of up to 0.91 (18, 19).

## Texture

As for beef, the texture of salmon is important for the consumer. Fifty-three farmed Atlantic salmons were measured on the inside of the fillets, using a NIR fiber-optic probe as described above (53) and a texture analyzer (26). The texture was measured by placing a $20 \times 20 \times 80$-mm$^3$ muscle in a five-blade Kramer shear force cell (HDP/KS5, Stable Micro systems, Surrey, UK) in duplicates. The area under the profile from the

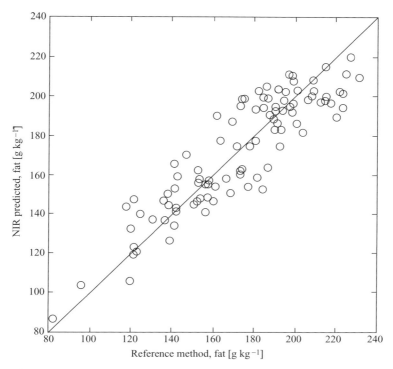

**Figure 7.3.11.** NIR predicted versus measured fat content in live Atlantic farmed salmon (from Ref. 42).

start to the maximum peak height was divided by the mass of the muscle cut, giving J/g units. The 53 salmon spanned a texture range of $2.16$–$7.33 \times 10^{-2}$ J/g and $1.48$–$2.24 \times 10^{-2}$ J/g for prerigor and postrigor fish, respectively. The cross-validated prediction errors were $0.88 \, (R = 0.76)$, $0.17 \, (R = 0.68)$, and $0.71 \, (R = 0.94) \times 10^{-2}$ J/g for prerigor, postrigor, and combined prerigor and postrigor fish, respectively. It was not possible to predict postrigor texture from a prerigor NIR calibration. These correlation coefficients are similar to NIR predictions of texture in beef.

## CONCLUSIONS

A wide range of NIR applications have been reported for meat and fish products. NIR spectroscopy is widely used in the meat and fish processing industries to determine different quality attributes. NIR spectroscopy gives good prediction results for a large variety of quality characteristics and can be used off-, at-, on-, and in-line. NIR spectroscopy can be used on both live and postmortem fish.

## REFERENCES

1. T. Arnalds, T. Fearn, G. Downey. Meat speciation using an hierarchical approach and logistic regression. *Proceedings of the 10th International Conference on Near Infrared Spectroscopy*, A. M. C. Davies, R. K. Cho, eds. NIR Publications, UK, 2002, p. 141–144.

2. H. Berg, K. Kolar. Evaluation of rapid moisture, fat, protein and hydroxyproline determination in beef and pork using the Infratec Food and Feed Analyser. *Fleischwirtschaft* **71**:787–789, 1991.

3. C. E. Byrne, G. Downey, D. J. Troy, D. J. Buckley. Non-destructive prediction of selected quality attributes of beef by near-infrared reflectance spectroscopy between 750 and 1098 nm. *Meat Sci* **49**:399–409, 1998.

4. D.I. Cazzolino, I. Murray, R. Paterson. Visable and near infrared spectroscopy of beef longissimus dorsi muscle as a means of discriminating between pasture and corn silage feeding regimes. *J Near Infrared Spectrosc* **10**:187–193, 2002.

5. G.B. Cornish. NIR analysis of sausage meat. In: *Proceedings of the Third International Conference on Near Infrared Spectroscopy*. ISBN 2/87286-001-0, 1990, p. 194–200.

6. J.L. DeBouver, B.G. Cottyn, L.O. Fiems, C.V. Boucque. Determination of chemical composition of beef by NIRS. In: *Bridging the Gap Between Data Analysis and NIR Applications*, K.I. Hildrum, T. Isaksson, T. Næs, A. Tandberg, eds. Ellis Horwood, 1992, p. 339–344.

7. G. Downey. Non-invasive and non-destructive percutaneous analysis of farmed salmon flesh by near infra-red spectroscopy. *Food Chem* **55**:305–311, 1996.

8. G. Downey, D. Beuchêne. Discrimination between fresh and frozen-then-thawed beef m. longissimus dorsi by combined visable-near infrared reflectance spectroscopy: a feasibility study. *Meat Sci* **45(3)**:353–363, 1997.

9. G. Downey, D. Beuchêne. Authentication of fresh vs frozen-then-thawed beef by near infrared reflectance spectroscopy of dried drip juice. Lebensmittel. *Wissenschaft Technol* **30**:721–726, 1997.

10. M. R. Ellekjær, T. Isaksson. Assessment of maximum cooking temperatures in previous heat treated meat. Part 1: near infrared spectroscopy. *J Sci Food Agric* **59**:335–343, 1992.

11. M. R. Ellekjær, T. Næs, K. I. Hildrum, T. Isaksson. Determination of NaCl of sausages by near infrared spectroscopy. *J Near Infrared Spectrosc* **1**:62–75, 1993.

12. M.R. Ellekjær, T. Isaksson, R. Solheim. Assessment of sensory quality of meat sausages using near infrared spectroscopy. *J Food Sci* **59(3)**:456–464, 1994.

13. P. Geladi, D. McDougall, H. Martens. Linearisation and scatter correction for near infrared reflectance spectra for meat. *Appl Spectrosc* **39**:491–500, 1985.

14. B. Gjerde, H. Martens. Predicting carcass composition of rainbow trout by near-infrared reflectance spectroscopy. *J Anim Breed Genet* **104**:137–148, 1987.

15. K.I. Hildrum, B.N. Nilsen, M. Mielnik, T. Næs. Prediction of sensory characteristics of beef by near-infrared spectroscopy. *Meat Sci* **38**:67–80, 1994.

16. K.I. Hildrum, T. Isaksson, T. Næs, B.N. Nilsen, P. Lea. Near infrared reflectance spectroscopy in the prediction of sensory properties of beef. *J Near Infrared Spectrosc* **3**:81–87, 1995.

17. Y. Huang, T.M. Rogers, M.A. Wenz, A.G. Cavinato, D.M. Mayes, G.E. Bledsoe, B.A. Rasco. Detection of sodium chloride in cured salmon by SW-NIR spectroscopy. *J Agric Food Chem* **49**:4161–4167, 2001.

18. Y. Huang, A.G. Cavinato, D.M. Mayes, G.E. Bledsoe, B.A. Rasco. Nondestructive prediction of moisture and sodium chloride in cold smoked Atlantic Salmon (*Salmo salar*). *J Food Sci* **67**:2543–2547, 2002.

19. Y. Huang, A.G. Cavinato, L.J. Mayes, D.M. Kangas, G.E. Bledsoe, B.A. Rasco. Nondestructive prediction of moisture and sodium chloride in cured Atlantic Salmon (*Salmo salar*) (Teijin) using short-wavelength near-infrared spectroscopy (SW-NIR). *J Food Sci* **68**:482–486, 2003.

20. T. Isaksson, M. H. R. Ellekjær, K. I. Hildrum. Determination of the previous maximum temperature of heat treated minced meat by near infrared reflectance spectroscopy. *J Sci Food Agric* **49**:385–387, 1989.

21. T. Isaksson, K.I. Hildrum. Near infrared transmittance (NIT) analysis of meat products. In: *Proceeding to the Third International Conference on Near Infrared Spectroscopy*. ISBN 2/87286-001-0, 1990, p. 202–206.

22. T. Isaksson, B. N. Nilsen. Noninvasive prediction of protein, fat and water in homogenized beef, vacuum packed in plastic bags. Proc. 5th Int. Conf. on Near Infrared Spectroscopy (Haugesund, Norway, 16-20 June 1992). ISBN 0-13-617416-7, 1992, p. 359-364.

23. T. Isaksson, C. E. Miller, T. Næs. Nondestructive NIR and NIT determination of protein, fat and water in plastic-wrapped, homogenized meat. *Appl Spectrosc* **46**:1685–1694, 1992.

24. T. Isaksson, G. Tøgersen, A. Iversen, K. I. Hildrum. Nondestructive determination of fat, moisture and protein in salmon filets by use of near infrared diffuse spectroscopy. *J Sci Food Agric* **69**:95–100, 1995.

25. T. Isaksson, B.N. Nilsen, G. Tøgersen, R.P Hammond, K.I. Hildrum. On-line, proximate analysis of ground beef directly at a meat grinder outlet. *Meat Sci* **43**:245–253, 1996.

26. T. Isaksson, L.P. Swendsen, R. Taylor, S.O. Fjæra, P.O. Skjervold. Non-destructive texture analysis of farmed Atlantic salmon using visual/near- infrared reflectance spectroscopy. *J Sci Food Agric* **82**:53–60, 2001.

27. E. Lanza. Determination of moisture, protein, fat , and calories in raw pork and beef by near infrared spectroscopy. *J Food Sci* **48**:471–474, 1983.

28. M.H. Lee, A.G. Calvanato, D.M. Mayes, B.A. Rasco. Non-invasive short-wavelength near-infrared spectroscopic method to estimate thr crude lipid content in muscle of intact rainbow trout. *J Agric Food Chem* **40**:2176–2181, 1992.

29. Y.L. Lui, B.G. Lyon, W.R. Windham, C.E. Realini, T.D.D. Pringle, S. Duckett. Prediction of color, texture, and sensory characteristics of beef steaks by visual and near infrared reflectance spectroscopy. A feasibility study. *Meat Sci* **65(3)**: 1107–1115, 2003.

30. B.G. Lyon, W.R. Windham, C.E. Lyon, F.E. Barton. Sensory characteristics and near-infrared spectroscopy of broiler breasts for various chill-storage regimes. *J Food Quality* **24(5)**:435–452, 2001.

31. J.A. Mathias, P.C. Williams, D.C. Sobering. The determination of lipid and protein in freshwater fish using near-infrared reflectance spectroscopy. *Aquaculture* **61**:303–311, 1987.

32. J. McElhinney, G. Downey, T. Fearn. Chemometric processing of visible and near infrared reflectance spectra for species identification in selected raw homogenised meats. *J Near Infrared Spectrosc* **7**:145–154, 1999.

33. M. Mitsumoto, S. Maeda, T. Mitsuhashi, S. Ozawa. Near-infrared spectroscopy determination of physical and chemical characteristics in beef cuts. *J Food Sci* **56**:1493–1496, 1991.

34. R. Nortvedt, O.J. Torrisen, S. Tuene. Application of near-infrared transmittance spectroscopy in the determination of fat, protein and dry matter in Atlantic halibut fillet. *Chemometrics Intell Lab Syst* **42**:199–207, 1998.

35. T. Næs, K.I. Hildrum. Comparison of multivariate calibration and discriminant analysis in evaluating NIR spectroscopy for determination meat tenderness. *Appl Spectrosc* **51**:350–357, 1997.

36. B. Park, Y.R. Chen, Hruschka, S.D. Shackelford, M. Kohmaraie. Near-infrared reflectance analysis for predicting beef Longissimus tenderness. *J Anim Sci* **76**:2115–2120, 1998.

37. H. Rannou, G. Downey. Discrimination of raw pork, chicken and turkey meat by spectroscopy in visible, near- and mid-infrared ranges. *Anal Commun* **34**:401–405, 1997.

38. B.A. Rasco, C.E. Miller, T.L. King. Utilization of NIR spectroscopy to estimate the proximate composition of trout muscle with minimal sample pretreatment. *J Agric Food Chem* **39**:67–72, 1991.

39. A. Ripoche, A.S. Guillard. Determination of fatty acid composition of pork fat by Fourier transform infrared spectroscopy. *Meat Sc* **58**:299–304, 2001.

40. R. Rødbotten, B.N. Nilsen, K.I. Hildrum. Prediction of beef quality attributes from early post mortem near infrared reflectance spectra. *Food Chem* **69**:427–436, 2000.

41. R. Rødbotten, B.H. Mevik, K.I. Hildrum. Prediction and classification of tenderness in beef from non-invasive diode array NIR spectra. *J Near Infrared Spectrosc* **9**:199–210, 2001.

42. C. Solberg, E. Saugen, L.P. Swensen, T. Bruun, T. Isaksson. Determination of fat in live farmed Atlantic salmon using non-invasive NIR techniques. *J Sci Food Agric* **83**:692–696, 2003.

43. H. Sollid, C. Solberg. Salmon fat content estimation by near infrared transmission spectroscopy. *J Food Sci* **57**:792–793, 1992.

44. K. Thyholt, T. Isaksson. Differentiation of frozen and unfrozen beef using near-infrared spectroscopy. *J Sci Food Agric* **73**:525–532, 1997.

45. K. Thyholt, U.G. Indahl, K.I. Hildrum, M.R. Ellekjær, T. Isaksson. Meat speciation by near infrared reflectance spectroscopy on dry extracts. *J Near Infrared Spectrosc* **5**:195–208, 1997.

46. K. Thyholt, G. Enersen, T. Isaksson. Determination of endpoint temperatures in previously heat treated beef using reflectance spectroscopy. *Meat Sci* **48**:49–63, 1998.

47. G. Tøgersen, T. Isaksson, B.N. Nilsen, E.A. Bakken, K.I. Hildrum. On-line NIR analysis of fat, water and protein in industrial scale ground meat batches. *Meat Sci* **51**:97–102, 1999.

48. G. Tøgersen, J.A. Arnesen, B.N. Nilsen, K.I. Hildrum. On-line prediction of chemical composition of semi-frozen ground beef by non-invasive NIR spectroscopy. *Meat Sci* **63(4)**:515–523, 2003.

49. E.V. Valdes, J.D. Summers. Determination of crude proteing and fat in carcass and breast muscle samples of poultry by near infrared reflectance spectroscopy. *Poultry Sci* **65**:485–490, 1986.

50. B.K. Watt, A.L. Merill. *Agricultural handbook No. 8*. U.S. Dept. of Agriculture, Washington, DC, US, 1975.

51. W.R. Windham, K.C. Lawrence, P.W. Feldner. Prediction of fat content in poultry meat by near-infrared transmission analysis. *J Appl Poultry Res* **12(1)**:69–73, 2003.

52. W.R. Windham, W.H. Morrison. Prediction of fatty acid content in beef neck lean by near infrared reflectance analysis. *J Near Infrared Spectrosc* **6**:229–234, 1998.

53. J. P. Wold, T. Isaksson. Non-destructive determination of fat and moisture in whole Atlantic salmon by near-infrared diffuse spectroscopy. *J Food Sci* **62**:734–736, 1997.

54. J.P. Wold, T. Jakobsen, L. Krane. Atlantic salmon average fat content estimated by near-infrared transmittance spectroscopy. *J Food Sci* **61**:74–77, 1996.

55. G. Xiccato, A. Trocin, F. Tulli, E. Tibaldi. Prediction of chemical composition and origion identification of European sea bass (*Dicentrarchus labrax L.*) by near infrared reflectance spectroscopy (NIRS). *Food Chem* **86**:275–281, 2004.

56. H.Z. Zhang, T.C. Lee. Rapid near-infrared spectroscopic method for the determination of free fatty acid in fish and its application in fish quality assessment. *J Agric Food Chem* **45**:3515–3521, 1997.

# CHAPTER 8

# APPLICATIONS TO FOODSTUFFS

# Flours and Breads

BRIAN G. OSBORNE

## INTRODUCTION

The flour milling and baking industry was one of the earliest adopters of NIR technology. This is not surprising because ground wheat and flour are near-ideal sample types for diffuse reflectance measurements. In addition, the need for rapid quality testing of wheat, flour, dough, and final products provided the motivation for the capital investment in NIR instruments. This investment, based on the application of NIR to the analysis of wheat for protein and moisture, prompted research into additional applications across a range of sample types. This chapter reviews the resulting varied and well-established applications of NIR to the analysis of wheat, flour, and bread. Particular emphasis is placed on on-line applications and their use in control systems.

## FLOUR MILLING

After the introduction of NIR for the determination of the protein content of wheat for payment, the flour milling industry was one of the first to adopt the same technology for quality testing of its main product (flour). A simplified diagram showing each of the main stages of the flour milling process and indicating the opportunities for NIR analysis is shown in Figure 8.1.1.

Figure 8.1.1 shows that the first point of NIR analysis in a typical mill is for the testing of wheat at intake. This was the first application of NIR in the flour milling industry. In addition, automated systems based on whole-grain NIR are available for the control of wheat blending. The next opportunity for a control system would be in the reduction system. Here, an NIR measurement of flour starch damage would provide the opportunity (not yet realized) to implement a feedback control of the

*Near-Infrared Spectroscopy in Food Science and Technology*, Edited by Yukihiro Ozaki, W. Fred McClure, and Alfred A. Christy.

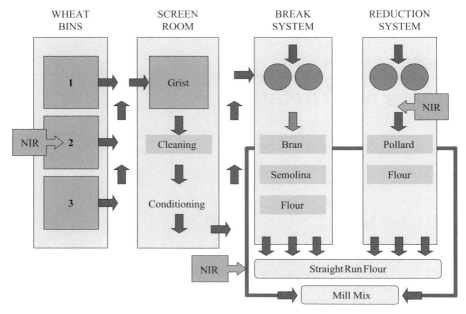

**Figure 8.1.1.** Schematic diagram of the flour milling process showing opportunities for NIR analysis.

reduction roll pressures. Finally, NIR is used on-line for the control of gluten addition and flour blending, and off-line for quality control of flour batches before shipment.

## WHEAT PURCHASE AND BLENDING

Wheat must be tested for quality to enable decisions to be made about acceptance, price, and binning. The key measurements are the protein and moisture contents and hardness.

Since the 1970s, NIR spectroscopy has provided the means to test wheat at delivery to handlers, merchants, or direct to the mill. For ground wheat, protein may be determined with three or four fixed filters: 2180, 2100 (1940 required for corrected moisture basis), and either 1680 or 2230 nm. In countries with wheat grading systems, such as Australia, Canada, and the US, segregation is carried out on the basis of load-by-load NIR testing of growers' deliveries at country silos. Nowadays, this is almost exclusively by whole grain instruments, the most popular of which is the FOSS Infratec. This type of instrument is calibrated with tens of thousands of samples and development of a model based on artificial neural networks (8, 9).

National Standard M8 "Pattern Approval Specifications for Protein Measuring Instruments for Grain," a legal standard for the measurement of grain protein content, has been introduced in Australia. It covers measurements "in use for trade," meaning those used to determine the consideration of a transaction—which includes determination of both the quantity of the commodity (the usual trade measurement focus) and the unit price. The Standard has two parts: Pattern Approval and Verification. Pattern Approval determines whether an instrument is capable of retaining its calibration over a range of environmental and operating conditions and ensures that the instrument is not capable of facilitating fraud. Verification is the process whereby individual instruments are examined to ensure that they meet the maximum permissible error (MPE). The MPE is the value of the maximum difference between the measurement of protein on a given instrument and the reference value, determined according to the Dumas method, for any sample used to verify the performance of that instrument. The MPEs are 0.4% for wheat and 0.5% for barley. Assuming a normal distribution of errors, these are equivalent to standard deviations of differences of 0.13% and 0.16%. An OIML Technical Committee is now considering international harmonisation of Standard M8 to smooth the introduction of legal standards around the world.

Moisture is routinely measured concurrently with protein. Measurement may be made on ground wheat with two or three fixed filters (1940, 2230, and/or 2310 nm) or on whole grain as described above. Under ideal laboratory conditions, the standard deviation of differences between NIR and reference moisture values should be 0.25% or better. It is essential to record the NIR spectral data and determine the oven moisture contents of the calibration samples as close as possible in time, ideally on the same day.

Wheat hardness has been measured in both meal (62) and whole grain (65). Because there is no international standard reference method for wheat hardness, a definition based on NIR spectra of ground grain was proposed (33). Subsequently, an algorithm was developed to correct the NIR hardness index for the effects of differences in moisture content (68). The NIR hardness index has also been found to be affected by protein content and growing season (7). In response to this situation, a systematic study was undertaken to define an NIR hardness index based on spectra of ground or whole grain that was insensitive to other factors (27–29). Hardness can also be measured on whole grain by transmittance in the region of 800 to 1050 nm or by reflectance in the region of 800 to 2500 nm. The standard deviation of differences for whole grain is of the same order of magnitude as that for ground grain (46, 65).

An empirical relationship can be established between the NIR spectrum of ground or whole wheat and the yield of flour that is obtained by laboratory milling. Monochromator or diode array instruments covering all or part of the region of 800 to 2500 nm have been used. The results of independent studies suggest that a standard deviation of differences between NIR and laboratory milling of approximately 1.2–1.4% for whole grain (6, 11–13) may be expected. Flour yield assessment is incomplete without taking account of the purity of the flour produced at a given yield. The CIE color space is often used for this purpose and the most useful properties are Minolta L* and b*. Minolta b* has been measured on whole grain to a standard deviation of differences between NIR and Minolta of 0.15 when NIR measurements were calibrated with reference Minolta measurements made on a flour-water paste (6) and 0.58 for dry Minolta reference values (11).

The results of independent studies based on the use of a monochromator in reflectance mode over the range of 1100 to 2500 nm suggest that a standard deviation of differences between NIR and Brabender Farinograph of 1.4–1.7% for whole grain (11, 46). Crosbie et al. (11–13) carried out a detailed validation that showed that the NIR calibration could predict water absorption beyond the effects of protein content and hardness, both known to exert a strong effect on the spectra. This was achieved by applying the calibration to a subset of samples for which water absorption had a low correlation with PSI and flour protein.

The first attempt to test wheat directly for dough strength and extensibility was described by Williams et al. (64) for ground wheat and later applied to whole wheat (6, 11, 46). Standard deviations of differences were typically of the order of 85 BU for maximum resistance and 1.5 cm for extensibility. However, when calibrations were put into practice, it was observed that there was a strong influence of environment on their performance (10). This is most likely due to the fact that extensibility is highly correlated with protein content. Therefore, if NIR calibrations for extensibility measure little or nothing beyond protein content the calibration will fail when the intercorrelation changes.

Glutenin and gliadin, the main fractions of functional protein in wheat, have some differences in their NIR spectra that enable them to be determined in mixtures with starch (60). Total glutenin, insoluble glutenin, and gliadin contents can also be measured in whole wheat by NIR when calibrated against HPLC (14, 52, 60). Extensibility is highly correlated with protein content, and this accounts for a high proportion of the correlation between NIR and extensibility. However, the measurement of glutenin and gliadin has been shown to account for an increased correlation between NIR and extensibility. Currently, this application requires a monochromator in reflectance mode over the range of 2000 to 2300 nm. The calibration accuracy is limited by the inherent variability in the Brabender Extensograph method. However, NIR can provide a means to rank wheat or flour for glutenin and gliadin contents independently of protein content and hence seasonal influences (58).

The key wheat starch components are amylose and amylopectin, the ratio of which is a key determinant of noodle quality. These are chains of glucose units that are essentially linear (amylose) or highly branched (amylopectin). Because the components contain the same glucose units, it is not surprising that their NIR spectra are very similar. However, the major difficulty in this case is that the reference method is not sufficiently accurate to enable the calculation of master curves. However, a workable calibration has been achieved with the default algorithm, partial least squares (59).

Recent progress in the prediction of flour yield, color, water absorption, and protein functionality from whole wheat NIR testing may provide millers with new blending opportunities (4).

## ANALYZING WHOLE GRAINS ON-LINE

NIR analysis of whole grain can also be accomplished on-line so as to monitor continuously the composition of the grist. An example of this type of equipment is the Infratec 1725 On Line Grain Analyzer. The optical component of the Infratec 1725 is identical to that of the laboratory Infratec 1225, and the same calibrations

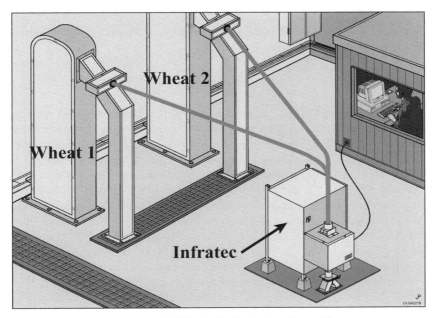

**Figure 8.1.2.** Infratec 1725 Whole Grain Analyzer in a mill screenroom.

can be employed. In the mill screenroom, wheat is sampled from two or more bins during blending and fed by gravity into the sampling station where the NIR analysis takes place. Figure 8.1.2 shows an example in which two wheat streams are being blended under the control of an Infratec. The system is controlled by Tecator Online Monitoring Software (TOMS). The target value and control limits are set for each constituent of interest; then TOMS will trigger an alarm if a predicted value for any of the constituents is outside the specified limits. The operator monitoring the system from the office can then take remedial action.

## FLOUR QUALITY TESTING

An essential aspect of mill monitoring and control is the maintenance of product quality, which can be accomplished by NIR in the laboratory, at-line or on-line. On-line NIR technology has enabled the analysis of intermediate and finished flours to be performed continuously. A discussion of commercially available NIR on-line samplers can be found in Sugden & Osborne (53). By careful selection of the sampling points, the miller can determine when and where adjustments should be made to ensure consistency of product quality. Many mills incorporate continuous flour blending to tailor-make products to each customer's order with a few base straight-run flours. This provides maximum flexibility in responding to customer requirements. The mixing is controlled by computerized loss-in-weight feeders. Any on-line NIR sampler can be used to monitor both the blending and outloading of the final product to provide quality assurance data (42).

## Protein

The oldest and best established NIR application is protein in ground wheat and flour. It is based on the absorption of NIR energy at specific wavelengths by peptide linkages between amino acids of protein molecules and at reference wavelengths. Inclusion of a measurement at a wavelength corresponding to an absorption by water enables the result to be corrected automatically to a standard moisture basis. The minimum instrument requirement is three or four fixed filters: 2180, 2100 (1940 required for corrected moisture basis), and either 1680 or 2230 nm. Under ideal laboratory conditions the standard deviation of differences between NIR and reference protein values should be 0.25% or better (1). A typical plot of Kjeldahl versus NIR flour protein, based on unpublished data, is shown in Figure 8.1.3. Large fluctuations in sample temperature require a correction factor of the order of +0.02% per °C.

Protein in flour can also be measured in the wavelength range of 800–1050 nm by means of the Flour Module accessory for the Infratec 1241 Whole Grain Analyzer (47).

NIR on-line samplers were first developed to measure the protein content of flour. This is still the most popular application and provides an excellent example of an NIR feedback control system. Dried gluten is commonly used, particularly in Europe, to replace wholly or partly the protein in flour that would otherwise be derived from high-protein wheat in the grist. The success of on-line NIR for monitoring flour protein content therefore led to its incorporation into a closed-loop control system

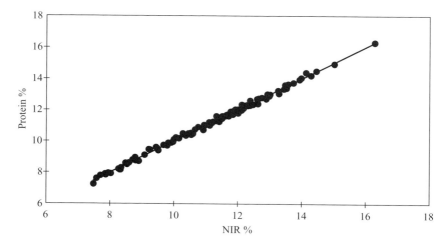

**Figure 8.1.3.** Plot of Kjeldahl versus NIR flour protein.

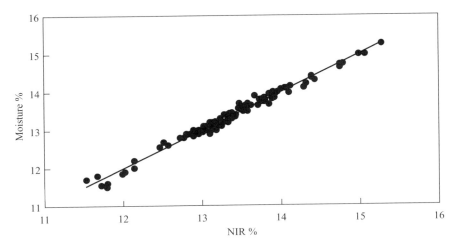

**Figure 8.1.4.** Plot of oven versus NIR flour moisture.

for gluten addition to flour (19). A mixing screw is installed between the gluten feed and the NIR sampler station from which a feedback signal controls the gluten feeder. The system has proved to be an efficient and accurate method of control of gluten addition to achieve a target protein content in the flour. The results of performance trials conducted with the NIROS system (the prototype of the Satake Nirotec SNIB) and the NDC Infrared Engineering MM55 gauge in two UK flour mills showed that with either system the protein levels of gluten-supplemented flours could be controlled to a standard deviation of less than 0.1%.

## Moisture

Measurement of the absorption of NIR energy at specific wavelengths by hydroxyl linkages in water and at reference wavelengths in flour requires two or three fixed filters: 1940, 2230 and/or 2310 nm or in the wavelength range of 800–1050 nm by means of the Flour Module accessory for the Infratec 1241 Whole Grain Analyzer. Under ideal laboratory conditions the standard deviation of differences between NIR and reference moisture values should be 0.20% or better (41). A typical plot of oven versus NIR flour moisture is shown in Figure 8.1.4.

## Particle Size (Granularity)

Particle size of flour can be measured by using only the first principal component calculated from its NIR spectrum (29). A plot of Malvern versus NIR flour particle size is shown in Figure 8.1.5.

Granulation is a particle size distribution in cumulative percentage of mass. It is used as a basis for information about roll flute condition, gap setting and sifter

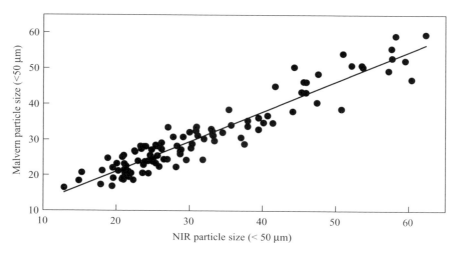

**Figure 8.1.5.** Plot of Malvern versus NIR particle size.

efficiency. Pasikatan et al. (43–45) have proposed an NIR on-line granulation sensor as a means of roller mill automation. The concept is to develop calibrations for the cumulative mass of sieved fractions to produce an NIR granulation curve as a basis for roll gap optimization. Initially, models were developed with six wheat classes but the models were later improved by subdividing into Hard Red Winter and Soft Red Winter subgroups.

## Starch Damage

Mechanical shear energy imparted by flour roller mills causes damage to the starch granules in the flour. This process is accompanied by cleavage of hydrogen bonds between starch molecules and between starch and bound water. Absorption of NIR energy in the region 1100 to 2500 nm due to free and hydrogen-bonded hydroxyl bonds in starch and water, hence starch damage, can be measured with a monochromator in reflectance mode over the range 1100 to 2500 nm (35). A standard deviation of differences between NIR and AACC Method 76–31 (Megazyme) of 0.4% is achievable as shown in Figure 8.1.6 (36). However, sample temperature must be controlled to within ±0.5°C or a correction factor applied.

## Water Absorption

The results of several independent studies based on the use of a monochromator in reflectance mode over the range 1100 to 2500 nm suggest that a standard deviation of differences between NIR and Brabender Farinograph of approximately 1% may be expected for flour (6, 15, 40, 42).

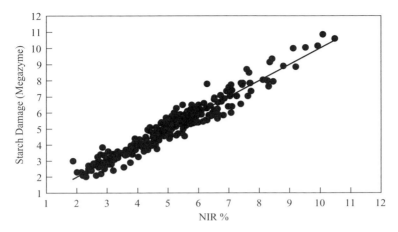

**Figure 8.1.6.** Plot of Megazyme versus NIR flour starch damage.

## Ash

There have been a number of reports in the literature of the measurement of ash content of flour by NIR, and most instrument manufacturers offer this application. There is complete agreement among experts that ash cannot be measured directly by NIR, and therefore indirect measurement has been proposed via cellulose (61) or oil and particle size (25, 66). Experience in commercial mills has shown that ash calibrations are matrix sensitive and frequent recalibration is required as the grist is changed. AACC Method 08–21 is based on the Perten Instruments 86 "Ash" Series instruments, based on patented wavelengths that measure color (50).

## Loaf Volume/Quality

There have been some reports in the literature of NIR calibrations for loaf volume or height but none for loaf quality. Rubenthaler & Pomeranz (51) and Williams et al. (64) obtained standard deviation of differences between NIR measurements on flour and loaf volume of 48 and 30 ml, respectively, whereas Delwiche and Weaver (15) reported a standard deviation of 0.11 cm for loaf height. These calibrations do not appear to have been adopted in routine use because the emphasis in more recent research has shifted to the measurement of biochemical parameters (5, 58).

## Qualitative Analysis

Rather than developing quantitative calibrations for a range of quality parameters, some researchers have proposed qualitative analysis to detect flours that are out of specification. Nielsen et al. (32) employed a new discriminant approach to the identification of flours that are out of specification with respect either to particle size or composition. Similarly, Gergely et al. (21) have used NIR discriminant analysis (polar qualification system) to separate, in quality space, flour fractions produced by two different mills. The discrimination focussed on NIR wavelengths around 1700 and 2300 nm with the aim of distinguishing starch-rich and bran-rich fractions. Kumagai et al. (26) have reported the discrimination of 15 kinds of noodle flour with principal components analysis. Williams and Dexter (63) recorded the standard deviation of optical density at 1978 nm as a measure of efficiency of flour mixing.

## DOUGH

Dough mixing is a critical stage of the breadmaking process. During mixing, energy is imparted to the dough to develop the gluten matrix. This matrix enables the gas produced during fermentation to be retained during proving and baking. Poor mixing is one of the major causes of bread faults. Craft bakers are able to judge optimum dough development by sight and touch, but in modern industrial bakeries an automated approach is required. The availability of diode array instruments, such as the Perten DA-7000, enabled the development of dynamic NIR spectroscopy and its patented application to the determination of the point of optimum dough mixing (57).

In its original implementation, shown in Figure 8.1.7, the DA-7000 instrument was placed above the mixer and reflected radiation from the moving dough mass was measured (56). The NIR absorption at 1160 nm decreases to the point of optimum

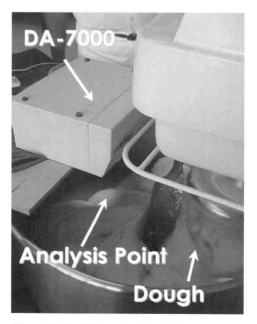

**Figure 8.1.7.** NIR monitoring of dough mixing.

dough mixing, as the water in the dough is bound to the protein and starch fractions and then increases as excess water is released. This provides a direct chemical measurement of the process that is independent of mixer design and speed. It has been shown to be consistent with the changes to the quality of the bread baked at different stages of mixing. Wesley et al. (55) have also employed a fiber-optic interactance probe to make the measurement in a 35-g Mixograph. Osborne et al. (39) demonstrated the use of a single instrument for the rapid quality testing of whole wheat, flour, and dough.

The use of NIR for real-time monitoring of dough mixing has been confirmed and extended by research groups in the UK (2, 3, 30, 31) and US (16, 23, 24, 48, 49). These groups have shown that the NIR method works for a variety of full-formula doughs in commercial-scale mixers and can be used for research into the relationships between the NIR mixing curves, flour quality, ingredient interactions, flour protein functionality, and final bread quality.

## BREAD

NIR has been used for the compositional analysis of bread. Calibrations have been reported for protein, fat, moisture, starch, and total sugars (22, 34, 38, 54). Two different approaches to sample presentation have been used: 1) drying and grinding the bread and 2) placing intact slices, cut into a disk of appropriate diameter using a pastry cutter, into a standard cell. There was no difference in the analytical performance of the NIR method with either approach. In two independent studies, calibrations for the analysis of dried bread (17) and bread disks (54) were found to be readily transferable between bread samples baked in different bakeries with widely different ingredients and processes.

If bread crumb is stored for a few hours or more, the process known as staling or retrogradation occurs. In the first report of the use of NIR for the study of reaction kinetics in food systems, second-derivative data at 1414 and 1465 nm were used to calculate rate constants for the retrogradation process in bread (37, 67). NIR spectra were recorded on disks sampled from bread slices taken from loaves during storage. A generic spectroscopic rationale for this application, as well as starch damage (36) and degree of cook (18) has been proposed by Osborne (35). Retrogradation occurs through the slow redevelopment of crystallinity in the amylopectin fraction of the starch. This crystallinity is a direct result of extensive hydrogen bonding, both intramolecularly and to water molecules, of the amylopectin molecules. It is these changes to the hydrogen bonding of the starch and water O-H bonds that are measured by the NIR spectra of the intact bread crumb (Fig. 8.1.8).

## SUMMARY

NIR is a mature technique for the rapid quality testing of whole wheat, ground wheat, and flour. Wheat protein, moisture, and hardness can be measured on ground grain

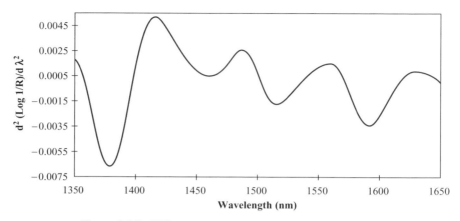

**Figure 8.1.8.** Difference spectrum for stored and fresh bread.

with a fixed-filter instrument or on whole grain. Indicative flour yield, color, water absorption, and protein functionality are also possible from whole grain NIR spectra recorded with a monochromator instrument. Flour protein, moisture, and particle size can be measured using the same fixed-filter instrument as for ground wheat, which makes this option popular with millers. The above flour measurements may be carried out on-line with fixed-filter instruments adapted for that purpose, but starch damage requires a monochromator instrument placed in a laboratory with accurate temperature control.

## REFERENCES

1. Anon. ICC Standard No. 159, Determination of protein by near infrared reflectance (NIR) spectroscopy, 1995.

2. J. M. Alava, S. J. Millar, S. E. Salmon. The determination of wheat breadmaking performance and bread dough mixing time by NIR spectroscopy for high speed mixers. *J Cereal Sci* **33**:71–81, 2001.

3. J. M. Alava, S. J. Millar, S. E. Salmon. The assessment of dough development during mixing using near infrared spectroscopy. *Spec Publ R Soc Chem* **261**: 439–441, 2001.

4. F. Bekes, O. Larroque, P. Hart, B. O'Riordan, D. Miskelly, M. Baczynski, B. Osborne, C. W. Wrigley. Non-linear behaviour of grain and flour blends made from dissimilar components. In: *Proceedings of Australian Cereal Chemistry Conference*, L. O'Brien, A. B. Blakeney, A. S. Ross, C. W. Wrigley, eds. RACI, Melbourne, 1998, p. 123–127.

5. D. G. Bhandari, S. J. Millar, C. N. G. Scotter. Prediction of wheat protein and HMW-glutenin contents by near infrared (NIR) spectroscopy. *Spec Publ R Soc Chem* **261**: 313–316, 2000.

6. C. K. Black, J. F. Panozzo. Whole grain quality evaluation in wheat based on near infrared spectroscopy. *Proceedings of 49th Australian Cereal Chemistry Conference*, J. F. Panozzo, M. Ratcliffe, M. Wootton, C. V. Wrigley, eds. RACI, Melbourne, 1999, p. 309–313.

7. G. L. Brown, P. S. Curtis, B. G. Osborne. Factors affecting the measurement of hardness by near infrared reflectance spectroscopy of ground wheat. *J Near Infrared Spectrosc* **1**: 147–152, 1993.

8. N. B. Buchmann, I. A. Cowe. Advantages of using artificial neural networks techniques for agricultural data. *Near Infrared Spectroscopy: Proceedings of the 10th International Conference*, A. M. C. Davies, R. K. Cho, eds. NIR Publications, Chichester, UK, 71–75, 2002.

9. N. B. Buchmann, H. Josefsson, I. A. Cowe. Performance of European artificial neural network (ANN) calibrations for moisture and protein in cereals using the Danish Near-Infrared Transmission (NIT) network. *Cereal Chem* **78**: 572–577, 2002.

10. P. M. Burridge, G. A. Palmer, G. J. Hollamby. An evaluation of the sensitivity of NIR calibrations to seasonal shifts in wheat quality. in *Proceedings of 45th Australian Cereal Chemistry Conference*, Y. A. Williams, C. A. Wrigley, eds. RACI, Melbourne, 1995, p. 275–276.

11. G. B. Crosbie, Y. Wang, B. G. Osborne, H. M. Allen, G. A. Palmer, C. Black, D. J. Mares. Collaborative development of NIR quality tests for application in wheat breeding. *Proceedings of 50th Australian Cereal Chemistry Conference*, M. Wootton, I. L. Batey, C. W. Wrigley, eds. RACI, Melbourne, 2000, p. 562–654.

12. G. B. Crosbie, Y. Wang, B. G. Osborne. Progress in the development of NIR calibrations for application in wheat breeding. *Proceedings of 51st Australian Cereal Chemistry Conference*, M. Wootton, I. L. Batey, C. W. Wrigley, eds. RACI, Melbourne, 2002, p. 100–104.

13. G. B. Crosbie, Y. Wang, B. G. Osborne, I. J. Wesley. Further development of NIR calibrations for application in wheat breeding. *Proceedings of 52*nd *Australian Cereal Chemistry Conference*, C. K. Black, J. F. Panozzo, C. W. Wrigley, I. L. Batey, N. Larsen, eds. RACI, Melbourne, 2002, p. 195–197.

14. S. R. Delwiche, R. A. Graybosch, C. J. Peterson. Predicting protein composition, biochemical properties, and dough-handling properties of hard red winter wheat flour by near-infrared reflectance. *Cereal Chem*, **75**: 412–416, 1998.

15. S. R. Delwiche, G. Weaver. Bread quality of wheat flour by near-infrared spectrophotometry: feasibility of modeling. *J Food Sci* **59**: 410–415, 1994.

16. R. E. Dempster, M. C. Olewnik, V. W. Smail. Determination of dough development using near-infrared reflectance. AACC Annual Meeting, Portland, Paper No. 39, 2003.

17. E. Duvenage. Measurement of fat and moisture in air-dried bread by near infrared spectrophotometry. *J Sci Food Agric* **37**: 384–386, 1986.

18. A. J. Evans, S. Huang, B. G Osborne, Z. Kotwal, I. J. Wesley. Near infrared on-line measurement of degree of cook in extrusion processing of wheat flour. *J Near Infrared Spectrosc* **7**: 77–84, 1999.

19. T. Fearn, P. I. Maris. An application of Box-Jenkins methodology to the control of gluten addition in a flour mill. *Appl Statist* **40**: 477–484, 1991.

20. P. L. Finney, J. E. Kinney, J. R. Donelson. Prediction of damaged starch in straight-grade flour by near-infrared reflectance analysis of whole ground wheat. *Cereal chem* **65**: 449–452, 1988.

21. S. Gergely, L. Handzel, A. Zoltan, A. Salgo. Near infrared spectroscopy – a tool for the evaluation of milling procedures. *Near Infrared Spectrocopy: Proceedings of 10th International Conference.* A. M. C. Davies, R. K. Cho, eds. NIR Publications, Chichester, UK, 2002, p. 33–37.

22. C. A. Groenewald. New applications of near infrared spectroscopy in the food industry. *Supplement to SA Food Review* 1984, 69–71.

23. W. N. Huang. Rheological dough and mixing behavior of food ingredients as studied by Labtron and NIR spectroscopy. Paper 28 AACC Annual Meeting, 14–18 October, Charlotte, NC, 2001, (*www.aaccnet.org/meetings/2001/abstracts*).

24. W. N. Huang, M. C. Olewnik, J. J. Psotka, R. E. Dempster. Measuring rheological dough and mixing properties of full formula ingredients and microingredients in real time using NIR. Paper 133 AACC Annual Meeting, 14–18 October, Charlotte, NC, 2001, (*www.aaccnet.org/meetings/2001/abstracts*).

25. M. Iwamoto, N. Kongseree, J. Uozomi, T. Suzuki. Determination of ash content in home-grown wheat flour in Japan by near-infrared diffuse reflectance analysis. *Nippon Shokuhin Kogyo Gakkaishi* **33**: 842–847, 1986.

26. M. Kumagai, K. Karube, T. Sato, N. Ohisa, T. Amano, R. Kikuchi, N. Ogawa. A near infrared spectroscopic discrimination of noodle flours using a principal-component analysis coupled with chemical information. *Anal Sci* **18**: 1145–1150, 2002.

27. M. Manley, A. E. J. McGill, B. G. Osborne. The effect of light scattering on NIR reflectance and transmittance spectra of wheat. *J Near Infrared Spectrosc* **2**: 93–99, 1994.

28. M. Manley, A. E. J. McGill, B. G. Osborne. Wheat hardness by NIR: new insights. in *Leaping Ahead with Near Infrared Spectroscopy*, G. D. Batten, P. C. Flinn, L. A. Welsh, A. B. Blakeney, eds. RACI, Melbourne, 1995, p. 178–180.

29. M. Manley, A. E. J. McGill, B. G. Osborne. Whole wheat grain hardness measurement by near infrared spectroscopy. in *Near Infrared Spectroscopy: The Future Waves*. AMC Davies, PC. Williams, eds. NIR Publications, Chichester, UK, 1996, 466–470.

30. S. J. Millar, J. M. Alava. Rapid assessment of dough properties. Paper 39, 87th AACC Annual Meeting, 2002, (*www.aaccnet.org/meetings/2002/abstracts*).

31. S. J. Millar, J. M. Alava, S. E. Salmon. NIR spectroscopy as a means of understanding changes occurring during dough mixing. Paper 186 AACC Annual Meeting, 14–18 October, Charlotte, NC, 2001, (*www.aaccnet.org/meetings/2001/abstracts*).

32. J. P Nielsen, D. Bertrand, E. Micklander, P. Courcoux, L. Munck. Study of NIR spectra, particle size distributions and chemical parameters of wheat flours: a multi-way approach. *J Near Infrared Spectrosc* **9**: 275–285, 2001.

33. K. H. Norris, W. R. Hruschka, M. M. Bean, D. C. Slaughter. A definition of wheat hardness using near infrared reflectance spectroscopy. *Cereal Foods World* **34**: 696–705, 1989.

34. B. G. Osborne. Applications of NIR in the baking industry. *Anal Proc* **20**: 79–83, 1983.

35. B. G. Osborne. Near infrared spectroscopic studies of starch and water in some processed cereal foods. *J Near Infrared Spectrosc* **4**: 195–200, 1996.

36. B. G. Osborne. Improved NIR prediction of flour starch damage. In: *Proceedings of Australian Cereal Chemistry Conference*, L. O'Brien, A. B. Blakeney, A. S. Ross, C. W. Wrigley, eds. RACI, Melbourne, 1998, p. 434–438.

37. B. G. Osborne. NIR measurement of the development of crystallinity in stored bread crumb. *Analusis* **26**: M39–M41, 1998.

38. B. G. Osborne, G. M. Barrett, S. P. Cauvain T. Fearn. The determination of protein, fat and moisture in bread by near infrared reflectance spectroscopy. *J Sci Food Agric* **35**: 99–105, 1984.

39. B. G. Osborne, P. Burridge, G. Palmer, G. Hollamby, J. A. Ronalds, I. J. Wesley, A. Laucke. Ultra-rapid quality testing of wheat, flour and dough using near infrared diode

array spectrometry. In: *Near Infrared Spectroscopy: Proceedings of the 9th Conference*, A. M. C. Davies, R. Giangiacomo, eds. NIR Publications, Chichester, UK, 2000, p. 835–839.

40. B. G. Osborne, S. Douglas, T. Fearn. Recent progress in the application of NIR to the measurement of quality parameters in flour. In: *Progress in Cereal Science and Technology*. J. Holas, J. Kratochvil, eds. Elsevier, 1983, p. 577–581.

41. B. G. Osborne, T. Fearn. Collaborative evaluation of universal calibrations for the measurement of protein and moisture in flour by near infrared reflectance. *J Food Technol* **18**: 453–460, 1983.

42. B. G. Osborne, T. Fearn, J. Blakeney. On line monitoring of flour starch damage by NIR. QWCRC Report No. 16, Quality Wheat Cooperative Research Centre, North Ryde, NSW 2113, Australia, 1998.

43. M. C. Pasikatan, E. Haque, J. L. Steele, C. K. Spillman, G. A. Milliken. Evaluation of a near-infrared reflectance spectrometer as a granulation sensor for first-break ground wheats: Studies with six wheat classes. *Cereal Chem* **78**: 730–736, 2001.

44. M. C. Pasikatan, E. Haque, C. K. Spillman, J. L. Steele, G. A. Milliken. Granulation sensing of first-break ground wheat using a near-infrared reflectance spectrometer: studies wirh soft red winter wheats. *J Sci Food Agric* **83**: 151–157, 2003.

45. M. C. Pasikatan, J. L. Steele, E. Haque, C. K. Spillman, G. A. Miliken. Evaluation of a near-infrared reflectance spectrometer as a granulation sensor for first-break ground wheats: studies with Hard Red Winter wheats. *Cereal Chem* **79**: 92–97, 2002.

46. T. Pawlinsky, P. C. Williams, Prediction of wheat bread-baking functionality in whole kernels, using near infrared reflectance spectroscopy. *J Near Infrared Spectrosc* **6**: 121–127, 1998.

47. J.-A. Persson, R. Sjodin. Efficiency all the way from grain to flour. *In Focus*, **28**: (1), 4–5, 2004.

48. J. Psotka, M. Olewnik. Using NIR for dough mixing—Diagnosing mixing problems in real time. Paper 303 AACC Annual Meeting, 5–9 November, Kansas City, MO, 2000.

49. J. Psotka, R. Chen, M. Olewnik. Monitoring dough development in real time using a near infrared spectrometer. Paper 293 AACC Annual Meeting, 31 October-3 November, Seattle, WA, 1999.

50. D. S. Reed, J. J. Psotka. Optical compositional analyzer apparatus and method for detection of ash in wheat and milled wheat products. *United States Patent 5,258,825*, 1993.

51. G. L. Rubenthaler, Y. Pomeranz. Near-infrared reflectance spectra of hard red winter wheats varying widely in protein content and breadmaking potential. *Cereal Chem* **64**: 407–411, 1987.

52. B. W. Seabourn, S. R. Bean, G. L. Lookhart, O. K. Chung. Prediction of gliadin and soluble/insoluble HMW glutenin fractions in whole kernel wheat by near-infrared reflectance spectroscopy. *Cereal Foods World*, **43**: 518, 1998.

53. T. D. Sugden, B. G. Osborne. Wheat flour milling. In: *Cereals and Cereal Products, Chemistry and Technology*. D. A. V. Dendy, B. J. Dobraszczyk, eds, 140–181. Aspen, Gaithersburg, MD, 2001.

54. K. Suzuki, C. E. McDonald, B. L. D'Appolonia. Near-infrared reflectance analysis of bread. *Cereal Chem* **63**: 320–325, 1986.

55. I. J. Wesley, P. W. Gras, N. Larsen, B. G. Osborne, J. H. Skerritt. A preliminary comparison of the 35g Mixograph and dynamic near infrared spectroscopy for studying

dough development. *Proceedings of 48th Australian Cereal Chemistry Conference*, L. O'Brien, A. B. Blakeney, A. S. Ross, C. W. Wrigley, eds. RACI, Melbourne, 1999, p. 447–450.

56. I. J. Wesley, N. Larsen, B. G. Osborne, J. H. Skerritt. Non-invasive monitoring of dough mixing by near infrared spectroscopy. *J Cereal Sci* **27**: 61–69, 1998.

57. I. J. Wesley, N. Larsen, B. G. Osborne, J. H. Skerritt. Monitoring of dough properties. United States Patent 6,342,259 B1, United States Patent Office, Washington DC, 2002.

58. I. J. Wesley, O. Larroque, B. G. Osborne, N. Azudin, H. Allen, J. H. Skerritt. Measurement of gliadin and glutenin content of flour by NIR spectroscopy. *J Cereal Sci* **34**: 125–133, 2001.

59. I. J. Wesley, B. G. Osborne, R. S. Anderssen, S. R. Delwiche, R. A. Graybosch. Chemometric localization approach to NIR measurement of apparent amylose content of ground wheat. *Cereal Chem* **80**: 462–467, 2003.

60. I. J. Wesley, S. Uthayakumaran, R. S. Anderssen, G. B. Cornish, F. Bekes, B. G. Osborne, J. H. Skerritt. A curve-fitting approach to the near infrared reflectance measurement of wheat flour proteins which influence dough quality. *J Near Infrared Spectrosc* **7**: 229–240, 1999.

61. D. Wetzel, H. Mark. Spectroscopic cellulose determination as a criterion of flour purity with respect to bran. *Cereal Foods World* **22**: 481, 1977.

62. P. C. Williams. Screening wheat for protein and hardness by near infrared reflectance spectroscopy. *Cereal Chem* **56**: 169–172, 1979.

63. P. C. Williams, J. E. Dexter. Application of near-infrared raw spectral data to determine efficiency of flour mixing. Paper 360, 87th AACC Annual Meeting, 2002, (*www.aaccnet.org/meetings/2002/abstracts*).

64. P. C. Williams, F. Jaby El-Haramein, G. Ortiz-Fereira, J. P. Srivastava. Preliminary observations on the determination of wheat strength by near-infrared reflectance. *Cereal Chem* **65**: 109–114, 1988.

65. P. C. Williams, D. C. Sobering. Comparison of commercial near infrared transmittance and reflectance instruments for analysis of whole grains and seeds. *J Near Infrared Spectroscopy*, **1**: 25–32, 1993.

66. P. C. Williams, B. N. Thompson, D. Wetzel, G. W. McKay, D. Loewen. Near-infrared instruments in flour mill quality control. *Cereal Foods World* **26**: 234–237, 1981.

67. R. H. Wilson, B. J. Goodfellow, P. S. Belton, B. G. Osborne, G. Oliver, P. L. Russell. Comparison of Fourier Transform mid infrared reflectance spectroscopy with differential scanning calorimetry for the study of the staling of bread. *J Sci Food Agric* **54**: 471–483, 1991.

68. W. R. Windham, C. S. Gaines, R. G. Leffler. Effect of wheat moisture content on hardness scores determined by near-infrared reflectance and on hardness score standardization. *Cereal Chem* **70**: 662–666, 1993.

# Cereal Foods

SANDRA E. KAYS and FRANKLIN E. BARTON, II

## INTRODUCTION

Cereals form the major staple crops worldwide, with production estimated at 2.25 GMt annually (14). Throughout the globe they are processed into hundreds of thousands of products with a wide range in quality, nutritional value, and acceptability and are widely consumed; for example, the per capita consumption of flour and cereal products in the U.S. is 200 pounds a year before adjustment for loss and spoilage (41). Cereals currently provide a substantial portion of the calories, carbohydrates, dietary fiber, iron, thiamine, riboflavin, niacin, vitamin $B_6$, magenesium and zinc intakes of humans where whole grain products are a significant portion of the foods consumed. In contrast to fresh produce, most cereals and cereal products, because of their lower moisture content, can be stored easily and can be processed by a wide variety of methods. Cereals typically have a low flavor impact that contributes tremendous versatility, allowing a diverse range of products to be processed for a wide range of palates and ethnicities. The nutritional value of the final product depends on the grain or combinations of grains, degree of refinement the method of processing, and the amount of sugar, fat, vitamins, minerals, and other additives included during preparation. Thus there is a wide range in nutritional quality and composition of the cereal food products available worldwide.

The ability of U.S. consumers to select healthy foods for their diets has been greatly facilitated by the 1990 Nutrition Labeling and Education Act (NLEA). As with similar legislation in other countries, this act requires food processors and manufacturers to state on a product's label the nutrients present in order of their magnitude (13); thus consistent and accurate analysis and monitoring of products or their ingredients is required. This can be laborious, time consuming, and expensive, depending on the parameter; for example, the approved methods for measurement of dietary fiber and

*Near-Infrared Spectroscopy in Food Science and Technology*, Edited by Yukihiro Ozaki, W. Fred McClure, and Alfred A. Christy.

total fat are labor intensive and take more than 24 hours to complete. New methods are needed and have been developed that are accurate, cost effective, and amenable to the measurement of not only large numbers of samples but also large numbers of components. This is the forte of near-infrared (NIR) technology.

Although the history of NIR spectroscopy actually begins with work by Herschel in 1800 (15–17), the emergence of NIR spectroscopy as an accepted technique in the analytical world began with the work of Karl Norris of the U.S. Department of Agriculture, Agricultural Research Service in the early 1960s focusing on grain, principally wheat (35, 37). The first report on the potential of NIR spectroscopy for the analysis of cereal foods was by Baker, Norris, and Li (10). Their research on the prediction of neutral detergent fiber in ready-to-eat breakfast cereals was based on the early success of Norris, Barnes, Moore, and Shenk (36) in predicting fiber content of forages. Subsequently, analysis of cereals and cereal foods by NIR has expanded to include properties such as color and many of the components that impact nutritional value, for example, protein, carbohydrate, fat, dietary fiber, and energy.

The application of NIR spectroscopy to the analysis of cereal food products has been reviewed by Osborne (39) and Kays (22). These reviews cover the use of NIR spectroscopy for the analysis of the composition of diverse cereal foods. The current chapter describes the approaches used in the development of NIR models for cereal foods. These approaches include cereal product sample preparation, sample conditioning, the importance of selecting the correct reference method where, in most cases, several reference methods are available, and the use of specific instruments to optimize sampling and performance. In addition, a table is presented with a comprehensive list of published research on the application of NIR spectroscopy to cereal food products from 1979 to the present. The application of NIR to the analysis of cereals processed into flour and baking products is reviewed in a prior section of this book.

The types of cereal food products included in this chapter are processed, packaged cereal products and certain bulk products available to consumers from retail sources, for example, granolas, ready-to-eat breakfast cereals, crackers, flours, pastas, bran, chips, snack bars, cookies, and some minimally processed, packaged whole grains (e.g., popcorn and rice). The processed products considered to be cereal foods are generally made up of wheat, maize (corn), oats, rice, rye, and millet; however, cereal products also include buckwheat, amaranth, and quinoa, which are handled as cereals by the food industry.

## ANALYSIS OF CEREAL FOOD PRODUCTS

A comprehensive list of research published on the application of NIR spectroscopy to the analysis of cereal food products is presented in Table 8.2.1. The table includes notes on the instrumentation and reference analyses used in the studies.

**TABLE 8.2.1. Research Published on Application of NIR Spectroscopy to Analysis of Cereal Food Products**

| Parameter | Analyte | Sample* | Instrument | Reference Method | Reference |
|---|---|---|---|---|---|
| Carbohydrate | Total carbohydrate | Snack foods | Neotec 6350 scanning spectrometer, reflectance | Summation of starch and NDF values | 8 |
| | Starch | Oat bran products | NIRSystems 6500 scanning spectrometer, dispersive, reflectance | Enzymatic-colorimetric | 8 |
| | Starch | Snack foods | Neotec 6350 scanning spectrometer, reflectance | Gas liquid chromatography of hydrolyzed starch | 8 |
| | Starch | Wheat bran mixes | Cary Model 14 prism-grating monochromator, reflectance | Ewers (Hungarian Standard 1953) | 18 |
| | Sucrose | Cake mixes | InfraAzer 400R, filter instrument, reflectance | HPLC | 38 |
| | Sugar | Breakfast cereals | Neotec 6350 scanning spectrometer, reflectance | Gas-liquid chromatography | 9 |
| | Dietary fiber and its components (hemicelluloses, pectin, cellulose, lignin) | Wheat bran mixes | Cary Model 14 prism-grating monochromator, reflectance | Modified enzymatic (dietary fiber) | 18 |
| | Total dietary fiber | Oat bran products | NIRSystems 6500 scanning spectrometer, reflectance | Enzymatic-gravimetric | 46 |

**TABLE 8.2.1.** *Continued*

| Parameter | Analyte | Sample* | Instrument | Reference Method | Reference |
|---|---|---|---|---|---|
| | Total dietary fiber | Diverse cereal products | NIRSystems 6500, scanning spectrometer, dispersive, reflectance; Nicolet Raman 950 | Enzymatic-gravimetric | 4, 5, 24, 30, 31, 48 |
| | Total dietary fiber | Diverse cereal products—intact | Perten Model DA7000 diode array spectrometer, reflectance | Enzymatic-gravimetric | 6 |
| | Insoluble dietary fiber/neutral detergent fiber (NDF) | Snack foods | Neotec 6350 scanning spectrometer, reflectance | Enzymatic-gravimetric | 8 |
| | Insoluble dietary fiber/neutral detergent fiber (NDF) | Breakfast cereals | Neotec GQA 41, filter instrument; Neotec 6350, scanning spectrometer, reflectance | Enzymatic-gravimetric | 7, 10 |
| | Insoluble dietary fiber/neutral detergent fiber (NDF) | Oat bran products | NIRSystems 6500 scanning spectrometer, reflectance | Enzymatic-gravimetric | 46 |
| | Insoluble dietary fiber | Diverse cereal products | NIRSystems 6500, scanning spectrometer, dispersive, reflectance | Enzymatic-gravimetric | 24, 25 |
| | Soluble dietary fiber | Oat bran products | NIRSystems 6500 scanning spectrometer, reflectance | Enzymatic-gravimetric | 46 |

| Constituent | Product | Instrument | Reference method | Reference |
|---|---|---|---|---|
| Soluble dietary fiber | Diverse cereal products | NIRSystems 6500, scanning spectrometer, dispersive, reflectance | Enzymatic-gravimetric | 25 |
| Color | Wheat, lentil, and barley flours | NIRSystems 6500, scanning spectrometer, dispersive, reflectance | Colorimeter | 11 |
| Energy | Diverse cereal products | NIRSystems 6500, scanning spectrometer, dispersive, reflectance | Bomb calorimetry (gross energy, energy from fat); calculation from gross energy, protein, and insoluble dietary fiber values (utilizable energy) | 26, 27 |
| Fat | Cake mixes | InfraAzer 400R, filter instrument, reflectance | Soxhlet solvent extraction-gravimetric | 38 |
| Fat | Pasta | Neotec 6450 RCA, reflectance; Cary Model 14 prism-grating monochromator, reflectance | Stoldt–Weibull; modified Lidner (acid hydrolysis-solvent extraction-gravimetric) | 20, 21 |
| Fat | Diverse cereal products | Pacific Scientific 6350, reflectance | Acid hydrolysis-solvent extraction-gravimetric | 40 |
| Fat | Diverse cereal products | NIR Systems 6500 dispersive spectrometer, reflectance mode | Solvent extraction-gravimetric | 29 |
| Fat | Diverse cereal products—intact | Perten Model DA7000 diode array spectrometer, reflectance | Solvent extraction/gravimetric | 23 |

*(Continued)*

**TABLE 8.2.1.** *Continued*

| Parameter | Analyte | Sample* | Instrument | Reference Method | Reference |
|---|---|---|---|---|---|
| | Fat (total fat) | Diverse cereal products | NIRSystems 6500 dispersive spectrometer, reflectance mode | Acid hydrolysis/solvent extraction/gas chromatographic | 45 |
| Protein | Protein | Wheat flour | Cary Model 14 prism-grating monochromator, reflectance | Kjeldahl; calculated from ingredients | 19 |
| | Protein | Oat bran products | NIRSystems 6500 scanning spectrometer, reflectance | Kjeldahl | 46 |
| | Protein | Pasta | Neotec 6450 RCA, reflectance; Cary Model 14 prism-grating monochromator, reflectance | Kjel-Foss instrument | 20, 21 |
| | Protein and nitrogen | Diverse cereal products | NIRSystems 6500, scanning spectrometer, dispersive, reflectance | Combustion analysis | 28 |
| | Nitrogen | Wheat bran mixes | Cary Model 14 prism-grating monochromator, reflectance | Kjel-Foss 162-10 instrument | 18 |
| | Egg content | Pasta | Neotec 6450 RCA, reflectance | Provided by preparer | 20 |
| Moisture | Moisture | Pasta | Cary Model 14 prism-grating monochromator, reflectance | Hungarian Standard 1972 | 21 |

*Cereal food products are ground unless specified.

## Sample Preparation and Sampling

Most studies on the application of NIR spectroscopy to the analysis of cereal food products have been conducted with ground or milled products. Ground samples have the advantage that they are more homogeneous in composition and particle size than intact products and have substantially reduced particle size. However, at grain elevators or during on-line processing, it is more practical and less time consuming if spectra can be obtained nondestructively on intact grains or intact products for assessment of quality. NIR spectroscopy is used successfully to predict protein and moisture content in intact wheat grains and has become a standard practice for the evaluation of grain quality (12, 47). With intact grain, NIR spectroscopy has been responsible for substantial cost savings, increased speed of analysis, and decreased chemical use.

The intact approach has also been applied to cereal food products (6, 23), which encompass a very wide range of particle sizes and shapes from flour to flakes, puffed grains, extruded products, crackers, and fines. A PLS regression model was developed to predict fat in intact cereal products. The model had a root mean squared standard error of performance of 11 g kg$^{-1}$ (range 1–205 g kg$^{-1}$) and coefficient of determination of 0.98 (23). When contrasted with models developed with similar products in a ground state, the latter had comparable standard errors of performance of 10–11 g kg$^{-1}$ fat and coefficients of determination of 0.98–0.99 (29, 40, 45).

Repacking the sampling device can be part of the sampling strategy to ensure representative spectra. The NIR reflectance model for prediction of total dietary fiber required repacking the sample six or seven times and averaging the spectra for optimum model performance (6). The need for repeated scanning to obtain representative spectra was attributed to the extreme heterogeneity of particle shapes and sizes within the data set. However, the model to predict crude fat in similar, intact cereal products did not require repeated repacking of the sample and averaging of spectra to improve NIR model performance (23). A single sample packing was as effective as several. This may be because the signal for fat is stronger than for dietary fiber and there is little interference from other constituents in fat prediction models (23, 45). Furthermore, cereal grains themselves contain very little fat; the bulk of the fat present is added during processing of the products and, therefore, may be more evenly distributed than fiber. Thus the requirement for repacking to obtain representative spectra for intact cereal products can vary depending on the parameter being tested.

Additional sampling strategies that have been used in developing NIR models for cereal food products are 1) conditioning of samples with relative humidity treatments resulting in wider ranges in moisture content and 2) treatment of intact samples by crushing, resulting in wider ranges in particle sizes (6, 23, 48). The former is to increase robustness of models to possible changes in ambient relative humidity during processing and concurrent product moisture contents and the latter to simulate problems arising from inadvertent crushing and breaking during product handling. For both total dietary fiber and fat, NIR models were expanded to include wider ranges in moisture content and particle sizes, with very little or no reduction in the accuracy over prediction of untreated test samples and with increased accuracy in prediction of relative humidity and particle size treated test samples (6, 23, 48).

## Instrumentation

Near-infrared instruments used to measure fiber and other properties of cereal products have evolved in tandem with the use of the technology. The early studies by Horváth et al. (18, 19), Norris et al. (36), and Kaffka et al. (20, 21) were accomplished with the Cary 14 instrument, which was really an ultraviolet/visible instrument whose range happened to extend into the NIR region. Although the Cary 14 was a reliable and stable instrument, it lacked the energy *throughput* of modern dispersive spectrometers. Baker looked at cereal products initially with Norris using the Cary 14 but later developed models for a simpler instrument with three fixed filters, the Neotec GQA 41, which was capable of monitoring three wavelengths (7–10). Baker also used the first commercial monochromator, the Neotec 6350, which was the precursor of the NIRSystems 6500 spectrometer later called the Foss 6500. Subsequent studies utilized some version of the Foss (formerly NIRSystems) dispersive monochrometer (Table 8.2.1). The Foss 6500 is distinguished by being a more robust and dependable instrument (some have been used for 15 years without the need for service). The signal-to-noise ratio (S/N) is excellent, and the instrument is very quiet, with noise levels measured in microabsorbance units. It has a spectral resolution (band pass) of 10 nm, and data are taken every 2 nm so there is a fivefold redundancy, which helps to increase the S/N and maintain a stable calibration performance. The 6500 has been replaced by the Foss XDS, which is currently the state-of-the-art dispersive instrument.

In the late 1990s a new type of spectrometer became available that had a unique sample geometry, the Perten DA 7000 diode array spectrometer. The instrument was used in cereal products research to measure total dietary fiber and fat in intact cereal products (6, 23). The DA 7000 used an eight-inch-wide sampling head that could be oriented to look either upward or down on a sample. A pour-through gate was devised for cereal products (6) and the DA 7000 mounted on its side so that a box of cereal could be poured through the instrument's light path and measured. The significance here is that total dietary fiber and other nutrients found on a product label could be measured as the box was filled; thus the nutrition label for each box potentially could be printed with data that truly reflected its contents.

Although NIR spectroscopy has been extremely successful in measuring the properties of cereal products, it is not the only spectroscopic technique available. Both mid-IR and Raman spectroscopy measure the fundamental frequencies of interest; however, the mid-IR instrument can only assess very small samples and the spectral region is more sensitive to moisture, thus creating sampling precision problems. Raman, in contrast, is not sensitive to moisture and can use the same sampling geometries as NIR spectroscopy. Archibald et al. (4, 5) have shown the advantages of using the combined NIR and Raman regions to measure total dietary fiber in cereal products. The results were obtained by selecting specific regions in the NIR with a Foss 6500 and the Raman with a Nicolet 950. The combined NIR-Raman models had a better SEP than either region by itself. The advantage is being able to obtain pertinent information from both the fundamental frequencies and the combination and overtones to yield a better model.

## Reference Method Selection

Several analytical methods are often available for the same parameter, and selection of the reference method can be critical, especially if the NIR technique is being developed for a specific purpose. For example, the U.S. Code of Federal Regulations (13) has specific definitions for nutrients appearing on the label, and it is important, if the NIR method is to be used for nutrition labeling or monitoring, that the reference method meet the definition of the component. The definition of a nutrient may also vary between countries, for example, as it does for dietary fiber, so regulations for the country of interest should also be considered.

The U.S. Code of Federal Regulations (13) defines total fat content of a food as "the sum of all fatty acids expressed as triglycerides" and all statements of total fat content on U.S. nutrition labels should reflect this definition. The accepted method of measurement of total fat and saturated fat in cereal products for nutrition labeling is AOAC Method 991.06 (3), which involves acid hydrolysis of the food sample (to release bound lipids), solvent extraction of the lipid material, saponification and esterification of the lipid extract, and capillary gas chromatographic analysis to quantify the fatty acids present. The Soxhlet solvent extraction-gravimetric method, for example, AOAC Method 945.16 (2), is not adequate for nutrition labeling in that some lipids in heat-treated cereal products are not easily extracted by the solvent alone, causing total fat to be underestimated (42, 49). Methods involving acid hydrolysis of the sample followed by solvent extraction and gravimetric analysis tend to overestimate total fat, particularly in products with large amounts of sugar, and are also not approved for nutrition labeling. All three of the above reference methods have been used to develop NIR models for the prediction of fat in cereal food products, and, surprisingly, all gave similar results in terms of residual standard errors of prediction and coefficients of determination for prediction of test samples (29, 40, 45).

There are several ways that the energy content of foods can be measured to comply with U.S. food labeling legislation. The inability of the human body to efficiently utilize protein and the lack of available energy from certain fractions of fiber in foods preclude the use of gross energy as a relevant measure of energy available in a food. However, calculation of the energy content from carbohydrate, protein, and fat with Atwater factors is an approved method, as are methods that involve the measurement of gross energy with an adjustment for the incomplete utilization of protein and/or the indigestibility of insoluble dietary fiber (13). In addition to total available calories, energy can be measured as the proportion of energy from fat, which is also defined in the U.S. Code of Federal Regulations (13). Near-infrared reflectance spectroscopy has excellent potential for predicting energy in cereal food products with these reference methods (26, 27).

The composition of cereal food products can be significantly modified by additives and some processing methods. For example, psyllium is often added to cereal products to increase the soluble fiber content; however, the accepted method for measuring total dietary fiber (AOAC Method 991.43, Ref. 1) is not accurate for psyllium-containing products and must be modified by the addition of a sonification step (33). Similarly, some low- or no-calorie sugar and fat substitutes and their carriers, used in certain snack products, contain short-chain resistant oligosaccharides in various quantities. These are included in the definition of dietary fiber for nutrition labeling but, because of their alcohol solubility, are not measured accurately by AOAC Method 991.03. Therefore, modifications of AOAC Method 991.43 are required for accurate reference measurement of total dietary fiber containing resistant oligosaccharides, inulin, and polydextroses (34).

## Model Development

Several approaches have been used to deal with interference when developing NIR models for the assessment of cereal food products. A high percentage of crystalline sugar is common among frosted breakfast cereals and some cookies. Because of characteristically sharp spectral features associated with absorption by OH groups (1434 and 2100 nm) in sugars, sugar can interfere in NIR calibrations (for examples, see spectra in Figure 8.2.1). Likewise, the spectra of cereal products with high fat content have unique features in regions of the spectrum associated with absorption of lipid alkyl groups (e.g., 1215, 1728, 1764, 2310, and 2346 nm), some of which can cause interference in the development of NIR models to predict fiber (see spectra in Figure 8.2.1). Several approaches have been used to combat this interference. In a reflectance model to predict neutral detergent fiber in breakfast cereals (10), extra high-fat and high-sugar cereal product samples were included in the calibration data set to minimize fat and sugar interference. This facilitated the prediction of neutral detergent fiber in high-fat and high-sugar breakfast cereals with the Cary Model 14 research instrument but not the smaller Neotec GQA 41 optimized for fiber.

To increase the robustness of a prediction model for total dietary fiber, high-fat and high-sugar samples were added to the calibration. Twenty-three high-sugar cereal samples ($>200$ g sugar kg$^{-1}$) and 17 high-fat samples ($>100$ g fat kg$^{-1}$) were added to the original calibration data set of 77 low-fat and low-sugar samples (30, 32). The high-fat and high-sugar samples added were selected from a larger group of high-fat and high-sugar samples with a selection algorithm (SELECT, Infrasoft International Software) (43, 44) to identify the new samples most uniquely different from the original calibration samples. The expanded models were tested on independent validation samples and successfully predicted total dietary fiber in high-fat samples, high-sugar samples (several with the spectral characteristics of crystallized sugar), and samples with both high fat and high sugar. Thus interference by fat and sugar was minimized by incorporating unique high-fat and high-sugar samples into the calibration.

Selection of wavelength variables can avoid interference and substantially improve performance of a calibration. In a model to predict neutral detergent fiber in cereal products with a Neotec 6350 instrument, it was found that the wavelengths chosen by the regression analysis avoided interference from fat and sugar (7). In a model to predict total dietary fiber in intact cereal products, a spectral window preprocessing technique was used that identified interference at 1370–1480 nm and established upper and lower limits for the spectral range (6). The window calculated by the algorithm corresponded to the region of interference from absorption by OH in crystallized sugar.

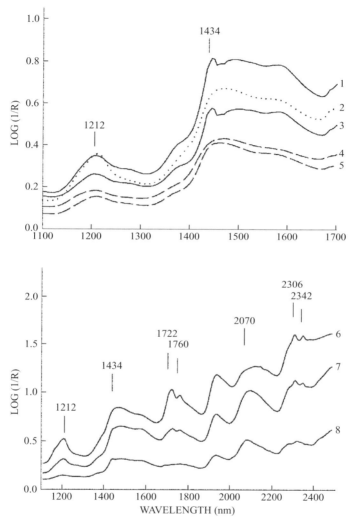

**Figure 8.2.1.** Spectral regions of cereal food samples obtained with the Perten DA7000 (upper panel) and Foss 6500 (lower panel) spectrometers. Spectra numbers 4 and 5 have low-sugar (<19.9%) and low-fat (<9.9%) content, spectra 1, 3, and 8 have high-sugar (>20%) and low-fat content, and spectra 2, 6, and 7 have high-fat (>10%) and low-sugar content.

## SUMMARY

This chapter provides a comprehensive list of published studies on the use of NIR spectroscopy for the assessment of nutritional quality and certain other attributes in cereal products (i.e., from the first work in 1979 to the present) and emphasizes the importance of several aspects of the technology with regard to accuracy. Most cereal food products fall under nutrition labeling regulations as the majority are processed and packaged. If this is the goal of the NIR method, it is important to be aware of the way in which nutrients are defined in the nutrition labeling regulations of each country to ensure that the reference method selected measures the parameter as defined. Sample preparation, for example, scanning of intact products versus ground products, can greatly affect the accuracy of the NIR method and determine whether the method is sufficiently accurate for screening, quality control, or nutrition labeling. In addition, it is advantageous to know as much as possible about the ingredients used in a product and the processing method, as both can affect the selection of the reference method and dictate any modifications required.

Several of the components of ground cereal products can be predicted within or approaching the accuracy required for nutrition labeling (i.e., total fat, protein, total dietary fiber, energy, and insoluble dietary fiber), and NIR spectroscopy models have been developed for screening for soluble dietary fiber. As improvements are made in instrument resolution, computer and software technology, and calibration transfer, the precision of NIR techniques for nutrition labeling will improve, enhancing their widespread acceptance by the food industry. Furthermore, the accuracy, speed, low cost, and environmental advantages of NIR spectroscopy for the evaluation of food components will lead to greatly increased utilization; thus the potential of NIR spectroscopy for monitoring nutritional quality of cereal products is excellent.

## REFERENCES

1. *AOAC Official Methods of Analysis*, 15th ed., 3rd Supplement. 991.43. AOAC, Arlington, VA, 1992.
2. *AOAC Official Methods of Analysis*, 16th ed., Method No. 945.16. AOAC, Arlington, VA, 1995.
3. *AOAC Official Methods of Analysis*, 17th ed., Method No. 996.01. AOAC, Arlington, VA, 1998.
4. D. D. Archibald, S. E. Kays, D. S. Himmelsbach, F. E. Barton II. Raman and NIR spectroscopic methods for determination of total dietary fiber in cereal foods: a comparative study. *Appl Spectrosc* **52**: 22–31, 1998a.
5. D. D. Archibald, S. E. Kays, D. S. Himmelsbach, F. E. Barton II. Raman and NIR spectroscopic methods for determination of total dietary fiber in cereal foods: utilizing model differences. *Appl Spectrosc* **52**: 32–41, 1998b.
6. D. D. Archibald, S. E. Kays. Determination of total dietary fiber of intact cereal food products by near-infrared reflectance. *J Agric Food Chem* **48**: 4477–4486, 2000.
7. D. Baker, The determination of fiber in processed cereal foods by near-infrared reflectance spectroscopy. *Cereal Chem* **60**: 217–219, 1983.
8. D. Baker, The determination of fiber, starch, and total carbohydrate in snack foods by near-infrared reflectance spectroscopy. *Cereal Foods World* **30**: 389–392, 1985.
9. D. Baker, K. H. Norris, Near-infrared reflectance measurement of total sugar content of breakfast cereals. *Applied Spectrosc* **39**: 618–621, l985.

10. D. Baker, K. H. Norris, B. W. Li. Food fiber analysis, Advances in methodology. *In* G.E. Inglett S.I. Falkenhag, eds, 67–78. *Dietary fibers: Chemistry and nutrition*, Academic Press, New York, NY, 1979.

11. C. K. Black, J. F. Panozzo, Accurate technique for measuring color values of grain and grain products using a visible-NIR instrument, *Cereal Chem* **81(4)**: 469–474, 2004.

12. M. Blanco, I. Villaroya, NIR spectroscopy: a rapid response analytical tool, *Trends Anal Chem* **21**: 240–250, 2002.

13. *Code of Federal Regulations*, F. D. A., H. H. S., 21, part 101. 9, 2004.

14. FAO, 2004, Production Statistics, January 6th, 2005, (*http://faostat.fao.org*).

15. W. Herschel. Investigation of the powers of the prismatic colors to heat and illuminate objects. *Phil Trans R Soc Lond* **90**: 255–283 and one plate, 1800a.

16. W. Herschel. Experiments on the refrangibility of the invisible rays of the sun. *Phil Trans R Soc Lond* **90**: 284–292 and one plate, 1800b.

17. W. Herschel. Experiments on the solar and the terrestrial rays that occasion heat. *Phil Trans R Soc Lond* **90**: 293–326 and five plates, 1800c.

18. L. Horváth, K. H. Norris, M. Horváth-Mosonyi, J. Rigó, E. Hegedüs-Völgye∏i. Study into determining dietary fiber of wheat bran by NIR-technique. *Acta Alimentaria* **13**: 355–382, 1984.

19. L. Horváth, K. H. Norris, and M. Horváth-Mosonyi. Comparative investigations into the determination of protein by the Kjeldahl method and NIR technique. *Acta Alimentaria* **14**: 113–124, 1985.

20. K. J. Kaffka, F. Kulcsár. Attempts to determine egg content in pastry products using the NIR technique. *Acta Alimentaria* **11**: 47–64, 1982.

21. K. J. Kaffka, K. H. Norris, M. Rosza-Kiss. Determining fat, protein and water content of pastry products by the NIR technique. *Acta Alimentaria* **11**: 199–217, 1982.

22. S. E. Kays. Application in the analysis of cereal food products. *In* C. A. Roberts, J. Workman, J. B. Reeves III. eds, 411–438. *Near-infrared Spectroscopy in Agriculture*, Agronomy Monograph No. 44., American Society of Agronomy Inc., Crop Science Society of America Inc., and Soil Science Society of America Inc., Madison, WI, 2004.

23. S. E. Kays, D. D. Archibald, M. Sohn. Prediction of fat in intact cereal food products using NIR reflectance spectroscopy. *J Sci Food Agric* **85**: 1596–1602, 2005.

24. S. E. Kays, F. E. Barton, II. The use of near-infrared reflectance spectroscopy to predict the insoluble dietary fibre fraction of cereal products. *J Near Infrared Spectrosc* **6**: 221–227, 1998.

25. S. E. Kays, F. E. Barton, II. Near-infrared analysis of soluble and insoluble dietary fiber fractions of cereal food products. *J Agric Food Chem* **50**: 3024–3029, 2002a.

26. S. E. Kays, F. E. Barton, II. Rapid prediction of gross energy and utilizable energy in cereal food products using near-infrared reflectance spectroscopy. *J Agric Food Chem* **50**: 1284–1289, 2002b.

27. S. E. Kays, F. E. Barton, II. Energy from fat determined by near-infrared reflectance spectroscopy. *J Agric Food Chem* **52**: 1669–1674, 2004.

28. S. E. Kays, F. E. Barton, II. W. R. Windham. Predicting protein content by near infrared reflectance spectroscopy in diverse cereal food products. *J Near Infrared Spectrosc* **8**: 33–43, 2000.

29. S. E. Kays, F. E. Barton, II. W. R. Windham. Analysis of nutritional components in cereal foods with near infrared reflectance spectroscopy. *In* A. M. C. Davies, R. Giangiacomo, eds,

813–816. Near Infrared Spectroscopy: Proceedings of the 9th International Conference, NIR Publications, Chichester, U.K., 2000.

30. S. E. Kays, F. E. Barton, II. W. R. Windham, D. S. Himmelsbach, Prediction of total dietary fiber by near-infrared reflectance spectroscopy in cereal products containing high sugar and crystalline sugar. *J Agric Food Chem* **45**: 3944–3951, 1997.

31. S. E. Kays, W. R. Windham, F. E. Barton, II. Prediction of total dietary fiber in cereal products using near-infrared reflectance spectroscopy. *J Agric Food Chem* **44**: 2266–2271, 1996.

32. S. E. Kays, W. R. Windham, F. E. Barton, II. Prediction of total dietary fiber by near-infrared reflectance spectroscopy in high-fat- and high-sugar-containing cereal products. *J Agric Food Chem* **46**: 854–861, 1998.

33. S. C. Lee, L. Prosky. Psyllium containing cereal products. *J AOAC Int* 1999.

34. B. V. McCleary. Measuring dietary fiber. *The World of Ingredients*, November/December, 1999.

35. K. H. Norris. Simple spectroradiometer for 0.4–1.2 micron region, *Trans ASAE* **7**: 240–242, 1964.

36. K. H. Norris, R. F. Barnes, J. E. Moore, J. S. Shenk. Predicting forage quality by infrared reflectance spectroscopy. *J An Sci* **43**: 889–897, 1976.

37. K. H. Norris, W. L. Butler. Techniques for obtaining absorption spectra on intact biological samples. *IRE Trans Biomed Electron* **8**: 153–157, 1961.

38. B. G. Osborne, T. Fearn, P. G. Randall. Measurement of fat and sucrose in dry cake mixes by near infrared reflectance spectroscopy. *J Food Technol* **18**: 651–656, 1983.

39. B. G. Osborne, NIR analysis of baked products. *In* D. A. Bums, E. W. Ciurczak, eds, 527–548. *Handbook of Near-Infrared Analysis*, Marcel Dekker, Inc., New York, NY, 1992.

40. B. G. Osborne, Determination of fat in a variety of cereal foods using NIR spectroscopy. *In* I. S. Crease, A. M. C. Davies, eds, 68–71. *Analytical applications of spectroscopy*, Royal Society of Chemists, London, 1988.

41. J. Putnam, J. Allshouse, L. S. Kantor, U. S. per capita food supply trends: more calories, refined carbohydrates, and fats. *Food Rev* **25**: 2–15, 2002.

42. G. S. Ranhotra, J. A. Gelroth, J. L. Vetter. Determination of fat in bakery products by three different techniques. *Cereal Foods World* **41**: 620–622, 1996.

43. J. S. Shenk, M. O. Westerhaus. Population structuring of near-infrared spectra and modified partial least squares regression. *Crop Sci* **31**: 1548–15, 1991.

44. J. S. Shenk, M. O. Wesrethaus. Population definition, sample selection, and calibration procedures for near-infrared reflectance spectroscopy. *Crop Sci* **31**: 469–474, 1991.

45. L. L. Vines, S. E. Kays, P. E. Koehler. A near-infrared reflectance model for the rapid prediction of total fat in cereal foods, *J Agric Food Chem* **53**: 1550–1555, 2005.

46. P. C. Williams, H. M. Cordeiro, M. F. T. Hamden. Analysis of oat bran products by near-infrared reflectance spectroscopy, *Cereal Foods World* **36**: 571–576, 1991.

47. P. C. Williams, D. C. Sobering. Comparison of commercial near infrared transmission and reflectance instruments for analysis of whole grains and seeds, *J. Near Infrared Spectros* **1**: 25–32, 1993.

48. W. R. Windham, S. E. Kays, F. E. Barton, II. Effect of cereal product residual moisture content on total dietary fiber determined by near-infrared reflectance spectroscopy, *J. Agric. Food Chem.* **45**: 140–144, 1997.

49. W. Zou, C. Lusk, D. Messer, R. Lane. Fat contents of cereal foods: comparison of classical with recently developed extraction techniques, *J AOAC Int* **82**: 141–150, 1999.

# Livestock Animal By-Products

D. COZZOLINO

## INTRODUCTION

Although Herschel discovered light in the near-infrared (NIR) region as early as 1800, spectroscopists of the first half of the last century ignored it, in the belief that it lacked of analytical interest (8, 26). The principle of NIR reflectance spectroscopy is that light of wavelength 1100–2500 nm, reflected off powdered solids, contains compositional information that can be unraveled by a computer to report multiple analyses almost instantaneously (4, 32, 37). NIR spectroscopy provides simultaneous, rapid, and nondestructive quantitation of major components in many organic substances (37). The technique is applicable to many foods and agricultural commodities and is widely used in the cereal, oilseed, dairy, and meat processing industries (49). NIR spectroscopy has been used to predict chemical composition of forages (e.g., crude protein, dry matter) with high accuracy (15, 36, 42, 43), but until recently published works (2, 13, 14, 16, 23, 50) it had not been widely used for concentrates, compound feeds, and by-products. It is well known that the use of rapid methods to measure the composition of foods and agricultural commodities increases efficiency and decreases the costs of quality control by allowing analysis to take part in management and decisions (31).

The combination of NIR spectroscopy and multivariate techniques (partial least squares, principal component regression) provides a powerful tool to analyze the raw materials used to manufacture compound feeds that are variable both in composition and in nutritional quality because of multiple factors (e.g., temperature, molds, contamination) (7, 29, 32). This variability is most important to the feed industry, where an uniform and consistent product is required (32). Analytical control is essential in order to assess raw materials, products, and by-products as well as to optimize the manufacturing process itself (32).

*Near-Infrared Spectroscopy in Food Science and Technology,* Edited by Yukihiro Ozaki, W. Fred McClure, and Alfred A. Christy.
Copyright © 2007 John Wiley & Sons, Inc.

The scope of this chapter is to review the applications of both quantitative and qualitative NIR spectroscopy to the analysis of livestock animal by-products.

## DEFINITIONS OF LIVESTOCK ANIMAL BY-PRODUCTS

A by-product is a substance that is unavoidably generated in the elaboration of a more valued product (28). It should be noted that this is an economic definition, so most of the effort of the manufacturer will concentrate on the quality of the product, while the by-product poses burdens of cost for storage, transport, and disposal. These costs must be borne by the product itself unless some market can be found for the by-product. A by-product for which no use has been found is a waste product that has to be disposed of, with costs incurred and environmental consequences. Of the farm animals grown and slaughtered for meat, only the muscles and a few internal organs are desirable butcher meat.

Viscera, gut contents, head, feet, skin, and bone are unpalatable but nutritionally rich by-products worth recycling, provided there is no risk of disease transmission. For example, a 100-kg live-weight pig yields a 75-kg carcass and 25 kg of offal and gut contents. The 75-kg carcass yields 57 kg of butcher meat, 4 kg of head, and 14 kg of skin and bones. The 57 kg of butcher meat is composed of 40 kg of lean meat and 17 kg of fatty tissue. Thus only 57% of a pig is edible butcher meat and 43% consists of slaughter by-products that have to be utilized as pet food or recycled into the food chain as protein supplements (see **Fig. 8.3.1**). If slaughter by-products are not recycled

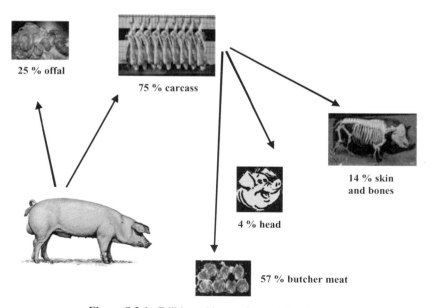

**25 % offal**

**75 % carcass**

**14 % skin and bones**

**4 % head**

**57 % butcher meat**

**Figure 8.3.1.** Edible and by products yield of a pig.

**TABLE 8.3.1. Definition of Livestock Animal By–Products**

| | |
|---|---|
| Meat meal,* CP, Oil | Product obtained by heating, drying, and grinding whole or parts of warm-blooded land animals from which the fat may have been partially extracted or physically removed. The product must be substantially free of hooves, horns, bristle, hair, and feathers, as well as digestive tract content. (Minimum crude protein content 500 g kg$^{-1}$ on a dry matter basis) |
| Meat and bone meal,* CP, Oil | Product obtained by heating, drying, and grinding whole or parts of warm-blooded land animals from which the fat may have been partially extracted or physically removed. The product must be substantially free of hooves, horns, bristle, hair, and feathers, as well as digestive tract content |
| Bone meal, CP | Product obtained by drying, heating, and finely grinding bones of warm-blooded land animals from which the fat has been largely extracted or physically removed. The product must be substantially free of hooves, horns, bristle, hair, and feathers, as well as digestive tract content |
| Greaves, CP | Residual product of the manufacture of tallow and other extracted or physically removed fats of animal origin |
| Poultry offal meal,* CP | Product obtained by drying and grinding waste from slaughtered poultry. The product must be substantially free of feathers |
| Feather meal, hydrolyzed, CP | Product obtained by hydrolyzing, drying, and grinding poultry feathers |
| Blood meal, CP | Product obtained by drying the blood of slaughtered warm-blooded animals. The product must be substantially free of foreign matter |
| Animal fat | Product composed of fat from warm-blooded land animals |

*Products containing more than 130 g kg$^{-1}$ fat in the dry matter must be named as "rich in fat." CP, crude protein (Adapted from Refs. 9, 25, 27, 28, 30).

they create an enormous hygiene and disposal problem (27, 28). Table 8.3.1 lists those tissues that are considered edible and those that are inedible slaughter by-products, which are stabilized by cooking, drying, and grinding to produce meat meal, meat and bone meal, bone meal, animal fat, blood meal, feather meal, or poultry offal meal (27, 30).

These materials are fed as protein concentrates to farm animals as a small proportion of the diet (9, 27, 28). They are not used primarily as sources of protein per se but to address deficiencies of certain indispensable amino acids, which may occur when nonruminants are fed cereal-based diets. In addition, they often make a significant

**TABLE 8.3.2. World Annual Production of Meat and Livestock By-Products (FAO 2002)**

|  | Production, Mt | As Feed, Mt |
|---|---|---|
| Meat | 247,576,769 | — |
| Bovine meat | 61,209,877 | — |
| Animal fat | 32,994,961 | 2,084,321 |
| Meat meal | — | 1,185,289 |
| Fish body oil | 957,819 | 384,580 |
| Fishmeal | 1,149,285 | 1,664,070 |
| Hides and skin | 11,444,003 | — |
| Offal | 16,035,692 | 924,380 |

*Source*: FAO statistics 2003 (*www.faostat.fao.org*)

contribution to the animal's mineral and vitamin requirements. A further reason why these products are fed in limited quantities to farm animals is that they are expensive, which makes their large-scale use uneconomic. Total usage of animal by-products in the compound feed industry in the UK during 1992 was 409,000 tons (11, 24). In the UK, meat and bone meal is defined as the product obtained by heating, drying, and grinding whole or parts of warm-blooded land animals from which the fat may have partially extracted or physically removed. The product must be substantially free of hooves, horn, bristle, hair, and feathers, as well as digestive tract content (9, 27, 28, 30). Meat meal generally contains from 660 to 700 g/kg of protein compared with about 450 to 550 g/kg for meat and bone meal. The fat content is variable, ranging from 30 to 130 g/kg (9, 27, 28, 30). It is important to point out that most of these by-products need a certain degree of process (drying, concentration) before storage and end use that make more complexity in this kind of product. Table 8.3.2 shows the world annual production of meat and livestock by-products for different species. The data show that the amount of livestock by-products produced annually around the world is about 61,000,000 Mt. It is interesting to note that less than 10% of animal by-products are reported to be used as feed.

## QUANTITATIVE ANALYSIS OF LIVESTOCK ANIMAL BY-PRODUCTS

Near-infrared reflectance spectroscopy has been traditionally developed in animal science to evaluate the chemical composition of forages (4, 5, 15). Compared with forages, little information is available on the suitability of NIR spectroscopy in the estimation of the chemical composition and nutritive value of by-product materials derived from livestock. The information available in the literature relates only to the use of NIR spectroscopy to determine gross chemical composition such as dry matter (DM), crude protein (CP), fat (oil), and amino acid content in a few livestock animal by-products. Examples of NIR applications in animal by-products are summarized in Tables 8.3.3 and 8.3.4. Good correlations were found for DM and CP in the different animal by-products. Although good correlations were found for some amino acids, Fontaine et al. (22) stated that the calibration for sulfur-containing amino acids are

**TABLE 8.3.3. Near-Infrared Cross-Validation Statistics for Crude Protein (CP) and Dry Matter (DM) in Livestock Animal By-Products Sourced from Different Authors**

|  |  | SECV | $R^2_{val}$ | SEP | Reference |
|---|---|---|---|---|---|
| Poultry meal byproduct | DM | 3.7 | 0.87 | n/a | 22 |
|  | CP | 16.6 | 0.92 | n/a |  |
| Meat meal | DM | 3.2 | 0.94 | 3.6 | 22 |
|  | CP | 12.1 | 0.96 | 15.5 |  |

SECV: standard error of cross-validation; DM: dry matter; CP: crude protein; $R^2_{val}$: coefficient of determination in cross validation; SEP: standard error of prediction, n/a: data non available (g kg$^{-1}$ DM basis).

less reliable than those for other amino acids susch as lysine. Similar results were reported by van Kempen (46) and van Kempen and Bodin (47).

## DIFFUSE REFLECTANCE: ANIMAL BY-PRODUCTS

An authentic food is one that is what it purports to be, in other words, one that conforms to the description provided by the producer or processor (10). Important aspects of this description may relate to the process history of a product (e.g., fresh meat opposed to frozen meat) or its geographic origin (e.g., Italian olive oil produced only with olives grown in Italy) (17–19). Food adulteration has been practiced since biblical times but has become more sophisticated in the recent past (11, 34). Foods or ingredients most likely to be targets for adulteration include those that are of high

**TABLE 8.3.4. Near-Infrared Cross-Validation Statistics for Methionine (Met), Met + Cys (Methionine + Cystine), Cystine (Cys), and Lysine (Lys) in Livestock Animal By-Products Sourced from Different Authors**

|  |  | SECV | $R^2_{val}$ | Reference |
|---|---|---|---|---|
| Poultry meal by-product | Met | 0.06 | 0.92 | 22 |
|  | Cys | 0.12 | 0.71 |  |
|  | Lys | 0.15 | 0.91 |  |
|  | Met + Cys | 0.12 | 0.90 |  |
| Meat meal | Met | 0.04 | 0.94 | 22 |
|  | Cys | 0.07 | 0.88 |  |
|  | Lys | 0.12 | 0.95 |  |
|  | Met + Cys | 0.07 | 0.93 |  |
| Meat and bone meal | Lys | 0.07 | 0.90 | 46, 48 |
|  | Met | 0.03 | 0.72 |  |
| Meat and bone meal | Dig. Lys | 0.09 | 0.77 | 46 |
|  | Dig. Met | 0.02 | 0.63 |  |

SECV: standard error of cross-validation in g/100 g; DM: dry matter; CP: crude protein; $R^2_{val}$: coefficient of determination in cross-validation; Dig: digestible amino acid.

value or are subject to the vagaries of weather during their growth or harvesting (18, 19, 44, 45). The practice of adulteration arises for two main reasons: First, it can be profitable, and second, adulterants can be easily mixed and are subsequently difficult to detect. To counter this problem manufacturers subject their raw material and by-products to a series of quality controls that includes high-performance liquid chromatography (HPLC), thin layer chromatography (TLC), enzymatic tests, and physical tests, to establish their authenticity and hence guarantee the quality of the products manufactured for the consumers (18, 19). Most uses of NIR spectroscopy have involved the development of NIR calibrations for the quantitative prediction of composition; this was a rational strategy to pursue during the initial stages of its application, given the type of equipment available, the state of development of the emerging discipline of chemometrics, and the overwhelming commercial interest in solving such problems (17, 19). More recently, there is a continuing demand for new, rapid, and cheaper methods of direct quality measurements in food and food ingredients. NIR spectroscopy has been examined to assess its suitability to detect adulteration in different types of products such as fat substitutes in sausages (20), milk fat (41), orange juice (45), fresh and frozen beef (44), and beef muscles sourced from different feeding regimes (12). It is well known that visual examination of the NIR spectra cannot discriminate between authentic and adulterated products (18, 19). Since the outbreak of bovine spongiform encephalopathy (BSE) in the UK, animal by-products have been banned from inclusion in the food chain (24). Increasing consumer concern also relates to meat speciation for religious reasons and also BSE and *Escherichia coli* 0157 infective risks (24, 34). This imposed great pressure on the food manufacturing industry to guarantee the safety of animal by-products. Tissue speciation of animal by-products, as well as speciation of animal-derived by-products fed to all classes of domestic animals, is now the most important uncertainty that the food industry must resolve to allay consumer concern (38, 39). Of five hundred and seventy meat products sampled in 1998 in the UK, 83 (14.6%) were found to contain species of meat other than those declared on the label (24). Of issues related to animal by-product adulteration, one of the most important will be the contamination of fishmeal with meat and bone meal, a by-product of the beef industry (11, 24, 34).

One of the advantages of NIR spectral technology is not only to assess chemical structures through the analysis of the molecular bonds in the NIR spectrum, but also to build an optical model characteristic of the samples that behaves like the "fingerprint" of the sample. This opens up the possibility of using spectra to determine complex attributes of organic materials that are in some way related to molecular chromophores. These may be attributes such as bioavailability, provenance, organoleptic scores, or classification of materials for authentication. In this context, the application of statistical packages like principal component (PCA) or discriminant analysis (e.g., discriminant partial least squares) opens the possibility of understanding the optical properties of the sample and allowing classification without chemical information (1, 35).

Murray et al. (34) demonstrated the potential of NIR spectroscopy to detect adulteration of fishmeal with meat and bone meal. More recently, Gizzi et al. (24) described the use of NIR spectroscopy that can be managed easily and can be used to rapidly detect suspect materials and provide "provisional declaration" for the gross volume of material traded. These authors state that NIR spectroscopy could be used as a first line of defense against accidental contamination (e.g., cross-contamination) and fraudulent practices. Discriminant equations successfully detected 44 of 45 fishmeal samples adulterated with meat and bone meal (34). The combination of the visible and NIR regions gave the best calibration and prediction models (SEC 0.85%) (34). Gizzi et al. (24) reported that at present the major drawback of the NIR technique is that the level of detection is higher than 1% and the method cannot be used alone as legal evidence. Nevertheless, NIR spectroscopic methods may provide initial screening in the food chain and enable more costly methods to be used more productively on suspect specimens and can be easily implemented in feed mills. Recently the use of NIR microscopy was reported to have a great success in authenticating and discriminating between different livestock animal by-products (3, 40).

## REFLECTANCE SPECTRA OF LIVESTOCK ANIMAL BY-PRODUCTS

**Figure 8.3.2** shows the mean spectra of meat and bone meal, poultry meal, and fishmeal. The most relevant spectra feaures were observed around 1500 nm (O-H and N-H overtones), mainly related with water (moisture) and protein content;

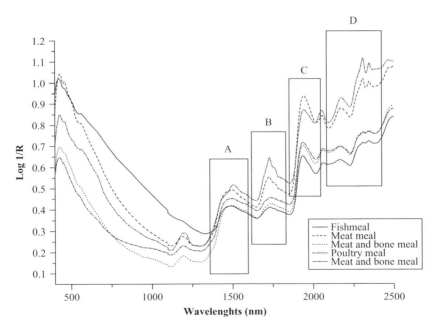

**Figure 8.3.2.** Mean spectra of animal and fish by-products. A = O-H and N-H overtones; B = C-H first overtones; C = O-H overtones; D = C-H combinations and C-$H_2$ overtones.

around 1700 nm (C-H first overtones) related with fat content and fatty acids; and around 1900 nm (O-H overtones) related with water. Both regions around 2200 nm and 2300 nm are related with C-H and C-$H_2$ overtones and combination bands, related with different chemical structures found in animal by-products such as fatty acids, amino acids, protein, and peptide groups (33).

## SUMMARY

NIR reflectance spectroscopy has proved highly successful for routine quality control (QC) analysis in livestock animal by-products by the animal feed industry. It is applicable both to raw material and processed products. Knowledge of the mass, moisture (dry matter) content, and composition (e.g., crude protein, fat, fatty acids) of raw material allows prediction of energy requirements by the feed industry as well as helping in management decisions to meet the specifications for the output generated.

NIR spectroscopy is not restricted to chemical analysis but also can perform a role in assessing attributes such as the animal species of the tissue being processed, the degree of thermal treatment achieved, and the likely destruction of potential pathogens or thermal damage to protein. NIR technology provides an alternative to other more sophisticated and labor-intensive analyses that would prove prohibitively costly to assess contamination or fraudulent adulteration of animal by-products.

Nowadays the importance of food quality in human health and well-being cannot be overemphasized. Consumer concern centers on the quality of the feed fed to farmed livestock whose tissues become our meat. Until now, too little effort has gone into surveillance of what enters the food chain. The consumer is less conserned with gross composition (e.g., crude protein or moisture) than with more sophisticated questions about the wholesomeness of food produced or derived from animal tissues or by-products; its freedom from zoonostic infection risk, hormone, and antibiotic residues, animal welfare issues; and honesty in production. Suspicions arise from exposed malpractice and fuel a campaign to reject intensive livestock animal production. The discovery of infective agents, which persist even after severe heat treatment, raises doubts about the recycling of animal by-products. The fact remains, however, that with only half of the live weight of an animal becoming meat, utilization of slaughter by-products is necessary for both economic and environmental reasons. NIR spectroscopy has a role in the production plant and for surveillance at critical points in the food chain.

## REFERENCES

1. M. J. Adams. Chemometrics in analytical spectroscopy. *Analytical Spectroscopy Monographs. Royal Society of Chemistry* 1995, *216 pp.*
2. J. Aufrere, D. Graviou, C. Demarquilly, J. M. Perez, J. Andrieu. Near infrared reflectance spectroscopy to predict energy value of compound feeds for swine and ruminants. *Anim Feed Sci Tech* **62**: 77–90, 1996.

3. V. Baeten, P. Dardenne. Spectroscopy: developments in instrumentation and analysis. *Grasas Aceites* **53**: 45–63, 2002.

4. G. Batten. Plant analysis using near infrared reflectance spectroscopy: the potential and limitations. *Aus J Exp Agric* **38**: 697–706, 1998.

5. D. Bertrand. La spectroscopie proche infrarouge et ses applicatins dans les industries de l'alimentation animale. *INRA Prod Anim* **15**: 202–219, 2002.

6. L. T. Black, A. C. Eldridge, M. E. Hockidge, W. F. Kwolek. Determination of texturized soybean flour in ground beef by near infrared reflectance spectroscopy. *J Agric Food Chem* **33**: 823–826, 1985.

7. M. Blanco, J. Pages. Classification and quantitation of finishing oils by near infrared spectroscopy. *Anal Chim Acta* **463**: 295–303, 2002.

8. M. Blanco, I. Villaroya. NIR spectroscopy: a rapid-response analytical tool. *Trends Anal Chem* **21**: 240–250, 2002.

9. D. C. Church. Supplementary protein sources. In: *Livestock feeds and feeding, 3rd. ed. Prentice Hall*, Chapter **8**: 133–164, 1991.

10. Ch. Cordella, I. Moussa, A. -C. Martel, N. Sbirrazzuoli, L. Lizzani-Cuvelier. Recent developments in food characterisation and adulteration detection: technique-oriented perspectives. *J Agric Food Chem* **50**: 1751–1764, 2002.

11. D. Cozzolino.*Visible and Near Infrared Reflectance Spectroscopy (NIRS) of Flesh Foods*. PhD Thesis. University of Aberdeen. 1998, 298 pp.

12. D. Cozzolino, D. Vaz Martins, I. Murray. Visible and near infrared spectroscopy of beef longissimus dorsi muscle as a means of discriminating between pasture and corn feeding regimes. *J Near Infrared Spec* **10**: 187–193, 2002.

13. D. Cozzolino. Use of near infrared reflectance spectroscopy (NIRS) to analyse animal feeds. *Agrociencia* **6**: 25–32, 2002a.

14. D. Cozzolino. Prediction of chemial composition of ruminat feeds by near infrared reflectane spectroscopy (NIRS) in Uruguay. *Rev Arg Prod Anim* **22**: 81–86, 2002b.

15. E. R. Deaville, P. C. Flinn. Near infrared (NIR) spectroscopy: an alternative approach for the estimation of forage quality and voluntary intake. In *Forage Evaluation in Ruminant Nutrition*, D. I. Givens, E. Owen, R. F. E. Axford, H. M. Omedi, eds. CABI Publishing UK, 301–320, 2000.

16. J. L. De Boever, B. G. Cottyn, J. M. Vanacker, Ch. V. Boucque. The use of NIRS to predict the chemical composition and the energy value of compound feeds for cattle. *Anim Feed Sci and Tech* **51**: 243–253, 1995.

17. G. Downey. Qualitative analysis in the near infrared region. *Analyst* **119**: 2367–2375, 1994.

18. G. Downey. Authentication of food and food ingredients by near infrared spectroscopy. *J Near Infrared Spectrosc* **4**: 47–61.

19. G. Downey. Food and food ingredient authentication by mid-infrared spectroscopy and chemometrics. *Trends Anal Chem* **17**: 418–424, 1998.

20. M. R. Ellekjaer, T. Naes, T. Isaksson, R. Solheim. Identification of sausages with fat–substitutes using near infrared spectroscopy. *In. Near Infrared spectroscopy: Bridging the gap between data analysis and NIR applications. Edited by K. I. Hildrum, T. Isaksson, T. Naes, A. Tandberg* 320–326, 1992.

21. FAO statistics 2003. *www.faostat.fao.org*

22. J. Fontaine, J. Horr, B. Schirmer. Near infrared reflectance spectroscopy enables the fast and accurate prediction of the essential amino acids content in soy, rapeseed meal, sunflower meal, peas, fishmeal, meat meal products and poultry meal. *J Agric Food Chem* **49**: 57–66, 2001.

23. M. Gerlach. NIR measuring technology for quality evaluation of feeds. *Kraftffuter*. **2**: 67–74, 1990.

24. G. Gizzi, L. W. D. van Raamsdonk, V. Baeten, I. Murray, G. Berben, G. Brambilla, C. von Holst. An overview of test for animal tissues in feeds applied in response to public health concerns regarding bovine spongiform encephalopathy. *Rev Sci Tech Off Int Epiz* **22**: 311–331, 2003.

25. R. A. Lawrie. *Meat Science*, 6th ed. Woodhead Publishing Ltd., UK (Cambridge), 1998.

26. F. W. McClure. 204 years of near infrared technology: 1800–2003. *J Near Infrared Spectrosc* **11**: 487–518, 2003.

27. P. McDonald, R. A. Edwards, J. F. D. Greenhalgh, C. A. Morgan. *Animal Nutrition*. 5th Edition, Longman (New York), 1995.

28. E. L. Miller, F. De Boer. By-products of animal origin. *Livest Prod Sci* **19**: 159–196, 1988.

29. Ch. E. Miller. Chemical principles of near infrared technology. In: *Near infrared technology in the agricultural and food industries*. P.C. Williams, K.H. Norris, eds., Second Edition. American Association of Cereal Chemists, 19–39, 2001.

30. F. B. Morrison. *Feeds and Feeding*. 21st ed. Morrison Publishing, Ithaca, NY, 1951.

31. L. Moya, A. Garrido, J. E. Guerrero, J. Lizaso, A. Gomez. Quality control of raw materials in the feed compound industry. In: *Leaping Ahead with Near Infrared Spectroscopy*. G. D. Batten, P. C. Flinn, L. A. Welsh, A. B. Blakeney, eds, 111–116, 1994.

32. I. Murray. The value of traditional analytical methods and near infrared (NIR) spectroscopy to the feed industry. In: *30th University of Nottingham Feed Manufacturers Conference*. 3–5 January, 1996 Manuscript.

33. I. Murray. The NIR spectra of homologous series of organic compounds. In: *NIR/NIT Conference*. J. Hollo, K. J. Kaffka, J. L. Gonczy, eds. *Akademiai Kiado, Budapest* 1986, pp. 13–28.

34. I. Murray, L. S. Aucott, I. H. Pike. Use of discriminant analysis on visible and near infrared refelctance spectra to detect adulteration of fish meal with meat and bone meal. *J Near Infrared Spectrosc* **9**: 297–311, 2001.

35. T. Naes, T. Isaksson, T. Fearn, T. Davies. *A User-Friendly Guide to Multivariate Calibration and Classification*. NIR Publications, Chichester, UK. 2002, 420 pp.

36. K. H. Norris, R. F. Barnes, J. E. Moore, J. S. Shenk. Predicting forage quality by infrared reflectance spectroscopy. *J Anim Sci* **43**: 889–97, 1976.

37. B. G. Osborne, T. Fearn, P. H. Hindle. Near Infrared Spectroscopy in Food Analysis, *2nd ed. Longman Scientific and Technical. 1993, 227 pp.*

38. R. L. S. Patterson. Some developments in analytical techniques relevant to the meat industry. In: *Chapter 10. Recent Advances in the chemistry of meat*. A. J. Bailey, eds. The Royal Society of Chemistry. 1983, pp. 193–205.

39. R. L. S. Patterson, S. J. Jones. Review of current techniques for the verification of the species origin of meat. *Analyst* **115**: 501–505, 1990.

40. F. Piraux, P. Dardenne. Microscopie-NIR appliquee aux aliments du betail. *Biotechnol Agron Soc Environ* **4**: 226–232, 2000.

41. T. Sato, S. Kawano, M. Iwamoto. Detection of foreign fat adulteration of milk by near infrared spectroscopic method. *J Dairy Sci* **73**: 3408–3413, 1990.

42. J. S. Shenk, M. O. Westerhaus, M. R. Hoover. Analysis of forages by infrared reflectance. *J Dairy Sci* **62**: 807–812, 1979.

43. J. S. Shenk, M. O. Westerhaus. Analysis of Agriculture and Food Products by Near Infrared Reflectance Spectroscopy. *Monograph. Infrasoft International. Port Matilda, PA.* 1993.

44. K. Thyoldt, T. Isaksson. Differentiation of frozen and unfrozen beef using near infrared spectroscopy. *J Sci Food Agric* **73**: 525–532, 1997.

45. M. Twomey, G. Downey, P. B. McNulty. The potential of NIR spectroscopy for the detection of the adulteration of orange juice. *J Sci Food Agric* **67**: 77–84, 1995.

46. T. van Kempen. NIR technology: can we measure amino acid digestibility and energy values? 12th *Annual Carolina Swine Nutrition Conference.* November 13th, 1996.

47. T. van Kempen, J. Ch. Bodin. Near infrared reflectance spectroscopy (NIRS) appears to be superior to nitrogen-based regression as a rapid tool in predicting the poultry digestibility amino acid content of commonly used feedstuffs. *Anim Feed Sci Tech* **76**: 139–147, 1998.

48. P. C. Williams, P. M. Starkey. Influence of feed ingredients upon the prediction of protein in animal feed—mixes by near infrared reflectance spectroscopy. *J Sci Food Agric* **31**: 1201–1213, 1980.

49. P. C. Williams. (2001). Implementation of near infrared technology. P. C. Williams, K. H. Norris, eds, 2nd. American Association of Cereal Chemist, USA. 145–171, 2001.

50. G. Xicatto, A. Trocino, A. Carazzolo, M. Meurens, L. Maertens, R. Carabaño. Nutritive evaluation and ingredient prediction of compound feeds for rabbits by near infrared reflectance spectroscopy (NIRS). *Anim Feed Sci and Tech* **77**: 201–212, 1998.

# Dairy Products

R. GIANGIACOMO and T.M.P. CATTANEO

## INTRODUCTION

According to the latest available statistics (28), world milk production accounts for an estimated 600 million tons, with an increase of 1.5% over the previous year. Cow milk represents 84%, water buffalo milk 12%, and goat and ewe milk 3%, with a major contribution to this growth provided by water buffalo milk, mainly produced on the Indian subcontinent. Gross income from dairy products in 2002 amounted to more than $US 92 billion. This figure explains why the dairy industry is interested in using simple and reliable quality control methods to produce high-quality products at the lowest possible cost.

Near-infrared (NIR) spectroscopy was initially used for measurements in low-moisture products. Its first application in dairy industry was mainly for the analysis of milk powders. Over the intervening years, developments both in hardware and in software have permitted analyses even of cheese and later of liquid milk. The transformation of milk into cheese or fermented milk is a very complex sequence of events, in which physical changes also take place, shifting from a basically liquid structure toward a solid or semisolid structure.

Applications of NIR spectroscopy to the dairy industry goes back to the late 1970s. Considerable work has been done since that time, justifying more than one review. Polesello and Giangiacomo in 1983 (43) summarized the first experiences in applying NIR to study the optical properties of dairy products and the use of fiber optics for determining the chemical composition of whole blue cheese. In 1994, Giangiacomo and Nzabonimpa (20) reported on the developments and applications of NIR technology to dairy products. Their review also cited the industrial applications and the main factors affecting measurements. Rodriguez-Otero et al. (48) in 1997 updated the 1994 review of the NIR analysis of dairy products, emphasizing recent data processing

*Near-Infrared Spectroscopy in Food Science and Technology*, Edited by Yukihiro Ozaki,
W. Fred McClure, and Alfred A. Christy.
Copyright © 2007 John Wiley & Sons, Inc.

techniques allowing for the analysis of the major components. They concluded that NIR spectroscopy was also useful as a classification tool for milk powders and a screening method to detect adulteration.

This chapter is a discussion of applications of NIR measurements throughout the entire milk production cycle, from its collection to final products, including analyses of powders obtained from milk, milk products, and waste, including literature through the year 2004.

## MILK

The composition of milk has been studied by several research groups, as an alternative to IR instruments. For example, Hall and Chan (23) reported the determination of protein, fat, and lactose in nonhomogenized raw bovine milk with NIR spectroscopy. Milk samples were analyzed by transmittance in a 0.5-mm quartz cuvette. Mathematical transformation and MLR were accomplished by NSAS software. Regression analyses were performed on the second-derivative data. For protein, after correction of the calibration equation for eliminating interference due to water absorption band, they obtained a SEC equal to 0.08%. The milk protein validation equation—on a new set of samples—showed a SEP value of 0.08%. They obtained a strong correlation between spectral data and fat content in the overtone and combination band regions. By linear least-squares regression analysis they found a correlation coefficient $r$ of $-0.998$ at 1724 nm with a SEC of 0.12%, a high value probably due to multiplicative scatter of the radiation, which varies as the number of scattering globules in the milk changes. These parameters also change the effective sample pathlength. To correct for these differences, absorption at 1724 nm was divided by absorption at 1610 nm. This model provided a strong reduction in SEC up to 0.04%. The validation regression equation for the fat determination confirmed a SEP of 0.04%.

Kamishikiryo-Yamashita et al. (25) reported peptide bond absorption at 2170 nm as the most suitable wavelength for determining protein content in presence of various food components, but this absorption was influenced by the presence of fat. Later (27), commercial milk samples in a fat range of 2–5% were analyzed from 1100 nm to 2500 nm in transmittance. The same calibration equation proposed for the determination of protein content in oil-water emulsion (26) was applied to milk samples for protein content determination, providing a SEC of 0.0897% and a SEP of 0.177%.

Tsenkova et al. (55) investigated the potential of NIR spectroscopy for measuring constituents in nonhomogenized milk for on-line milk analysis. They compared on-line measurements with a restricted-wavelength fiber-optic probe to that of an instrument that scanned from 1100 to 2400 nm and studied the influence of sample thickness on the accuracy of milk composition determination. Transmittance spectra with path lengths of 1, 4, and 10 mm were obtained. The accuracy of total protein and of fat content determination depended strongly on the spectral region and the pathlength. The best results for the region from 1100 to 2400 nm were obtained with a 1-mm sample thickness. The accuracy of prediction was close to the accuracy of the mid-infrared method used as reference. For the spectral region from 700 to 1100 nm the best results were obtained for samples of 10-mm thickness. Laporte and Paquin (30) evaluated the feasibility of NIR for determining crude protein (CP), true protein (TP), casein, and fat in nonhomogenized cow's milk in the range from 1100 to 2500 nm. Calibrations were performed by MPLS and scattering correction standard normal variate (SNV). The statistical data for the calibration of each nitrogen component, with 7 PLS factors, showed that the calibration models fit the reference data well, with a SEP of 0.09%, 0.12%, and 0.07% for crude protein (CP), true protein (TP), and casein, respectively. In the same paper, the authors evaluated the homogenization effects on NIR determination of CP, TP, and casein. For calibration purposes, 20 homogenized samples were added to the previous calibration set. The addition of new samples accounted more efficiently for the larger spectral variations of the nitrogen components and improved the robustness of calibration. The new SEPs met the required upper limit of the International Dairy Federation (IDF) Standard (52) of 0.06% for mid-infrared instruments. Using the overall calibration, with the added homogenized samples, a decreased accuracy for fat determination was observed, with a SEP of 0.07%. This may be explained by the increase in spectral variability due to the addition of homogenized samples to the calibration set, characterized by a different mean size of fat globules that modified the scattering effects. As a consequence, the authors suggested that to develop an accurate and robust NIR calibration including homogenized and nonhomogenized samples would necessitate over 150 samples.

Purnomoadi et al. (45) investigated whether different feed sources affected the accuracy of NIR spectroscopy. The rations containing nitrogen components, included also in the rations of the cows, used for the calibration set did not negatively influence the prediction accuracy. Rations containing additional components modified the protein fractions in milk proteins, affecting prediction accuracy. The authors suggested that, as there was an influence of diet on composition of milk protein, a wide range of milk from various rations would be needed to develop an applicable calibration. They also studied the influence of feed source on determination of fat. The calibration equation for milk fat was evaluated as suitable for prediction on the basis of the same level of correlation coefficients obtained in the calibration ($r = 0.971$) and in the validation ($r = 0.970$). The predictions for milk obtained from cattle fed with supplemented rations were predicted accurately, because the supplementation of different protein sources did not affect fat intakes except for one. The fat intake in these diets was similar to the ration for the cows used for the calibration purpose.

Diaz-Carrillo et al. (13) studied the potential of NIR spectroscopy for quantifying protein, casein, casein subfractions, and fat contents in goat's milk. A total of 2000 individual milk samples were collected. A sample pretreatment step, using glass fiber filters, was introduced in order to eliminate the water absorption before NIR analyses. Samples were scanned in transmission mode, using a blank glass fiber filter as reference. A first-derivative mathematical transformation of the original spectra was applied to solve standardization problems. Stepwise MLR and PLS were used to obtain the calibration equations for each nitrogen component studied. Linear regression gave the best calibration and validation parameters for total protein and total casein. The PLS method using nine factors was the best method for calibration of casein subfractions ($\alpha_s$-, $\beta$-, and $\kappa$-cn), and to determine fat content the best calibration was obtained using a PLS regression method with eight factors. The authors suggested that NIR spectroscopy could be an useful tool in routine analysis of a large number of goat's milk samples and even more suitable for dairy goat recording schemes.

Perez-Marin et al. (42) significantly improved these results by using MPLS for regression purposes and SNV and detrending treatments to correct for scatter. Nuñez-Sanchez et al. (36) compared the accuracy of folded transmission in liquid milk and reflectance of dried milk NIR calibration equations to predict quality parameters of ewe's milk. The results obtained for each constituent in both NIR analysis modes showed excellent capacity for quantitative analysis with $r^2$ higher than 0.9. Also, the equation obtained for somatic cell count (SCC) by both methods had adequate accuracy, similar to that obtained for goat's milk by Perez-Marin et al. (42) and much higher than the model reported by Tsenkova et al. (54).

Tsenkova et al. (53) examined the feasibility of the NIR range for simultaneous milk composition analysis and mastitis diagnosis at very early stage with high re-peatability with individual cow's milk. For qualitative analysis, to detect mastitis, foremilk spectra of the udder quarters of the examined cows were compared. To highlight the spectra differences when one or more quarters of the same udder were mastitic, a spectral function (F) was created, using mathematical treatment of raw absorbance spectral data at four wavelengths in the near-NIR range. The examined milk samples were defined as mastitic samples when their somatic cell count (SCC) was higher than 500,000/ml. The authors observed changes in the foremilk spectra when SCC increased compared with the other quarters of cows with stable and low SCC. By comparing the spectral function F with SCC content, the accuracy of NIR mastitis diagnosis was 95%, with the same repeatability as SCC but higher sensitivity. From these observations the authors concluded that NIR could be a suitable method for mastitis diagnosis even if a threshold of the spectral function F, corresponding to the standard SCC of 500,000/ml, needed to be determined.

Later, Tsenkova and Atanassova (56) used SIMCA models based on principal components to diagnose cow's mastitis (Fig. 8.4.1). The best calibration results were obtained with first-derivative spectra transformation. The success of NIR mastitis diagnostics was defined by the effect on NIR spectra of milk compositional changes caused by mastitis.

## COAGULATION OF MILK

Several studies have been carried out with the aim of understanding the mechanisms of milk coagulation and of developing sensors able to define the optimal curd cutting time after curd formation. Ustunol et al. (57) monitored the NIR diffuse reflectance profiles at 950 nm of coagulating whole pasteurized milk over 60 min, by using eight commercially available milk-clotting enzyme preparations and comparing several coagulation parameters. The wavelength at 950 nm was chosen on the basis of the detected changes in reflectivity of energy at $950 \pm 5$ nm most likely due to a change in bound water held by casein during coagulation.

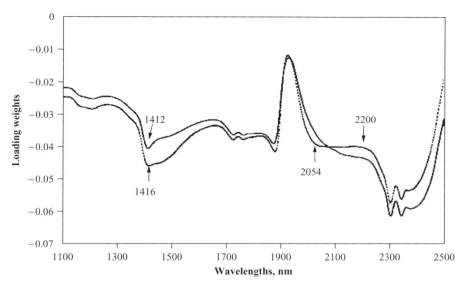

**Figure 8.4.1.** NIR milk spectra: first principal component loading plots of SIMCA models for healthy cows (class 1) and mastitic cows (class 2).—class 1; ... class 2.
*Source*: R. Tsenkova, S. Atanassova. p 123–128. [In Davies A. M. C. and Cho R. K. eds. Near Infrared Spectroscopy, Proceedings of 10th International Conference NIR Publications, Chichester, UK, 2002, **Fig. 2**].

Payne et al. (41) measured diffuse reflectance changes during milk coagulation to develop a cutting time prediction equation with a fiber-optic probe, which utilized a photodiode light source emitting a center wavelength at 940 nm with an half-power bandwidth of 50 nm. This wavelength range was different than that used by Banon and Hardy (2), which, by a turbidimetry method, used a single-wavelength instrument emitting a radiation at 860 nm. Because diffuse reflectance changes during milk coagulation over a typical range of fat, protein, added calcium, pH, temperature, and enzyme concentrations, Payne et al. (41) used a fractional factorial design to test these six factors at three levels each for two enzyme types.

Duffy et al. (15) evaluated the performance of an NIR sensor over a range of process variations normally encountered in Irish cheese making and developed cutting time prediction equations based on comparison to a rheological instrument.

Laporte et al. (29) used a NIR reflectance optic probe for monitoring rennet coagulation of milk from 52 individual cows in comparison with the coagulometer device of INRA (Patent # 8800803) for monitoring Cheddar cheese-type coagulations. Spectral variations were then analyzed by PLS in order to verify the feasibility of NIRS in monitoring milk coagulation, expressed as percent of coagulation. Qualitative evaluation of the original spectra and their first derivative collected during renneting demonstrated that there was a well-shown trend in the NIR reflectance spectra during rennet coagulation.

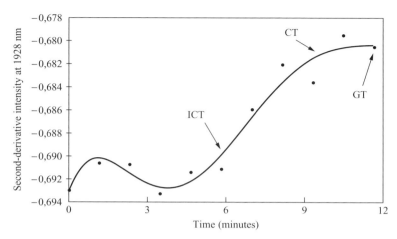

**Figure 8.4.2.** NIR second-derivative absorbance at 1928 nm during coagulation of reconstituted whole milk.
[*Source*: Giangiacomo et al. (1998) *J Near Infrared Spectrosc* **6**: 205–212, **Fig. 12**].

Giangiacomo et al. (21) studied the possibility of detecting the principal critical events related to the primary milk clotting phase by NIR in comparison with other methods based on the interaction of light and material that provide a quick response (Fig. 8.4.2). Because water is the major component in milk and water is involved in the physico-chemistry of coagulation, an attempt was made to use NIR to monitor an evolving process and to detect eventually the relationship between water and the other constituents during the phases changes. It was assessed that NIR was monitoring, along the coagulation process, some rearrangements in water structure according to a specific path in the time domain. It was possible to correctly assign on NIR second derivative vs. time curves the flocks formation time (CT) and the clotting time (GT) measured by visual observation and by Formagraph, respectively.

## CHEESE

Frank and Birth (17) determined the water content of thirty commercial cheese samples. Cheeses were grated and part of them freeze-dried. They observed that water absorption band in the wavelength region over 1850 nm was so great that the reflectance signal was inadequate for useful measurements. Consequently, they used freeze-dried samples to obtain reflectance data over a wide spectral range. The high correlation of reflectance data with moisture content suggested to the authors the potential use of reflectance spectroscopy in measuring cheese composition after appropriate selection of wavelengths for different varieties of cheese with samples representative of the population. They showed evidence of direct correlation between reflectance measurements and protein and fat contents, with some variations from the regression line for composition and variability of cheese varieties.

Frankhuizen and Van Der Veen (18) analyzed 145 commercial samples of cheese consisting of three varieties. The results suggested that the most accurate predictions of composition were obtained if the instrument was calibrated for each specific variety of product. They obtained a lower correlation for individual cheese types than for the global calibration, where the fat content range was much higher. However, even if correlation was higher for the global set of samples, standard error was lower for individual cheeses, in accordance with the suggestions of Frank and Birth (17).

Wehling and Pierce (59) determined the Cheddar cheese moisture content, with a ripeness interval between 3 and 6 months. They concluded that reflectance measurements from grated samples allowed reliable determination of the constituents.

Rodriguez-Otero et al. (46) tried to measure cheese composition by NIR in cheese. They tried to analyze cheese without any prior sample manipulation (not even grating) on the basis of increased knowledge in calibration techniques based on multivariate analysis. Repeatability of NIR moisture determination was approximately double in comparison with the reference method. Repeatability of determination of protein by reference (Kjeldahl) and NIR methods was higher for the NIR spectroscopy, probably because of the large sample size rather than to a lack of method precision. The repeatability of fat determination by reference (gravimetric extraction) and NIR methods was 0.31% for reference and 0.40% for NIRS.

Lee et al. (32) investigated the potential for utilizing NIRS to rapidly determine composition of curds during cheese making. The authors reported that an acceptable linear relationship was found between chemical and NIRS analysis and that the standard deviation for NIRS measurements was lower than that for the chemical values for all the constituents.

Sorensen and Jepsen (50) used both NIR reflectance and NIR transmittance in the quality control of semihard Danbo cheese production with the aim of analyzing ungrated cheese samples. They reported that in the case of NIR reflectance spectroscopy for moisture determination, no significant improvement was obtained by data pretreatment. In the case of NIR transmittance spectroscopy, small improvements were obtained in SECV by data pretreatment, especially with MSC. Concerning protein determination by reflectance spectroscopy, the best result obtained was a SECV of 0.41% in the full range of wavelengths and 0.40% in the segmented range. The same result was achieved by different data pretreatment. By NIR transmittance spectroscopy the result was significantly better, with a SECV of 0.21%. The results were slightly different with different data pretreatments but were the same for the full and segmented spectra. Just as for protein, NIR transmittance spectroscopy provided significantly better results in determining fat content. Significant differences were also found for different data pretreatments.

Wittrup and Norgaard (60) recorded NIR spectra of 2-cm-thick slices of 107 samples of semihard Danbo cheese without pretreatments. The best calibration equation for dry matter was achieved by using four PLS factors and standardized data obtained by dividing each wavelength in the spectra by the standard deviation of this wavelength based on all the calibration samples. The best calibration equation for nitrogen content was obtained by using eight PLS factors and the multiplicative signal correction (MSC). The spectral range used when fat was modeled was different from that used for the other components. The area from 1918 to 2298 nm was excluded because it affected the results. The results of truncating area and using MSC as data pretreatment showed a significant improvement of SEP.

Nuñez-Sanchez et al. (37) evaluated NIR calibration equations for the main constituents of ewe's cheese under two different sample preparation methods (homogenized and intact) and under reflectance and fiber-optic probe. The SECV values obtained for the homogenized cheeses and for both analysis modes were comparable for fat, protein, and dry matter. The calibration statistics for the intact cheese analyzed by fiber-optic probe were higher than those obtained with homogenized cheese.

## FERMENTED MILK PRODUCTS

Rodriguez-Otero and Hermida (47) used NIRS for analysis of total solids content in fermented milk products. They analyzed 141 commercial fermented milk products, unflavored and flavored, made from whole milk and skim milk. Samples were divided into four homogeneous subsets, and reflectance spectra were collected in the range from 400 to 2498 nm. The authors attempted a global calibration using the whole set of data. For total solids the best results were obtained by using first-derivative spectra, with a SEC of 0.083% and a SEP of 0.25%. Statistical evaluation for specific calibration and validation subsets exhibited slightly worse results than when global calibration was used. The reason could be the greater number of samples in global calibration, which allowed the use of more terms without risk of overfitting.

A protein calibration was performed by using a modified PLS (MPLS) regression of first- and second-derivative spectra. In the validation set a standard error of 0.14% was obtained. This value is considered to be quite high, probably due to the variable nonprotein nitrogen (NPN) contents of samples. The error in protein calibration may occur partly because the reference method includes true protein and NPN, whereas in NIRS NPN behaves differently from true protein.

Fat calibration was performed by MPLS regression of first and second derivatives of the spectra. For fat calibration the authors observed no appreciable differences between the statistical indicators with two different data treatments and between the two scatter correction treatments. They recommended the use of the first derivative with any of the two scatter correction treatments because of the better signal-to-noise ratio.

## MILK POWDER

DeVilder and Bossuyt (12) determined water content in industrial skim milk and whole milk powders obtained by low, medium, and high thermal treatment of milk before spray drying. They also used two different technological processes for drying. They concluded that the accuracy of the method, even if very reliable, could be improved by calibrating for each factory or, preferably, for each production line. They determined protein content with a residual standard deviation amounting to 0.18%. Validation of calibration was made on 51 samples of different kinds and origin, with a SEP of 0.18%. In all cases, the authors recorded deviating values for some straight-through powders. They concluded that the kind of powder affected the determination of protein content in whole milk powder and suggested the use of separate calibrations for straight-through or instant granulated powders. Furthermore, the authors reported that the kind of powder affected the fat determination in whole milk powders, suggesting, as in the case for proteins, development to separate calibrations for straight-through or instant-type powders.

At the same time, Baer et al. (1) determined water content, protein, and fat in commercial skim milk powders and in the laboratory prepared from nonfat dry milk (NFDM) samples. They used 44 commercial low-, medium-, and high-heat samples and 38 laboratory prepared samples, to increase the composition range. Data indicated that prediction of moisture content by NIR was linear throughout the tested range (2.85–9.70%). They reported that usable predictions of composition could be obtained even when physical factors such as particle size were not constant. They also concluded that NIR calibrations at individual dairy plants probably result in lower SEP values because of the greater physical homogeneity of samples.

A large number of wavelengths (filters) were needed to determine protein in skim milk powders. This may be because of a partial masking of protein absorption by lactose in wavelength bands at which the protein wavelengths are located. SEC values of 0.254% (compared to micro-Kjeldahl) and 0.352% (compared to dye binding) were found, indicating the influence of accuracy of the reference method in calibration. A set of 22 samples was used for validation, and SEP of 0.438% and 0.509% were found, respectively. These results showed standard errors higher than those reported by De Vilder and Bossuyt (12) with the same instrumentation, probably because of the smaller number of analyzed samples. Fat determination on 18 different (from the calibration) samples produced a SEC of 0.81% and a SEP of 0.99%.

Frankhuizen and Van Der Veen (18), using a filter instrument, estimated moisture in about 300 samples of skim milk powders from different sources and processing technologies. The reference method was the oven drying method. They obtained a SEC of 0.12% in the range 2.9–5.8% and a SEP of 0.08% in the range 3.3–4.7%. These results were comparable to those of De Vilder and Bossuyt (12), suggesting the influence of the reference method for NIR calibration. Concerning protein determination, a high correlation was observed between protein contents estimated by the NIRS and those determined by the Kjeldahl method. The SEC of 0.27% was acceptable compared with the standard deviation of the reference method of about 0.15% for milk powder. The authors also noted that the differences are probably due to the NPN content variability, which differs in NIR absorption from the true protein and could result in an increased standard error in calibration. A prediction set of 19 samples showed a SEP value of 0.20%, comparable with that obtained in calibration. They concluded that this calibration was robust and reliable for total protein content determination of skim milk powders. In the fat concentration range 0.8–2.8, a SEC value of 0.09% was found. Prediction, carried out on 19 samples, confirmed the suitability of the calibration developed, which gave a SEP of 0.09%.

Vuataz (58) studied the possibility of detecting the moisture content in dietetic milk powders during storage. A total of 146 samples were analyzed by a B+L InfraAlyzer 500 in the range of 1400–2400 nm. The milk powder samples were stored up to 12 months at different temperatures in the range 22–65°C.

Barabassy and Kaffka (3) determined a quantitative relationship between water content and NIR spectra for mixtures of multicomponent milk powder products. The water content in a range 0.0–4.8% gave a standard error in calibration of 0.098% with five wavelengths. They also determined a quantitative relationship between the protein content and the NIR spectra for mixtures of multicomponent milk powder products. Reflectance spectra were collected by a scanning PMC Spectralyzer 1025 in the range 1000–2500 nm. The protein content in a range 0.0–80.3% gave a standard error in calibration of 1.89% with five wavelengths. The authors commented on this high value as a consequence of the large calibration range and of possible lactose alterations in the drying process. The fat content in a range 0.0–2.8% gave a standard error in calibration of 0.128% with five wavelengths. On the same set of samples they compared the performance of two NIR instruments: quite different wavelengths were selected in the best equations for the determination of the various constituents.

Barabassy and Turza (4), using the same set of samples, studied the influence of the lyophilization process on spectral response, by taking reflectance spectra of the mixed powders as prepared and after reconstitution and lyophilization. The SEC increased by more than three times after lyophilization, also showing a calibration range double that before lyophilization, most likely because of residual water in the freeze-dried products. In protein determination the SEC increased from 1.89% to 2.772% after lyophilization, and for fat the SEC increased from 0.128% to 0.166%. The authors concluded that the accuracy obtained was acceptable for industrial practice. The use of NIR spectroscopy in milk powder evaluation could be extended also to applications for classification purposes.

Downey et al. (14) used a statistical approach to classify commercial skim milk powders according to heat treatment. They used 66 samples of commercially produced skim milk powder including high-heat, medium-heat, and low-heat powders. Principal component analysis (PCA) was applied to the normalized spectral data, with the use of wavelengths as principal variables and "class" values as supplementary variables. Factorial discriminant analysis (FDA) was performed on the PC scores. Ten components were needed to correctly classify all samples in the calibration development set; 91% of those in the evaluation set were correctly identified. Three samples of the medium-heat class were incorrectly classified, but the authors pointed out difficulties in the exact definition of the heat treatment classes, particularly the medium-heat class.

A further aspect of NIR analysis of milk powders concerns authenticity. Giangiacomo et al. (19) investigated the possibility of using NIR reflectance spectroscopy to develop a quantitative method for determination of whey powder added to milk powder. With a calibration set of 61 samples prepared by mixing whey powder and milk powder to obtain a range of whey addition between 0 and 30%, the best performance from multiple linear regression was obtained with a five-wavelength equation, with a SEC of 0.71%. In validation, good results were obtained using the same type of milk powder as in calibration. When the type of milk powder was changed, the SEP was doubled. A spectral analysis of milk dried by different technological processes indicated that areas of the spectra where major differences among the various milk types were found overlapped the region where differences between whey and milk powder were most noticeable. This could explain the difficulties in obtaining a good prediction on samples of mixtures containing milk powders different than the one used in calibration. Furthermore, the authors tried the potential of discriminant analysis in revealing whey powder addition. A calibration step was carried out on 31 samples of powder mixtures and of milk powders dried in different conditions. A validation set represented by 37 mixtures and 4 milk powders was correctly classified, except for 1 milk powder classified as a mixture.

Maraboli et al. (33) tried to develop a fast and reliable method for detecting vegetable proteins added to milk powder. One hundred and fifty-five samples of genuine and adulterated skim milk powder containing up to 5% of selected vegetable isolates were used and scanned in reflectance mode. Different data pretreatments were applied to obtain the best performance. Pretreated data were processed by using PLSR, PCR, and MLR on a limited number of wavelengths previously selected among those explaining major differences in the set of samples. The best relationship between NIR data and the quantifiable sample property was obtained by applying MLR, which is characterized by a good $R^2$ value, a low SEC value, and a minimum number of outliers, to the first derivative of NIR absorbance values. A performance similar to MLR was obtained by Cattaneo et al. (9) by using PLSR applied to calibrations of individual vegetable isolates (Fig. 8.4.3). The authors in both cases conclude that NIR prediction

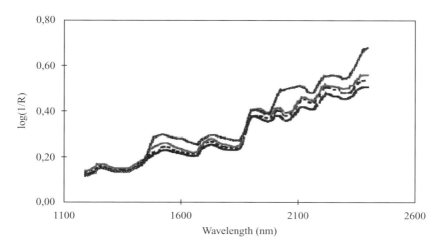

**Figure 8.4.3.** Examples of NIR spectra for genuine skim milk powder (pasteurized . . . ), soy isolate (—), pea isolate (– – –); wheat isolate (—). [Source: Cattaneo T. M. P. et al. (2002), p. 161–166, In Davies A. M. C. and Cho R. K. eds. *Near Infrared Spectroscopy Proceedings of 10th International Conference*, NIR Publications, Chichester, UK, **Fig. 1**].

of adulteration of milk powders by vegetable isolates provided better results than other analytical techniques and NIR could be a suitable method for screening purposes.

## ON-LINE APPLICATIONS

Hall and De Thomas (24) described the development of spectroscopic models for the on-line analysis of fat in milk and cheese and moisture in butter and cheese. For fluid milk analysis, the process interface consisted of an interactance immersion probe inserted into a pipe. This probe was useful for minimally scattering fluid streams as it reflected light that passes through the sample back to the fibers, providing a "double pass" through the sample. For butter analysis, the process interface consisted of two probes mounted at 180° to each other welded to the pipe at the end of the creamer. For process cheese analysis, transmission probes were positioned in a process pipe with sapphire windows.

Parker (40) reported the application of on-line NIR analyses for monitoring continuous-process butter production. The point chosen for NIR sampling was the pipe before the packing line. Two probes, a transmitter and a receiver, were installed in the butter pipe. Fiber-optic cables ran from the probes to the instrument. The laminar flow of the butter in the pipe gave a very reproducible sample presentation with very good spectral data. The instrument was calibrated for moisture in unsalted, salted, and cultured butter and for salt. The same author reported the on-line NIR application in the control of the milk powder spray-drying process.

Pouliot et al. (44) explored, obtaining modest results, potential applications of NIR in evaluating protein denaturation during heat treatment of whey, following up chemical composition (total nitrogen, nonprotein nitrogen, lactose) of whey protein concentrate (WPC) during its production, and evaluating the degree of hydrolysis (DH) during tryptic hydrolysis of whey proteins.

## TWO-DIMENSIONAL CORRELATION SPECTROSCOPY

Experience gained in the last 30 years in 2D NMR appears to lend support to the idea that spreading NIR data yields additional quantitative and qualitative information. This approach has gained popularity and has found application in several areas [Barton et al. (6), Barton et al. (7), Murayama et al. (34), Wu et al. (61), Segtnan et al. (51), Barton et al. (8), Sasic et al. (49), Czarnik-Matusewicz et al. (11), Estephan et al. (16)]. The technique as applied to NIR spectroscopy is called two-dimensional correlation spectroscopy (or 2D-CORR). The theoretical and/or background for 2D-CORR can be found elsewhere [Ozaki et al. (38)]. Spreading spectral peaks over a second dimension can simplify the visualization of complex spectra consisting of many overlapped bands, as in the case of NIR spectra. The "generalized 2D-CORR" method developed by Noda (35) can handle signals fluctuating as an arbitrary function of time, or any other physical variables such as temperature, pressure, or concentration. One of the main advantages, described by Ozaki et al. (39), is the possibility of investigating in detail various inter- and intramolecular interactions through selective correlation of peaks. Barton et al. (5) reported that the primary use of 2D-CORR across the NIR and MIR regions was to interpret and explain the NIR spectra and to provide additional confidence in the analytical models developed with empirical data.

Czarnick-Matusewicz et al. (10) demonstrated the potential of generalized 2D-CORR in analyzing protein and fat concentration-dependent NIR spectral changes of milk after MSC and smoothing. The synchronous spectrum constructed from the fat concentration-dependent NIR spectral changes developed autopeaks at 2311 and 2346 nm, whereas the corresponding spectrum from the protein concentration-dependent variations showed autopeaks at 2047, 2075, 2100, 2351, and 2375 nm. The appearances of the autopeaks meant that the intensities of those bands vary most significantly with the increase of the concentrations of fats and proteins in milk. Power spectra along the diagonal line on the synchronous spectra indicated that generalized 2D-CORR can select bands due to fat and proteins separately from the complicated NIR spectra of milk. An asynchronous 2D NIR correlation map generated from the same set of spectra and a slice of this map at a single wavelength revealed a number of peaks missing in the corresponding synchronous spectrum. This meant that, in general, asynchronous correlation spectra have more powerful deconvolution ability for highly overlapped bands.

Laporte et al. (31) applied two-dimensional spectroscopy to investigate whether denaturation and aggregation phenomenon during processing of whey protein isolates (WPI) could be observed and interpreted in both mid- and near-infrared regions. Synchronous 2D mid-IR and NIR correlation contour maps of spectra of WPI heated at various temperatures highlighted that denaturation processes, related to protein unfolding events, were described by disappearance of several bands and formation of new bands for intermolecular hydrogen bonds. On the basis of these results the authors concluded that 2D-CORR provided a new insight into the study of thermal denaturation and aggregation phenomenon of WPI and led to the assignment of NIR regions highly correlated with well-characterized mid-IR denaturation and aggregation bands.

Giardina et al. (22) used 2D-NIR-COSS to study water molecules rearrangements during the process of milk rennet coagulation (Fig. 8.4.4). They analyzed the water combination band at 1930 nm by two-dimensional analysis of NIR/NIR spectra. Absorptions above 1924 nm suggested that the process involves in particular H-bonded water molecules, rather than free molecules, and more specifically molecules with one H-bond. When $\kappa$-casein, which is the hydrophilic part of the micelle, starts to release the caseino-macropeptide, micelles reorganize and consequently water finds a new equilibrium, by breaking the old bonds and creating new ones. This mechanism involves in particular, the S1 form (molecules with 1 H-bond). Two-dimensional maps permitted the interpretation of secondary interactions of water and water constituents. When studied through water rearrangements and modifications, structural rearrangements of constituents could be monitored long before the apparent system destabilization assessed by classic methods.

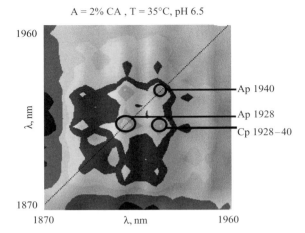

A = 2% CA , T = 35°C, pH 6.5

**Figure 8.4.4.** Contour map of the synchronous 2D NIR correlation spectrum of standard coagulation process. [Source: Giardina C. et al. (2004), p. 187–190, In Davies A.M.C. & Garrido-Varo A. Eds. *Proceedings of 11th International Conference on Near Infrared Spectroscopy.* NIR Publications, Chichester, UK, **Fig. 2**].

## CONCLUSIONS

The literature cited in this chapter and the various aspects discussed represent the most important advances in the use of spectroscopy applied to milk and dairy products. The subjects and the papers cited are not inclusive of all applications, yet it is clear that this NIR technology has enormous potential both now and in the future, particularly with the recent advances in hardware, software, fiber optic, chemometry, and two-dimensional analyses of spectral data. These developments could open up further applications for NIR spectroscopy monitoring of both production processes and final product quality.

## REFERENCES

1. R. J. Baer, J. F. Frank, M. Loewenstein. Compositional analysis of nonfat dry milk by using near infrared diffuse reflectance spectroscopy. *J Assoc Off Anal Chem* **66**(4): 858–863, 1983.
2. S. Banon, J. Hardy. Study of acid milk coagulation by an optical method using light reflection. *J Dairy Res* **58**: 75–83, 1991.
3. S. Barabassy, K. Kaffka. The application possibilities of the near infrared technique in the non destructive investigation of mixed milk powder products. *J Food Phys* **57**: 39–43, 1993.
4. S. P. Barabassy, S. Turza. Investigation of the lyophilization effect on mixed milk powder products by near infrared spectroscopy. In: *Near Infrared Spectroscopy: The Future Waves*, A. M. C. Davies, Ph. Williams, eds. NIR Publications, Chichester, UK, 1996, p. 611–616.

5. F. E. Barton, II, D. S. Himmelsbach, J. H. Duckworth, M. J. Smith. Two-dimensional vibrational spectroscopy: correlation of mid- and near infrared regions. *Appl Spectrosc* **46**: 420, 1992.

6. F. E. Barton, II, D. S. Himmelsbach, D. E. Akin, A. Sethuraman, K. -E. L. Eriksson. Two-dimensional vibrational spectroscopy III: interpretation of the degradation of plant cell walls by white rot fungi. *J Near Infrared Spectrosc* **3**: 25–34, 1995.

7. F. E. Barton, II, D. S. Himmelsbach, D. D. Archibald. Two-dimensional vibrational spectroscopy V: Correlation of mid- and near infrared of hard red winter and spring wheats. *J Near Infrared Spectrosc* **4**: 139–152, 1996.

8. F. E. Barton, II, D. S. Himmelsbach, A. M. McClung, E. L. Champagne. Two-dimensional vibrational spectroscopy of rice quality and cooking. *Cereal Chem* **79(1)**: 143–147, 2002.

9. T. M. P. Cattaneo, A. Maraboli, S. Barzaghi, R. Giangiacomo. Detection of soy, pea and wheat proteins in milk powder by near infrared spectroscopy. In: *Near Infrared Spectroscopy: Proceedings 10th Internation Conference*, A. M. C. Davies, R. K. Cho, eds. NIR Publications, Chichester, UK, 2002, p. 161–166.

10. B. Czarnick-Matusewicz, K. Murayama, R. Tsenkova, Y. Ozaki. Comparison of two-dimensional correlation analysis and chemometrics in near infrared spectroscopy: protein and fat concentration-dependent spectral changes. In: *Near Infrared Spectroscopy: Proceedings of 9th International Conference*, A. M. C. Davies, R. Giangiacomo, eds. NIR Publications, Chichester, UK, 2000, p. 17–23.

11. B. Czarnik-Matusewicz, K. Murayama, Y. Wu, Y. Ozaki. Protein-water interaction monitored by analysis of near infrared spectra. In: *Near Infrared Spectroscopy: Proceedings of 11th International Conference*, A. M. C. Davies, A. Garrido-Varo, eds. NIR Publications, Chichester, UK, 2004, p. 913–918.

12. J. De Vilder, R. Bossuyt. Practical experiences with an InfraAlyzer 400 in determining the water, protein and fat content of milk powder. *Milchwissenschaft* **38(2)**: 65–69, 1983.

13. E. Díaz-Carrillo, A. Muñoz-Serrano, A. Alonso-Moraga, J. M. Serradilla-Manrique. Near infrared calibrations for goat's milk components: protein, total casein, $\alpha_s$-, $\beta$-, and $\kappa$-caseins, fat and lactose. *J Near Infrared Spectrosc* **1**: 141–146, 1993.

14. G. Downey, P. Robert, D. Bertrand, P. M. Kelly. Classification of commercial skim milk powders according to heat treatment using factorial discriminant analysis of near-infrared reflectance spectra. *Appl Spectrosc* **44(1)**: 150–155, 1990.

15. A. P. Duffy, D. J. O'Callaghan, E. P. Mulholland, J. Fitzpatrick, C. P. O'Donnell, F. A. Payne. Evaluation of an on-line NIR sensor to predict renneted cheese curd firmness during coagulation. In: *Engineering & Food at ICEF7*, R. Jowitt, eds. Part 1. Brighton (UK), Sheffield Academic Press Ltd, Sheffield, UK, 1997, p. A49–A52.

16. N. Estephan, A. Barros, I. Delgadillo, D.N. Rutledge. Characterisation and classification of vegetable oils by combining near and mid infrared signals. In: *Near Infrared Spectroscopy: Proceedings of 11th International Conference:* A. M. C. Davies, A. Garrido-Varo, eds. NIR Publications, Chichester, UK, 2004, p. 583–588.

17. J. F. Frank, G. S. Birth. Application of near infrared reflectance spectroscopy to cheese analysis. *J Dairy Sci* **65**: 1110–1116, 1982.

18. R. Frankhuizen, N. G. Van der Veen. Determination of major and minor constituents in milk powders and cheese by near infra-red reflectance spectroscopy. *Milk Dairy J* **39**: 191–207, 1985.

19. R. Giangiacomo, F. Braga, C. Galliena. Use of NIR spectroscopy to detect whey powder mixed with milk powder. In: *Near Infrared Spectroscopy*, *Making Light Work*: *Advances in* I. Murray, I. A. Cowe, eds. VCH Weinheim-New York-Basel-Cambridge, 1991, p. 399–407.

20. R. Giangiacomo, R. Nzabonimpa. Approach to near infrared spectroscopy. *Bull Int Dairy Fed* **298**: 37–42, 1994.

21. R. Giangiacomo, R. Lizzano, S. Barzaghi, T. M. P. Cattaneo, A. S. Barros. NIR and other luminometric methods to monitor the primary clotting phase of milk. *J Near Infrared Spectrosc.* **6**: 205–212, 1998.

22. C. Giardina, N. Sinelli, T. M. P. Cattaneo, R. Giangiacomo. 2D-IR-COSS as a tool in understanding milk rennet coagulation processes. In: *Near Infrared Spectroscopy: Proceedings of 11th International Conference*, A.M.C. Davies, A. Garrido-Varo, eds. NIR Publications, Chichester, UK, 2004, p. 187–190.

23. J. W. Hall, K. Chan. Near-infrared spectroscopic analysis of bovine milk for fat, protein and lactose. In: *Proceedings of Cheese Yield & Factors Affecting its Control*, IDF Seminar, Cork, Ireland – April 1993, International Dairy Federation Publishing, Bruxelles, Belgium, 1993, p. 230–239.

24. J. W. Hall, F.A. De Thomas. Near-infrared spectroscopic analysis of bovine milk for fat, protein and lactose. In: *Proceedings of Cheese Yield & Factors Affecting its Control*, IDF Seminar, Cork, Ireland – April 1993, International Dairy Federation Publishing, Bruxelles, Belgium. 1993, p. 222–229.

25. H. Kamishikiryo, K. Hasegawa, T. Matoba. Stability of 2179 nm as a key wavelength for protein analysis by near-infrared spectroscopy. *J Jpn Food Sci Technol* **38**: 850–857, 1991.

26. H. Kamishikiryo, K. Hasegawa, H. Takamura, T. Matoba. Near-infrared spectroscopic measurement of protein content in oil/water emulsions. *J Food Sci* **57**: 1239–1241, 1992.

27. H. Kamishikiryo, Y. Oritani, H. Takamura, T. Matoba. Protein content in milk by near-infrared spectroscopy. *J Food Sci* **59**: 313–315, 1994.

28. IDF World Dairy Situation. Bulletin N°384/2003, International Dairy Federation Publishing, Bruxelles, Belgium, 2003, p. 5–8.

29. M. F. Laporte, R. Martel, P. Paquin. The near-infrared optic probe for monitoring rennet coagulation in cow's milk. *Int Dairy J* **8**: 659–666, 1998.

30. M. F. Laporte, P. Paquin. Near-Infrared analysis of fat, protein and casein in cow's milk. *J Agric Food Chem* **47**: 2600–2605, 1999.

31. M. F. Laporte, M. Subirade, P. Paquin. Two-dimensional mid infrared and near infrared correlation spectroscopy: a useful tool for study and quantification of heat denaturation and aggregation of whey protein isolates. In: *Near Infrared Spectroscopy: Proceeding of 9th International Conference*, A. M. C. Davies, R. Giangiacomo, eds. NIR Publications, Chichester, UK. 2000, p. 9–15.

32. S. J. Lee, I. J. Jeon, L. H. Harbers. Near-infrared reflectance spectroscopy for rapid analysis of curds during Cheddar cheese making. *J Food Sci* **62(1)**: 53–56, 1997.

33. A. Maraboli, T. M. P. Cattaneo, R. Giangiacomo. Detection of vegetable proteins from soy, pea and wheat isolates in milk powder by near infrared spectroscvopy. *J Near Infrared Spectrosc* **10**: 63–69, 2002.

34. K. Murayama, B. Czarnik-Matusewicz, Y. Wu, R. Tsenkova, Y. Ozaki. Comparison between conventional spectral analysis methods, chemometrics, and Two-dimensional

Correlation Spectroscopy in the analysis of near-infrared spectra of protein. *Appl. Spectrosc* **54(7)**: 978, 2000.

35. I. Noda. Two-Dimensional infrared (2D-IR) spectroscopy: theory and applications. *Appl Spectrosc* **44(4)**: 550–561, 1990.

36. N. Nuñez-Sanchez, A. Garrido-Varo, J. Serradilla, L. Ares. Near infrared analysis of liquid and dried ewe milk. In: *Near Infrared Spectroscopy: Proceedings 10th International Conference*, A. M. C. Davies, R. K. Cho, eds. NIR Publications, Chichester, UK, p. 179–182, 2002.

37. N. Nuñez-Sanchez, A. Garrido-Varo, J. Serradilla–Manrique, L. Ares-Cea. Reflectance versus interactance reflectance near infrared analysis of homogenised and intact ewe cheese. In: *Near Infrared Spectroscopy: Proceeding 9th International Conference*, A. M. C. Davies, R. Giangiacomo, eds. NIR Publications, Chichester, UK, 2000, p. 135–138.

38. Y. Ozaki, S. Šašic, T. Tanaka, I. Noda. Two-Dimensional Correlation Spectroscopy: Principle and Recent Theoretical Development *Bull Chem Soc Jpn* **74**: 1–17. 2001.

39. Y .Ozaki, S. Šašic, J. Jiang. How can we unreveal complicated near infrared spectra? Recent progress in spectral analysis methods for resolution enhancement and band assignments in the near infrared region. *J Near Infrared Spectrosc* **9**: 63–77, 2001.

40. E. F. Parker. NIR analysis of dairy products in the New Zealand dairy industry. In: *Near Infrared Spectroscopy Leaping Ahead With*, G. D. Batten, P. C. Flinn, L. A. Welsh, A. B. Blakeney, eds. NIR Spectroscopy Group, Royal Australian Chemical Institute Publishing, North Melbourne, Victoria, Australia. 1995, p. 282–286.

41. F. A. Payne, C. L. Hicks, Pao-Sheng Shen. Predicting optimal cutting time of coagulating milk using diffuse reflectance. *J Dairy Sci* **76**: 48–61, 1993.

42. M. D. Perez-Marin, A. Garrido-Varo, J. M. Serradilla, N. Nuñez , J. L. Ares, J. Sanchez. Chemical and microbiological analysis of goat's milk, cheese and whey by near infrared spectroscopy. In: *Near Infrared Spectroscopy: Proceeding 10th International Conference*, A. M. C. Davies, R. K. Cho, eds. NIR Publications, Chichester, UK, 2002, p. 225–228.

43. A. Polesello, R. Giangiacomo. Application of near infrared spectrophotometry to the nondestructive analysis of foods: a review of experimental results. *Crit Rev Food Sci Nutr* **18**: 203–237, 1983.

44. M. Pouliot, P. Paquin, R. Martel, S. F. Gauthier Y. Pouliot. Whey changes during processing determined by near infrared spectroscopy. *J Food Sci* **62(3)**: 475–479, 1997.

45. A. Purnomoadi, K. K. Batajoo, K. Ueda, F. Terada. Influence of feed source on determination of fat and protein in milk by near-infrared spectroscopy. *Int Dairy J* **9**: 447–452, 1999.

46. J. L. Rodriguez-Otero, M. Hermida, A. Cepeda. Determination of fat, protein, and total solids in cheese by near-infrared reflectance spectroscopy. *J AOAC Int* **78(3)**: 802–806, 1995.

47. J. L. Rodriguez-Otero, M. Hermida. Analysis of fermented milk products by near-infrared reflectance spectroscopy. *J AOAC Int* **79(3)**: 817–821, 1996.

48. J. L. Rodriguez-Otero, M. Hermida, J. Centeno. Analysis of dairy products by near-infrared spectroscopy: a review. *J Agric Food Chem* **45(8)**: 2815–2819, 1997.

49. S. Šašić, Y. Katsumoto , H. Sato, Y. Ozaki. Applications of moving window Two-Dimensional correlation spectroscopy to analysis of phase transitions and spectra classification. *Anal Chem* **75**: 4010–4018, 2003.

50. L. K. Sørensen, R. Jepsen. Comparison of near infrared spectroscopic techniques for determination of semi-hard cheese constituents. *Milchwissenschaft* **53(5)**: 263–267, 1998.

51. V. H. Segtnan, S. Šašić , T. Isaksson, Y. Ozaki. Studies on the structure of water using two-dimensional near-infrared correlation spectroscopy and principal component analysis. *Anal Chem* **73**: 3153–3161, 2001.

52. Standard 141B, FIL-IDF. Whole milk: Determination of milk fat, protein and lactose content – Guide for the operation of mid-infrared instruments. International Dairy Federation, Brussels, Belgium, 1996.

53. R. Tsenkova, K. I. Yordanov, K. Itoh , Y. Shinde, J. Nishibu. Near infrared spectroscopy of individual cow milk as a means for automated monitoring of udder health and milk quality. In: *Dairy Systems for the 21st Century*, Proceedings of the third International Dairy Housing Conference R. Bucklin, ed. Orlando, Florida, February 2–5, 1994, ASAE, Michigan, USA, 1994, p. 82–91.

54. R. Tsenkova, K. Itoh, J. Himoto, K. Asahida. NIR spectroscopy analysis of unhomogenized milk for automated monitoring in dairy husbandry. In: *Near Infrared Spectroscopy: Leaping Ahead With*, G.D. Batten, P. C. Flinn, L. A. Welsh, A. B. Blakeney, eds. NIR Spectroscopy Group, Royal Australian Chemical Institute Publishing, North Melbourne, Victoria, Australia, 1995, p. 329–333.

55. R. Tsenkova, S. Atanassova, K. Toyoda, Y. Ozaki, K. Itoh, T. Fearn. Near-Infrared Spectroscopy for dairy management: measurement of unhomogenized milk composition. *J Dairy Sci* **82**: 2344–2351, 1999.

56. R. Tsenkova, S. Atanassova. Mastitis diagnostics by near infrared spectra of cow's milk, blood and urine using soft independent modelling of class analogy classification. In: *Near Infrared Spectroscopy: Proceedings 10th International Conference*, A. M. C. Davies, R. K. Cho, eds. NIR Publications, Chichester, UK, 2002, p. 123–128.

57. Z. Ustunol, C. L. Hicks, F. A. Payne. Diffuse reflectance profiles of eight milk-clotting enzyme preparations. *J Food Sci* **56(2)**: 411–415, 1991.

58. G. Vuataz. Some NIR observations of lactose crystallization in milk powders during storage. In: *Near-Infrared Spectroscopy: Proceedings 3rd Int Conf*, R. Biston, N. Bartiaux-Thill eds. Agricultural Research Centre Publishing, Gembloux, Belgium, 1990, p. 218–234.

59. R. L. Wehling, M. M. Pierce. Determination of moisture content in cheddar cheese by near infrared reflectance spectroscopy. *J Assoc Off Anal Chem* **71**: 571–574, 1988.

60. C. Wittrup, L. Nørgaard. Rapid near infrared spectroscopic screening of chemical parameters in semi-hard cheese using chemometrics. *J Dairy Sci* **81**: 1803–1809, 1998.

61. Y. Wu, K. Murayama, Y. Ozaki. Two-Dimensional infrared spectroscopy and Principal Component Analysis studies of the secondary structure and kinetics of Hydrogen-Deuterium exchange of Human Serum Albumin. *J Phys Chem* **105**: 6251–6259, 2001.

# CHAPTER 9

# OTHER TOPICS

# Fermentation Engineering

TAKUO YANO

## INTRODUCTION

Near-infrared spectroscopy (NIR), which is a nondestructive analytical technique, has been employed for the simultaneous prediction of the concentrations of several substrates, products, and constituents in the mixture sampled from fermentation process. In this chapter, applications of NIR to monitoring of the various fermentation processes are introduced. The fermentation processes mentioned here are wine, beer, Japanese sake, miso (soybean paste), soy sauce, rice vinegar, alcohol, lactic acid, glutamic acid, mushroom, enzymatic saccharification, biosurfactant, penicillin, and compost. The analysis of molasses, which is a raw material of fermentation, with NIR is also introduced. These studies indicate that NIR is a useful method for monitoring and control of fermentation process.

## FERMENTATION PROCESS

From ancient times, various foods, bread, yogurt, cheese, vinegar, and alcoholic beverages have been produced by fermentation. Amino acids (glutamic acid, lysine, etc.), organic acids (citric acid, lactic acid, etc.), cell masses (yeast, single cell protein, chlorella, etc.), nucleic acids (inosine monophosphate, guanosine monophosphate, etc.), antibiotics (penicillin, streptomycin, etc), enzymes (amylase, cellulase, lipase, protease, etc.), lipids, polysaccharides, vitamins, antibodies, and anticanceragents have been also produced with microorganisms, animal cells, or plant cells.

A bioreactor, also called a fermentor, used in aerobic fermentation has an aeration and an agitation system to supply oxygen into the culture broth. When nutrients are supplied to the culture broth suitably in addition of the supply of oxygen, organisms can grow until the product(s) inhibits the growth. High-density culture can be achieved

*Near-Infrared Spectroscopy in Food Science and Technology*, Edited by Yukihiro Ozaki, W. Fred McClure, and Alfred A. Christy.

in some cases. To operate a fermentation process with high efficiency, it is very important that the concentration of nutrients and the environmental factors are maintained in good condition. Primary and secondary measurement items in the fermentation process include inner-pressure of the fermentor measured with a diaphragm pressure gauge, aeration rate, agitation speed, dissolved oxygen concentration of the culture broth, partial pressures of oxygen and carbon dioxide in exhaust gas and respiratory quotient calculated from these partial pressures, optical density of the culture broth, temperature, pH of the culture broth, and amount of acid or base added to control pH. To measure the items mentioned above, the same instruments and sensors used in the other industry process monitoring are applied to the fermentation process and these monitoring systems satisfy the needs of fermentation process.

The concentrations of nutrients such as carbon and nitrogen sources and mineral ions are very important operational indicators, and the measurement of these factors accurately is essential for fermentation with high efficiency. Measurement of the concentration of the product, which inhibits the growth and the production of the main product, is also a very important factor for high-efficiency fermentation. Liquid and gas chromatographs are used to measure the concentrations of carbon source and product(s). However, a long operational time, from a few to several dozen minutes, is usually required in these chromatographs. For fermentation process management, an easy, speedy, aseptic, and nondestructive monitoring system is favorable and has been desired for a long time.

The fermentation process is operated under aseptic and optimum environmental conditions and ordinary temperature and pressure. In addition to the multicomponent culture broth, the concentration of each component contained in the culture broth is changed over the cultivation time. Therefore, heat resistance for steam sterilization and measurement specificity for the target compound are required for a monitoring system applied to fermentation process.

## NEAR-INFRARED SPECTROSCOPY

NIR is widely used for rapid and nondestructive analysis in industries such as agriculture, food, pharmaceuticals, textiles, cosmetics, and polymer production. NIR has several advantages. In addition to nondestructive and rapid analysis, several components can be assayed simultaneously. The sample is supplied to the photometer without pretreatment. Both dry and wet materials can be accepted as a sample. Therefore, both aqueous and solid fermented samples can be accepted. NIR is an assay technique with high precision, and on-line measurement is available with fiber optics. These characteristics of NIR are an excellent match for the requirements of monitoring systems for the fermentation process. NIR will be increasingly apt for fermentation process monitoring.

## APPLICATION OF NIR TO FERMENTATION PROCESS

In this section, application of NIR to fermentation process is introduced. NIR has been applied to the measurement of the concentrations of constituents in wine, beer, sake, miso, soy sauce, vinegar, alcohol fermentation, lactic acid fermentation, mushroom production, enzymatic saccharification, and biosurfactant, penicillin, and compost fermentation.

## Wine

Wine is brewed by the addition of yeast to grape juice. In a natural wine, all the ethanol present has been produced by fermentation. Wines are distinguished by color, flavor, bouquet or aroma, and ethanol content. Wines are red, white, or rose (depending on the grape used and the amount of time the skins have been left to ferment in the juice). Wines are also classified as dry or sweet, according to whether the grape sugar is allowed to ferment completely into ethanol (dry) or some residual sugar has been left (sweet).

To prepare the grape juice called must, the grapes are gathered when fully ripe and expressed mechanically. For red wines, the must is fermented with the skins and pips, from which the newly formed alcohol extracts coloring matter and tannin. Fermentation starts when wine yeasts (*Saccharomyces* sp.) existing on the skins of ripe grapes come in contact with the must. It may take from a few days to several weeks, according to the temperature (ordinarily 15°C) and the amount of yeast present or inoculated. When the new wine has become still and fairly clear, it is run off into large casks, where it undergoes a complicated series of chemical processes including oxidation, precipitation of proteins, and formation of esters that create a characteristic bouquet. The wine is periodically clarified and then racked into smaller casks. After some months or several years, the wine is ripe for bottling.

The concentrations of ethanol, fructose, and tartaric acid in white and red wines have been measured with NIR (1). The optical density (transmittance) of a 1-mm thickness of the wine sample was recorded in the 800-to 2400-nm-wavelength region by 0.8-nm steps. Six wavelengths, 1072, 1450, 1696, 1816, 2150, and 2232 nm, were found to be the most characteristic for ethanol, fructose, and tartaric acid. SEP was less than 0.1 v/v% for ethanol, 2 $g \cdot l^{-1}$ for fructose, and 1 $g \cdot l^{-1}$ for tartaric acid.

Wine quality grading by NIR has been studied (2). Good correlations between NIR spectra and testing data were obtained with dry red wines. Turbid samples were clarified by centrifugation. The samples were scanned in transmission mode with a Foss NIRSystems 6500 spectrometer, over the wavelength range 400–2500 nm, using a 1-mm path length. The calibration used partial least-squares (PLS) factor loadings in the wavelengths related to anthocyanins, ethanol, and possibly phenolic compounds. For dry white wines, calibrations were less significant and appeared to be more dependent on the ethanol-related regions. The tawny port wine quality calibrations indicated strongest loadings in the water regions of the spectrum, suggesting that the concentration of the wines was important, and the visible and alcohol regions of the spectrum also featured as important factors. The tawny port wines were aged for long periods in wooden barrels, The water is lost through the barrels, and the wines are concentrated during aging.

Soft independent modeling of class analogy (SIMCA) has been applied on FT-NIR spectra of must and wine to discriminate between the samples in terms of free amino nitrogen values, the status of the malolactic fermentation, and the level of ethyl carbamate present (3). A high recognition rate was obtained for each term.

The important red wine grape quality parameters, total anthocyanin concentration, TSS (total soluble solids, mainly sugars, in Brix), and pH, have been measured by NIR (4). Homogenized samples were scanned in diffuse reflectance mode on a Foss NIRSystems 6500 spectrometer. The optimal total anthocyanin calibrations used two to four PLS factors and the pH calibrations four to five factors. For anthocyanins, large "global" calibrations across multiple cultivation regions, grape varieties, and seasons ($n = 909$) provided useful degrees of measurement precision with 0.90 multiple correlation coefficient $R^2$ and 0.14 standard error of calibaration (SEC). The calibrations further refined to particular regions, varieties, and seasons provided significantly greater precision. For pH ($n = 912$) and TSS ($n = 909$), good results were also obtained with 0.81 and 0.98 $R^2$ and 0.08 and 0.33 SEC, respectively.

## Beer

In beer production process, barley is transformed into malt by steeping the grains in water and allowing them to germinate to break the complex molecules of starch, cellulose, and protein inside each grain so they can be used in the brewing process. This reduces the enzyme activity but increases color and flavor, and small quantities of these darker malts are used to provide the color and flavor of the final beer.

The crushed malt is mixed or mashed with hot water. This dissolves the starch and allows the enzymes in the malt to convert the starch to sugars. When this process is complete, the sugary liquid or wort is separated from the grain residue and boiled. Hops are added at the start of the boil to provide bitterness and may also be added near the end of the boil to add aroma and flavor. The wort is transferred to a fermentation vessel. Fermentation begins with the addition of yeast to the wort. During fermentation, the yeast converts the sugars in the wort to ethanol, carbon dioxide, and other subtle flavor compounds. As the sugars are used up, the fermentation is stopped. This rough beer is then filtered to remove the yeast and produce bright beer, which is carbonated with carbon dioxide, pasteurized, and packaged.

In the study of Coventry et al. (5), beer samples from different countries covered all major types. Color varied from dark Guiness-type stouts to light lagers, and ethanol content ranged from 0% in the alcohol-free lagers to 11% in the high-original gravity barley wines. All samples were degassed by filtration and scanned from 1100 to 2500 nm with a Pacific Scientific Gardner/Neotec 6350 Mark II spectrocomputer configured for transmission measurement. The influence of color and the brewing process on NIR was investigated, and no significant difference was found. An important wavelength for ethanol was 1672 nm, and the performance of the first-derivative equation is vastly superior to that of the raw spectral data.

Both transmission and transflectance NIR have been evaluated for the rapid analysis of beer components (6). The sample set consisted of 52 ales, 34 lagers, 37 light ales, 21 brown ales, 14 stouts, 14 barley wines, 3 mild ales, and 11 diet lagers. Each beer was degassed by stirring at ambient temperature until the form collapsed (5–10 min). NIR analysis was performed with two instruments, transmission work on a Neotec

6350 Mark II and transflectance work on a Neotec 6350 Mark I. The transmission data were measured with a cuvette of 3-mm light path length and the transflectance data with one of 0.25-mm light path length. Data were recorded over the wavelength range 1100 to 2500 nm.

For ethanol analysis by transmission NIR spectroscopy, good results were obtained, with SEC = 0.16%, simple correlation coefficient ($r$) = 0.998, and standard error of prediction (SEP) = 0.16% with first-derivative values of the optical data obtained at the wavelength of 1672 nm. The ethanol content of low-alcohol beers defined as <1 v/v% could also be measured by NIR to the same accuracy, but this might not be sufficient for alcohol-free beers defined as <0.05%, although, for ethanol analysis by transflectance NIR, good results were also obtained with SEC = 0.08%, $R$ = 0.999, and SEP = 0.16% with first-derivative values of the optical data obtained at the wavelengths of 1672 and 1212 nm. Ethanol can be measured accurately and precisely in beer by both transmission and transflectance NIR methods.

For gravity analysis by transmission NIR spectroscopy, good results were obtained, with SEC = 0.94, $R$ = 0.999, and SEP = 1.27 degrees with first-derivative values of the optical data obtained at the wavelengths of 1542 and 1426 nm, while, for gravity analysis by transflectance NIR, good results were also obtained, with SEC = 1.21 degrees, $R$ = 0.998, and SEP = 1.15 degrees with first-derivative values of the optical data obtained at the wavelengths of 1562, 1412, and 1216 nm. The SEC and SEP values at the transflectance NIR were all slightly worse than the corresponding figures with transmission.

However, it was not possible to construct a reasonable calibration equation for either total soluble nitrogen or bitterness with either transmission or transflectance NIR. With transmission NIR, the best SEC obtained for total soluble nitrogen was 140 mg·l$^{-1}$ and the correlation coefficient was 0.94, and for bitterness was 8 B.U. and correlation coefficient was 0.72, which were far too great to be of practical use. These constituents are found in beer in only ppm concentrations, and NIR methods may be not capable of reliable measurement at these levels.

## Japanese Sake

The brewing process of Japanese sake is more complex than that employed for beer and wine. Brown rice is polished until 20–70% of the periphery of the rice is removed. The amount of polishing greatly influences the taste of sake. Polishing must be done gently so as not to generate too much heat, which adversely affects water absorption, and not to crack the rice kernels, which is not good for the fermentation process. A greater amount of the periphery of the rice is removed for higher grades of sake product.

The next process is washing and soaking. These operations are carried out to wash away the white powder left on the rice after polishing and to attain a certain water content deemed optimal for steaming. The conditions of these processes make a significant difference in the final quality of the steamed rice. The more polished rice causes faster water absorption and shorter soaking time.

Next the rice is steamed to give the rice a slightly harder outside surface and a softer center. The steaming process is also an important process in sake fermentation.

Making rice malt (Koji) is the heart of sake brewing. Koji mold, *Aspergillus oryzae*, is sprinkled on the steamed rice after cooling the rice. Koji making is carried out in a special room controlled for humidity and temperature. Yeast starter is prepared with rice malt, steamed rice, yeast, and water.

The main alcohol fermentation is started by moving yeast starter and adding water, steamed rice, and rice malt to a larger tank. The temperature and other factors of the alcohol fermentation process are measured and adjusted to create the flavor and taste profiles of the sake product. After the fermentation, the mash is pressed with a filter press to remove the white lees and unfermented solids, and clear sake is obtained.

Application of NIR to the sake brewing process was reviewed by Wakai (7). As mentioned above, the moisture and protein contents of polished rice are very important factors in sake brewing. These contents of rice were measured with a NIR photometer, InfraAlyzer 400LR (8). Good results were obtained: The values of $R$ were 0.9452 and 0.9868 for moisture and protein contents, respectively. However, the wavelengths used were not mentioned.

NIR spectra of polished rice have been measured by a spectrophotometer, Comp-scan 3000 (9). The true polishing ratio of rice could be estimated accurately and easily by NIR. Although values of absorbance and true polishing ratio were most affected by the moisture content of rice, these influences could be avoided by measuring at a wavelength of 2171 nm and by adjusting the moisture content of samples to from 2% to 6%. Good results of $R = 0.934$ and SEP $= 2.81\%$ were obtained with second-derivative values of four wavelengths, 2170, 2274, 2103, and 2300 nm. Optical data at these four wavelengths were affected by the contents of crude fat and crude protein.

Making rice malt (Koji) is the heart of sake brewing. After ethanol dehydration and grinding rice malt, NIR spectra of the rice malt have been measured from the wavelength of 1900 to 2320 nm by a spectrophotometer, Compscan 3000 (10). Good results were obtained, but the details were not clear.

Without pretreatment of rice malt, NIR spectra of rice malt were measured from the wavelength of 1100 to 2500 nm by a spectrophotometer and estimation of $N$-acetylglucosamine content was attempted (11). Good result was obtained, with $R = 0.99$.

Kojima et al. applied NIR to determination of mycelial weight of *A. oryzae* in Koji (12). The mycelial weight in Koji is the most important factor in Japanese sake processing because the reaction rate from starch to sugar depends on the content of enzymes produced by *A. oryzae*. NIR measurements of ground Koji rice were made over the 1100-to 2500-nm region. A calibration equation for mycelial weight in Koji was formulated with the NIR spectrum data at 1730, 1738, 2348, and 2360 nm, and the values of $R$ and SEC were 0.98 and 0.56 mg·g$^{-1}$, respectively. A good calibration equation was obtained by MLR based on the NIR spectral data and the mycelial weight measured by enzymatic method. Aramaki et al. also formulated a calibration equation for mycelial weight in Koji (13). NIR data at 2348 nm were also used in this calibration equation. The assignment of the absorbance at 2348 nm may be caused by lipid in *A. oryzae*.

Concentrations of various properties of sake, ethanol, nihonshu-do, acid content, amino acid content, and total sugar content have been measured with a NIR photometer, InfraAlyzer 400LR (14). Although the wavelengths used were not clear, good results were obtained, $R = 0.9999, 0.9945, 0.9982, 0.9897$, and $0.9949$ for ethanol, nihonshu-do, acid content, amino acid content, and total sugar content, respectively. Here, nihonshu-do is a measure of the density of the sake relative to water. It is a very general reference to the sweetness or dryness of sake.

## Miso (Soybean Paste)

Fermented soybean pastes, which are called miso, are made from soybean, salt, and rice malt, barley malt, or soybean malt. Soybean is soaked in water for a while, after being washed in water. After soaking, it is mashed roughly after boiling in heat water of about 110°C for about 10 minutes and cooling. Rice and wheat are steamed after washing and soaking in water. The spore of "Koji-mold," which is the base of the delicious miso, is mixed into soybean, rice, or barley to make malt. Soybean pastes are called rice miso, barley miso, or bean miso depending on the type of malt used. The mixture of soybean, malt, salt, and water is fermented.

NIR has been applied to the analysis of the moisture and salt content in shikomi miso and pH, protease, and amylase contents in Koji (15). The optical data in the 1100- to 2500-nm wavelength region by 2-nm intervals were measured with a spectrophotometer, InfraAlyzer 500. For moisture of miso, $R = 0.941$, SEC $= 0.756\%$, and SEP $= 0.506\%$ when optical data at six wavelengths, 1412, 1488, 1880, 2204, 2212, and 2244 nm, were used. For salt content of miso, $R = 0.964$, SEC $= 0.219\%$, and SEP $= 0.178\%$ when optical data at six wavelengths, 1256, 1268, 1288, 1784, 1808, and 1848 nm, were used. For moisture of Koji, $R = 0.980$, SEC $= 0.885\%$, and SEP $= 0.851\%$ when optical data at six wavelengths, 1176, 1312, 1272, 2276, 2312, and 2332 nm, were used. No good results were obtained for measurements of pH and activities of protease and amylase.

## Soy Sauce

Soy sauce is made only from soybean, wheat, and salt. The protein in soybean changes to components that produce flavor and color unique to soy sauce by microorganisms. Wheat, rich in starch, is mainly used to create the aroma unique to soy sauce. At first, soybeans are steamed. Wheat is roasted and ground into small pieces. After that, the soybeans and wheat are mixed together.

Koji production is one of the most important processes of soy sauce production. Equal volumes of steamed soybeans and crushed roasted wheat are mixed, seed Koji-mold is added to the mixture, and the mixture is kept for 3 days.

Salt is dissolved into water, which is then mixed with Koji and poured into a large tank. The mixture is called Moromi (mash of unmatured soy sauce). Moromi continues to change in the tank for 6 months. During this period, enzymes, yeast, and microorganisms such as lactic acid bacilli work actively. Decomposition of ingredients during fermentation process makes Moromi tasty by producing various important components for soy sauce flavor, color, and aroma.

Next, Moromi is wrapped in a cloth and squeezed slowly for as long as 3 days. The squeezed raw soy sauce is heated up to a certain temperature. This is not only for sterilization but for standardizing aroma and color and increasing stability of soy sauce by stopping enzyme activity. This process is an important process of soy sauce production. After heating process, the soy sauce is checked by inspectors for not only chemical composition but also color, aroma, and taste by using their eyes, nose, and tongue. Inspected soy sauce is bottled in the clean room.

NIR has been applied to the determination of the ethanol content of soy sauce (16). Ethanol is a factor in aroma and antibacterial activity of soy sauce. NIR spectra of soy sauce samples were measured with InfraAlyzer 400, equipped with 19 filters to select wavelength, or with InfraAlyzer 500. Gas chromatograph was used as a conventional method. A calibration equation for ethanol was formulated with the optical data at 2270 and 2310 nm of wavelength; $R = 0.997$, and SEC $= 0.07\%$. Good validation results were obtained, with $r = 0.996$ and SEP $= 0.12\%$.

## Rice Vinegar Fermentation

In the process of rice vinegar fermentation, Japanese sake is made from rice by alcohol fermentation, and then rice vinegar is produced from sake by acetic acid fermentation. The compositions of sake, raw material of acetic acid fermentation, and rice vinegar are very complex. It is not surprising that both sake and rice vinegar have very complex compositions because these are products obtained by fermentation. All substances contained in sake and rice vinegar are contained in the broth of acetic acid fermentation. The concentrations of ethanol and acetic acid in the culture broth are high, whereas the concentrations of the other constituents are very low.

Yano et al. applied NIR to the measurement of the concentrations of the constituents in rice vinegar fermentation broth (17). Samples of the culture broth of a rice vinegar fermentation were removed from a 23-m$^3$ fermentor set up in a vinegar brewery. The samples drawn from several runs of the fermentation contained ethanol and acetic acid ranging in concentrations from 5.7 to 34.8 g·l$^{-1}$ and from 66.9 to 109 g·l$^{-1}$, respectively. A set of 42 samples was used as the calibration sample set, and the set of the remaining 21 samples was used as the prediction sample set. The culture broth was incubated in a water bath to heat it to the required temperature and was then placed in a cuvette with a 2-mm light path length. After the cuvette was placed in the cell holder of the NIR spectrophotometer (NIRS6500SPL), the transmittance values at wavelengths ranging from 400 to 2500 nm were measured at 2-nm intervals.

Multiple linear regression (MLR) was conducted on second-derivative spectra and the actual concentration for ethanol or acetic acid in the culture broth. In the calibration equation for ethanol using the wavelengths of 1686 and 1738 nm, $R$ and SEC were 0.999 and 0.374 g·l$^{-1}$, respectively. In the equation for acetic acid using the wavelengths of 1674 and 1718 nm, the values of $R$ and SEC were 0.940 and 0.387 g·l$^{-1}$, respectively. Good validation results were also obtained. The culture broth of the rice vinegar fermentation also contains several other constituents such as organic acids. MLR was conducted on NIR spectra and the actual values for several kinds of the organic acids such as gluconic, lactic, 2-ketogulutaric, succinic, and propionic acids in the culture broth. MLR for glucose, reducing sugar, and total sugar were also attempted. For calibration and validation of gluconic and lactic acids, good results were obtained. However, calibration of other organic acids and sugars obtained no good results. However, the assignments of these wavelengths in the calibration for gluconic, lactic, 2-ketogulutaric, succinic, and propionic acids were not clear.

## Alcohol Fermentation

Ethanol production has gotten a lot of attention recently in view of its potential as an energy source. Although various substrates are studied as a carbon source of ethanol fermentation, molasses, which is a residue of sugar factories, is used as a carbon source of ethanol and amino acid fermentation. NIR has been used to predict the ethanol content in fermented molasses prepared by the addition of baker's yeast to the diluted molasses medium (18). Before NIR measurement, the samples were centrifuged to remove excess stabilizer and yeast. NIR measurements were made with an InfraAlyzer 400 equipped with a 1-mm cuvette and a set of 19 pre-selected filters in the wavelength range of 1445 to 2348 nm. The calibration equation for ethanol was formulated with the optical values at the wavelengths of 2270, 2230, and 2139 nm, and good results were observed: SEC $= 0.0766\%$, $R = 0.997$, and SEP $= 0.117\%$ in the range 5.9% to 9.5% v/v ethanol.

The yeast, *Saccharomyces cerevisiae*, was grown anaerobically on the medium with glucose as the carbon source in a reactor. Short-wavelength NIR spectroscopy, which lies between 700 and 1100 nm, was carried out by means of a Hewlett-Packard 8452A photodiode array-based spectrophotometer with NIR option (19). Spectra were taken at half-hour intervals for a 30-h fermentation period and then first smoothed, and a second-derivative transformation was calculated. A decrease in relative baseline absorbance was observed with development of the fermentation. A linear regression model was constructed by using a single wavelength, 905 nm, of the second-derivative spectra. Correlation between the gas chromatograph and NIR resulted in SEC $= 0.19\%$ and $r = 0.993$.

## Lactic Acid Fermentation

Vaccari et al. also applied NIR to the control of the lactic acid fermentation process (20). To get a chemical parameter such as the concentrations of the substrate, nutrients, and biomass, an on-line NIR system has been introduced to the quantitative determination of glucose, lactic acid, and biomass in real time during fermentation. The spectrophotometer, InfraAlyzer 450 (Bran-Luebbe Co., Germany) equipped with a cuvette for liquid was used. Lactic acid and glucose concentrations in the broth were measured with HPLC as a conventional method. The cell (biomass) concentration was measured conventionally by a drying method. A set of 45 samples of broth was used as the calibration sample set to create the calibration equation. *Lactobacillus casei* was cultivated in a fermentor with working volume of 3 l. The calibration equations for lactic acid, glucose, and biomass in the culture broth were formulated. As the validation results, the values of $r$ were 0.9988, 0.9971, and 0.9870 for lactic acid, glucose, and biomass, respectively.

Glucose (main carbon source), lactic and acetic acids (main metabolites), and biomass in the culture broth of the lactic acid fermentation with *Staphylococcus sp.* and *Lacrtobacillus sp.* have been measured with an optical fiber probe (2-mm optical path provided by 1 mm of the probe slit) connected to the NIR spectrometer (Foss NIRSystems 6500)(21). With the second-derivative spectra (from wavelengths of 700 to 1800 nm) of the culture broth of *Staphylococcus sp.*, good results were obtained with 0.968, 0.939, 0.901, and 0.954 $R^2$ and 2.91, 1.367, 0.628, and 0.853 $g \cdot l^{-1}$ of SEP for glucose, lactic and acetic acids, and biomass, respectively. In the effect of the agitation speed on the NIR spectrum, a higher absorbance value was observed at higher agitation speed, and the effect was significant at lower biomass concentration rather than higher. These results may be due to the effect of the population of air bubbles suspended in the culture broth.

NIR has been applied to prediction of the concentrations of glucose and lactic acid in peritoneal dialysis solution, which is a medical product and not a fermentation broth (22). The peritoneal dialysis solution is introduced into the peritoneal cavity of renal failure patients, and waste materials in the blood are dialyzed into the solution through the peritoneum. MLR was used to obtain calibration equations relating the NIR spectral data and the glucose and lactic acid concentrations of a calibration sample set obtained by enzymatic methods. A calibration equation for glucose in peritoneal dialysis solution was formulated with second-derivative NIR spectral data at 2270 nm, and the values of $r$ and SEC were 0.996 and 2.03 $g \cdot l^{-1}$, respectively. A calibration equation for lactic acid in peritoneal dialysis solution was formulated with the second-derivative NIR spectral data at 1688 and 1268 nm, and the values of $R$ and SEC were 0.997 and 0.178 $g \cdot l^{-1}$, respectively. In the validation results of the calibration equations, excellent agreement between the results of the enzymatic method and the NIR method was also observed for these constituents. The values of $r$ for glucose and lactic acid in the peritoneal dialysis solution were 0.996 and 0.996, respectively.

## Glutamic Acid Fermentation

Okayasu et al. introduced an on-line NIR analyzing system of the L-glutamic acid fermentation process (23). The on-line sampling system with a deforming apparatus was used as a pretreatment of NIR analysis. The simultaneous determination of the main constituents of the L-glutamic acid fermentation broth, such as sugar, cell, L-glutamic acid, and $NH_4^+$ concentration was attempted. However, this had some problems that must be resolved to realize the on-line monitoring of the fermentation process by NIR. At first, bubbles in the culture broth influenced the measurement value of the NIR absorption. Second, adhesion of the scale in the cuvette also influenced the measurement values of the NIR. These are common problems in the fermentation processes. In this NIR system, deforming apparatus of the cyclone type and U-shaped pipe were set onto the outlet stream from the fermentor and onto the inlet stream to a NIR cuvette. To avoid adhesion of scale of the NIR cuvette, the cuvette was washed after analysis, using a detergent with an enzyme. A calibration equation for sugar in the culture broth of L-glutamic fermentation was formulated with the raw NIR spectral data at 1445, 1722, 2100, and 2180 nm, and the value of $R$ was 0.992. Calibration equations for cell, L-glutamic acid, and $NH_4^+$ concentrations were formulated, and the values of $R$ were 0.999, 0.996, and 0.966, respectively. The measurement of time courses of sugar, cell, and L-glutamic acid was successfully obtained on the NIR system. The system is partially used in the industrial fermentation processes; however, improvement of analytical accuracy in change of raw material in each lot is necessary.

## Mushroom Production

Mushroom is one of the important materials for foods and medical products. Solid cultivation has been used for the commercial production of mushroom. Measurement of the cell mass, growth rate of mycelium, and constituents in the solid media are very hard because the analysis is time consuming and laborious. NIR has been applied to the prediction of the cell growth rate of mushroom, *Ganoderma lucidum*, in a solid culture (24). Cell mass is conventionally measured by analyzing the concentration of glucosamine, which is a component of the cell wall. The glucosamine concentrations in the culture materials were predicted by six-wavelength regression analysis. Values of the first-derivative spectra values at 1203, 1635, 1751, 2103, 2375, and 2431 nm were used to develop a calibration equation. The values of $R$ and SEC of the calibration sample set ($n = 30$) were 0.969 and 0.622 mg·g$^{-1}$, respectively. The value predicted by NIR was in fairly good agreement with that obtained by the conventional method. The $r$ and SEP of the validation sample set ($n = 11$) were 0.992 and 0.346 mg·g$^{-1}$, respectively. Specific growth rate of the cell could be calculated by using NIR.

Simultaneous determination of the water and rice bran contents in solid media material was also attempted (25). To obtain a calibration equation for water content in the mushroom media material, a simple linear regression was carried out on the NIR spectral data at 1450 nm and on the water content of calibration sample set ($n = 113$) obtained by a drying method. The values of $r$ and SEC were 0.995 and 1.33%, respectively. On the basis of the result of the NIR on the content of rice bran in the solid media, a calibration equation using the second-derivative reflectance data at the wavelengths of 672 and 2100 nm was obtained with values for $R$ and SEC of 0.978 and 1.73%, respectively. To validate the calibration equations obtained, water and rice bran content in the prediction sample set ($n = 56$) were calculated using the calibration equations. For both the water and rice bran contents, excellent agreement was observed between the results of the conventional method and those of the NIR method. The $r$ and SEP were 0.997 and 1.33% for water content and 0.975 and 1.84% for rice bran content, respectively. The NIR method is a useful method for rapid measurement of the cell mass, growth rate of mycelium, and constituent content in the solid media of mushroom cultivation.

## Enzymatic Saccharification

Enzymatic hydrolysis has been used extensively in saccharification of starch. To control the enzymatic hydrolysis automatically, it is necessary to measure the concentration of substrate and of products in the reactor in real time. The applicability of NIR for the on-line determination of produced glucose content in starch hydrolysis by glucoamylase (EC3.2.1.3) has been examined to develop automatic control of the reaction in a bioreactor (26). The reaction mixture was circulated from the reactor to an NIR flow-cell with a peristaltic pump. The transreflectance spectrum of the solution was measured over a wavelength range of 1000 to 2500 nm in 2-nm increments, using a NIR InfraAlyzer 500C. The absorbance at 2008 nm gave a high correlation between glucose content measured by NIR and HPLC. SEP was 0.077% for the calibration with two wavelengths, 2008 and 2148 nm. The 2000-nm wavelength has been attributed to $2 \times$ O-H deformation plus C-O deformation in starch.

## Biosurfactant Fermentation

Yeast, *Kurtzumanomyces sp. I-11*, produces mannosyl erythritol lipid (MEL) from soybean oil. MEL is a kind of biosurfactant and is classified by a glycolipid. MEL is the typical amphiphilic compound included in both lipophilic and hydrophilic moieties in the molecule. MEL is composed of mannose, erythritol, and fatty acids. Biosurfactants such as MEL have special properties over their chemically synthesized counterparts. The properties include low-toxicity, biodegradability, biological activity, a wide variety of possible structures, and ease of synthesis from inexpensive, renewable resources. Consequently, biosurfactants provide new possibilities for a wide range of industrial applications, especially the food, cosmetic, and pharmaceutical fields and chemicals for biotechnology.

A measurement system for the concentrations of MEL and soybean oil in the fermentation process has been developed with NIR (27). MEL and soybean oil in the culture broth were extracted with ethyl acetate. NIR transmittance spectra of the ethyl acetate extract were measured. The absorption caused by MEL was observed at 1436, 1920, and 2052 nm. To obtain a calibration equation, MLR was carried out between the second-derivative NIR spectral data at 2040 and 1312 nm and MEL concentrations obtained with a thin layer chromatography with a flame-ionization detector (TLC/FID) method. The values of $R$ and SEC were 0.994 and 0.48 g·l$^{-1}$, respectively. The absorption caused by soybean oil was observed at 1208, 1716, 1766, 2182, and 2302 nm. A calibration equation for soybean oil was formulated with the second-derivative NIR spectral data at 2178 and 2090 nm. The values of $R$ and SEC were 0.974 and 0.77 g·l$^{-1}$, respectively. For the results of the validation of the calibration equation, good agreement was observed between the results of the TLC/FID method and those of the NIR method for both constituents. The values of $r$ and SEP for MEL were 0.994 and 0.45 g·l$^{-1}$, respectively. The values of $r$ and SEP for soybean oil were 0.979 and 0.56 g·l$^{-1}$, respectively.

## Penicillin Fermentation

In a submerged fermentation of *Penicillium chrysogenum*, mycelial biomass has been monitored with NIR (28). The spectra of the culture broth were acquired with a Model 6500 NIR spectrophotometer (Foss NIRSystems) in the transmittance mode with a cuvette of 1-mm path length. The second derivatives were used with a segment size of 10 and a gap size of 2. For MLS models, SEC was $0.820$–$1.060$ $g \cdot l^{-1}$ and $R^2$ was $0.954$–$0.979$ when three wavelength terms, 1724, 1644, and 1684 nm, were used. The wavelength term 1724 nm was traced to the contribution from biomass, and the other two wavelength terms, 1644 and 1684 nm, were traced to the contribution from the culture broth (sugars, ammonium at early stage of the fermentation, and penicillin at the later stages). For the PLS models with the optical data from the wavelength of 1600 to 1800 nm, a factor size of 4 was found to be optimal, SEC was $0.620$–$0.740$ $g \cdot l^{-1}$, and $R^2$ was $0.978$–$0.988$.

## Compost Fermentation

Compost fermentation is one of the key technologies for waste treatment and recycling of the residue from food processing. In general, compost fermentation process is classified into three stages, primary fermentation, secondary fermentation and maturation. Easily decomposable organic compounds such as protein, carbohydrate, and lipid in the compost material are decomposed rapidly by thermophilic bacteria at high temperature (over $60°C$) under aerobic conditions. This process is called thermophilic composting (primary fermentation), and the term of the process is about 1 week. The fermented product of this stage is not suitable for compost because slowly decomposable organic compounds such as cellulose, hemicellulose, and lignin in the compost material are not decomposed. Therefore, the product of the thermophilic composting process is often transferred to secondary fermentation and maturation.

On the point of a rapid treatment of organic waste, monitoring and control at the optimal level of the fermentation conditions such as temperature, aeration rate, pH, and moisture content of the compost are very important to management of the thermophilic composting process. Organic compounds in the compost material are changed to $CO_2$ and $NH_3$ by microorganisms. Characterization of the compost fermentation could be measured as the time courses of the carbon and nitrogen contents and C-to-N ratio (29). Furthermore, change of the lipid content in the compost could be one of the indicators to determine the end point of the fermentation (30).

NIR was applied to determine the moisture, carbon, nitrogen, and lipid contents and C-to-N ratio in a compost sample during the tofu (soybean curd) refuse compost fermentation (29–31). Control of the moisture content of the compost fermentation process and simultaneous measurement of the carbon, nitrogen, and lipid contents and C-to-N ratio in the compost were also attempted. The compost sampled from the composter was put into a polyethylene bag, and it was placed in a sample holder. The reflectance values were measured with a NIR spectrophotometer (NIRS6500SPL).

With the second-derivative values at the wavelength of 960 nm, the best calibration equation for moisture was obtained. The values of the moisture content of the compost measured and controlled by the NIR method were in good agreement with those obtained by the conventional method.

In the spectra of compost, prominent negative peaks were also observed at 1210, 1360, 1584, 1730, 1820, and 2174 nm. The peaks at 1360 and 1584 nm may be related to carbon compounds in the compost. These absorptions may be caused by

C-H stretching and C-H deformation in the carbon compounds of the compost, such as lipid, protein, and cellulose of soybean. It should be noted, however, that most wavelengths were common to both carbon and nitrogen.

The first wavelength used to formulate a calibration equation of carbon content should be selected from 922, 1360, 1584, 1718, and 1830 nm. These absorptions may be caused by the structure of carbon compounds in the compost. The first wavelength used to formulate a calibration equation of nitrogen content should be selected from 900, 1060, 1570, and 2174 nm. Because of the absorptions in the NIR spectrum caused by nitrogen-containing structure, four wavelengths were chosen. A calibration equation for carbon and nitrogen contents was developed from the various first and second wavelength combinations, and the best combination (best values of $R$ and SEC) was chosen. To determine the lipid content of compost, 1208 and 1712 nm were selected as the first and second wavelengths. Good calibration and validation results were obtained.

C-to-N ratio of the compost calculated based on the values of the carbon and nitrogen contents of the compost is also an important indicator of compost fermentation. The predicted values of C-to-N ratio with NIR were in good agreement with the values obtained by the conventional method.

Masui et al. reported that the time course of the lipid content of the compost during the compost fermentation was coupled to the decrease of the total dry weight of the compost in the composter and the change of the lipid content in the compost could be one of the indicators to determination of the end point of the thermophilic compost fermentation (30). Lipid content was predicted with NIR. Good agreement between the values obtained by the conventional method and those obtained by NIR was observed during the compost fermentation.

Reliability of NIR in the evaluation of compost quality, contents of total nitrogen, total phosphorus, organic matter, total carbon, moisture, copper, potassium, and sodium, has been investigated (32). NIR spectra of dried and ground compost samples were measured in an InfraAlyzer 500 diffuse reflectance NIR spectrophotometer from 1100 to 2500 nm at 2-nm intervals. MLR and PLS were conducted between raw and first-derivative optical data and the concentrations of each comonent in the compost. Good results were obtained for the contents of total nitrogen (SEP = 0.085%), total carbon (1.12%), moisture (0.11%), and copper (0.048%).

## Molasses (Raw Material of Fermentation)

Molasses, the residue of sugar factories, was used as a carbon source of ethanol and amino acid fermentation. Beet molasses consists of 17 w/w% water, 53% sugars, and 30% other substances such as partclies in suspension and soluble components. The latter consist of organic nonsugars (19%) (nitrogenous materials, glutamic acid, betaine, acids, soluble gummy substances) and inorganic constituents (11%). The sugars are mainly sucrose (51%), invert sugar (1%), and raffinose (1%). Analyzing the sugar content of molasses used to the fermentation is very important to manage a good fermentation.

NIR has been used to predict the total sugar content in aqueous beet molasses (18). Before NIR measurement, the samples were centrifuged to remove excess stabilizer.

NIR measurements were made with an InfraAlyzer 400 equipped with a 1-mm cuvette and a set of 19 preselected filters in the wavelength range 1445 to 2348 nm. The calibration equation for sugar was formulated with the optical values at the wavelengths of 2348, 2270, 2180, 2139, 1778, and 1680 nm, and good results were observed: SEC $= 0.72 g \cdot l^{-1}$, $R = 0.999$, and SEP $= 3.45 g \cdot l^{-1}$ in the range 99–167 $g \cdot l^{-1}$ sugar.

Cane molasses was also used as a carbon source of fermentation because it usually contains 30–40% sugar. Total sugar and reducing sugar contents in molasses were estimated with NIR diffused reflectance spectrosocopy (33). For both kinds of sugar, good results were obatained when the calibration was carried out with the wavelength range from 2100–2500 nm.

## BRIGHT FUTURE FOR NIR

Rapid measurement of the main constituents of the fermentation process with the NIR method has been introduced in this chapter. For management of the fermentation process, NIR is a useful method in the various fermentation processes. The operational procedure of NIR is very simple, and the time required for the measurement (full range scanning of 32 times) is only 1.5 min. The time required will be able to be shortened more drastically if only absorptions in the wavelength used in the calibration equation are measured. In addition, it is possible to transport the fermented sample back from the spectrophotometer to the fermentor after the analysis, because NIR is a nondestructive method. On-line monitoring with an optical fiber probe also might be possible instead of packing the fermented sample into a cuvette or a polyethylene bag before the measurement because the fermented sample was not pretreated and the measurement of these constituents in the sample were possible with transmitted or reflected rays. To make an efficient fermentation process, computerized control systems for process management using NIR will be developed in the future.

However, several significant problems must be resolved. The fermentation sample contains various constituents such as proteins, lipids, and sugars. These compounds easily attach on the surface of a probe and a cuvette and cause the problem of shift of the optical density. It becomes a problem when the cell density increases to a hundred times with development of the fermentation. The increase causes the shift of the baseline of the NIR spectrum. Furthermore, the attachment of the bubbles generated by the fermentation on the surface of a probe and a cuvette also causes shift of the optical value measured. These problems should be studied as soon as possible to make sure the bright future of NIR in fermentation engineering.

## REFERENCES

1. K. J. Kaffka, K. H. Norris. Rapid instrumental analysis of composition of wine. *Acta Alimenta* **5**: 267–279, 1976.
2. R. G. Dambergs, A. Kambouris, N. Schumacher, I. L. Francis, M. B. Esler, M. Gishen. Wine quality grading by near infrared spectroscopy. In: *Near Infrared Spectroscopy*, A. M. C. Davies, R. K. Cho, eds. NIR Publications, UK, 2002, p. 187–189.

3. A. Zyl, M. Manley, E. E. H. Wolf. The application of Fourier transform near infrared spectroscopy in the wine industry of South Africa. In: *Near Infrared Spectroscopy*. A. M. C. Davies, R. K. Cho, eds. NIR Publications, UK, 2002, p. 203–205.

4. M. B. Esler, M. Gishen, I. L. Francis, R. G. Dambergs, A. Kambouris, W. U. Cynkar, D. R. Boehm. Effects of variety and region on near infrared reflectance spectroscopic analysis of quality parameters in red wine grapes. In: *Near Infrared Spectroscopy*. A. M. C. Davies, R. K. Cho, eds. NIR Publications, UK, 2002, p. 249–253.

5. A. G. Coventry, M. J. Hunston. Application of near infrared spectroscopy to analysis of beer samples. *Cereal Foods World* **29**: 715–718, 1984.

6. S. A. Halsey. The use of transmission and transflectance near infrared spectroscopy for the analysis of beer. *J Inst Brew* **91**: 306–312, 1985.

7. Y. Wakai. Outline of near infrared spectroscopy and applications in sake brewing process. *J Brew Soc Japan* **87(7)**: 492–496, 1992. (in Japanese)

8. Y. Wakai, Y. Inoue, Y. Nishikawa, J. Murata, T. Miura. Determination of various properties of sake and rice using the near infrared reflectance analyzer. *J Brew Soc Japan* **79(6)**: 445–446, 1984. (in Japanese)

9. N. Okazaki, K. Fukuda, Y. Kizaki, S. Kobayashi. Estimation of true polishing ratio of rice by near infrared reflectance spectrophotometric analysis. *J Brew Soc Japan* **86(4)**: 299–303, 1991. (in Japanese)

10. N. Okazaki, K. Fukuda, Y. Kizaki, S. Kobayashi. Estimation of polished rice by near infrared reflectance spectrophotometric analysis. *Nippon Nogeikagaku Kaishi* **64(3)**: 431, 1990. (in Japanese)

11. Y. Kojima, Y. Nakamura, Y. Hata, E. Ichikawa, A. Kawado, A. Imayasu. Measurement of cell density of rice malt using near infrared spectroscopy. *Nippon Nogeikagaku Kaishi* **66(3)**: 544, 1992. (in Japanese)

12. Y. Kojima, Y. Asai, Y. Hata, E. Ichikawa, A. Kawato, S. Imayasu. Estimation of mycelial weight in rice Koji by near infrared reflectance spectroscopy. *Nippon Nogeikagaku Kaishi* **68(4)**: 801–807, 1994. (in Japanese)

13. I. Aramaki, K. Fukuda, T. Hashimoto, T. Ishikawa, Y. Kizaki, N. Okazaki. *Near infrared diffuse reflectance* spectrophotometric analysis of mycelial weight in rice Koji and search for characteristic wavelenght for mycelia. *Seibutu Kougaku Kaishi* **73(1)**: 33–36, 1995. (in Japanese)

14. Y. Wakai, Y. Inoue, Y. Nishikawa, J. Murata, T. Miura. Determination of various properties of sake and rice using the near infrared reflactance analyzer. *J Brew Soc Japan* **79(6)**: 445–446, 1984. (Japanese)

15. Y. Kitamura, H. Yasuhira. Analysis of shikomi miso and koji by NIR. Report of the shinshu-miso research institute, **33**, 81–84, 1992. (in Japanese)

16. K. Kobayashi, K. Iizuka-Yamada, T. Okada, H. Hashimoto. Application of NIR analysis in quality control of soy sauce (part 3). *Syoyu Kenkyu* **19(6)**: 307–312, 1993. (in Japanese)

17. T. Yano, T. Aimi, Y. Nakano, M. Tamai. Prediction of the concentrations of ethanol and acetic acid in the culture broth of a rice vinegar fermentation using near-infrared spectroscopy. *J Ferment Bioeng* **84(5)**: 461–465, 1997.

18. E. D. Dumoulin, B. P. Azais, J. T. Guerain. Determination of sugar and ethanol content in aqueous products of molasses distilleries by near infrared spectrophotometry. *J Food Sci* **52(3)**: 626–630, 1987.

19. A. G. Cavinato, D. M. Mayes, Z. Ge, J. B. Callis. Noninvasive method for monitoring ethanol in fermentation processes using fiber-optic near infrared spectroscopy. *Anal Chem* **62**: 1977–1982, 1990.

20. G. Vaccari, E. Dosi, A. L. Campi, A. Gonzales-Vara, D. Matteuzzi, G. Mantovani. A near-infrared spectrocopy technique for control of fermentation processes. An application to lactic acid fermentation. *Biotech Bioeng* **43**: 913–917, 1994.

21. E. Tamburini, G. Vaccari, S. Tosi, A. Trilli. Application of in-line near infrared spectroscopy to the control of industrial fermentation process. In *Near Infrared Spectroscopy*. A. M. C. Davies, R. K. Cho, eds. NIR Publications, UK, 2002, p. 447–451.

22. T. Yano, H. Matsushige, K. Suehara, Y. Nakano. Measurement of the concentrations of glucose and lactic acid in the peritoneal dialysis solutions using near-infrared spectroscopy. *J Biosci Bioeng* **90(5)**: 540–544, 2000.

23. S. Okayasu, M. Katayama, S. Miyashiro. Application of near infrared spectrometer to management of the operation of fermentation plant. *Kagakukougaku ronnbunnsyu* **21**: 847–851, 1995. (in Japanese)

24. K. Suehara, Y. Nakano, T. Yano. Application of near infrared spectroscopy to the measurement of cell mass in solid cultures of mushroom. *J Near Infrared Spectrosc* **6**: 273–277, 1998.

25. T. Yano, K. Suehara, Y. Nakano. Determination of the content of water and rice bran in solid media used for mushroom cultivation using near-infrared spectroscopy. *J Ferment Bioeng* **86(5)**: 472–476, 1998.

26. K. Nishinari, R. K. Cho, M. Iwamoto. Near Infra-red monitoring of enzymatic hydrolysis of starch Starch **41(3)**: 110–112, 1989.

27. K. Nakamichi, K. Suehara, Y. Nakano, K. Kakugawa, M. Tamai, T. Yano. Measurement of the concentrations of mannosyl erythritol lipid and soya oil in the glycolipid fermentation process using near-infrared spectroscopy. *J Near Infrared Spectrosc* **10(1)**: 53–61, 2002.

28. S. Vaidyanathan, G. Macaloney, B. McNeil. Measurement of mycelial biomass in a submerged bioprocess using near infrared spectroscopy. In: *Near-Infrared Spectroscopy*. A. M. C. Davies, R. Giangiacomo, eds. NIR Publications, UK, 2000, p. 429–433.

29. K. Suehara, Y. Nakano, T. Yano. Simultaneous measurement of the carbon and nitrogen content of compost using near-Infrared Spectroscopy. *J Near Infrared Spectrosc* **9(1)**: 35–41, 2001.

30. D. Masui, K. Suehara, Y. Nakano, T. Yano. Measurement of lipid content of compost fermentation using near infrared spectroscopy. *Near Infrared Analysis* **2(1)**, 37–42, 2001.

31. K. Suehara, Y. Ohta, Y. Nakano, T. Yano. Rapid measurement and control of the moisture content of compost using near-infrared spectroscopy. *J Biosci Bioeng* **87(6)**: 769–774, 1999.

32. J. Nam, K. Jung, S. Lee. Non-destructive analysis for quality control of compost by near infrared spectroscopy. In: *Near-Infrared Spectroscopy*. A. M. C. Davies, R. Giangiacomo, eds. NIR Publications, UK, 2000, p. 613–615.

33. R. Mehrota, A. Gupta, J. Tewari, S. P. Varma. Estimation of total sugar and reducing sugar in molasses using near infrared diffused reflectance spectroscopy. In: *Near Infrared Spectroscopy*. A. M. C. Davies, R. K. Cho, eds, NIR Publications, UK, 2002, 207–211.

# On-Line Analysis in Food Engineering

KATHRYN A. LEE

## INTRODUCTION

Economic, safety, quality, and environmental issues motivate the development of real-time process analytical chemistry. Near-infrared (NIR) technology provides many of the key attributes that are needed for rapid process control. NIR allows non-destructive multiconstituent chemical analysis in just a few seconds, and the instrumentation is rugged, intrinsically safe, and easy to integrate into a process.

Continuous and batch processes require different strategies for implementation of process analyzers. For continuous processes, startup must be controlled to quickly reach the optimal process conditions. Constant monitoring is needed for both continuous and batch processes to detect when the process is going out of control so that corrective action can be taken. For batch processes, end point detection is an important control parameter. For all analyses, the simplest method that performs the desired analysis is best. As the complexity of a method increases, changes in the instrument, materials, process, and measurement conditions can cause problems (1).

An overlooked drawback is that NIR spectroscopy appears easier to implement than it is. One reason for this is sample variability, which can be particularly important for food applications. Natural products can have significant seasonal and genetic variation and often contain water and other ingredients that are sensitive to environmental factors (such as temperature) during measurement. Food can have several ingredients, and when only a few are of interest, it is easy to overlook the other ingredients and their effects on the measurements. Finally, the fact that NIR relies on correlations for calibrations (even with noise) and the ease of running calibration software programs combine to make feasibility studies that appear to succeed, but applications that fail.

Samples selected for developing a NIR calibration not only must cover the range of analyte levels expected in the process, but must also include all of the variation in the other components in the matrix, including physical variables (e.g., particle size).

*Near-Infrared Spectroscopy in Food Science and Technology*, Edited by Yukihiro Ozaki, W. Fred McClure, and Alfred A. Christy.

A commonly used method development strategy is to develop "starter" calibrations that can be used to find unrepresented sample and process variation (1).

Calibration methods used for process NIR analysis include multiple linear regression (MLR), which uses measurements at a few wavelengths, and partial least squares (PLS) or principle component analysis (PCA), which use spectral regions. Generally, the simplest model that provides adequate information should be used; a more complicated model may provide greater accuracy for the current set of samples but not provide the best long-term reliability or tolerance to variability. Validation depends on the modeling approach, the materials, and the process, and ideally validation samples should be collected over several manufacturing cycles or batches. The standard error of prediction (SEP) should closely match the standard error of calibration (SEC). If the SEP is much larger than the SEC, that could indicate too many wavelengths in MLR models, too many factors in full-spectrum regression models, or that the model does not truly represent the analyte or property being modeled (1).

Segtnan analyzed temperature-dependent NIR spectra of pure water and found that temperature-induced concentration shifts in hydrogen bonding in water could dominate the spectral dynamics of water-based samples. Thus, in experiments involving NIR spectra of high-moisture samples, the sample temperature ideally should be held constant, or at least measured and included in the calibration (2).

Specialists should be relied on for development of NIR process analysis applications. Routine analyzer and NIR method maintenance are also critical components in the success of the implementation of process NIR methods. Either the instrument can drift or the samples can vary. It is preferable to perform analyzer and method maintenance procedures automatically, for example, through control charting, rather than requiring a specialist (1).

The first step in developing a new NIR method is to do a feasibility test. Ideally, feasibility tests include all the variation that is expected in the final application as well as the variation in the conditions of analysis. Actual feasibility studies often cover only a small subset of the variations, reduced to the samples at hand under current conditions. Because food and food ingredients are often, though not always, based on natural products, with potentially wide variation in content that can vary with nurture, nature, time, and current conditions, feasibility studies succeed far more often than implemented applications. Feasibility studies start with one sample and one question, "Is it possible to get a NIR spectrum?", and move to two samples and a new question, "Are the spectra different for different samples?", to three samples and the ultimate question, "Is there a reliable correlation?" To answer this question, proper experimental design and many more samples are required. Sample presentation to the instrument also may be crucial to the success of the analysis. Thus sample variability, composition, condition, environment, and proprietary nature result in many more reports of feasibility studies than reports of actual applications in the literature. For example, using NIR for monitoring blended food products would be a good application, but the ingredient list or proportions may be too secret for publication. Consequently, there are not many reports in the literature. NIR analysis of food has been reported starting in the late 1960s, and there have been many reviews (3). Science builds on the work of others, so much of the recent literature of NIR applications to on-line food analysis consists of applications that are similar to what has already been reported but include new technology or data analysis that yield improved results. Some previous reviews of NIR used on-line for food analysis include the following: O'Sullivan on milk and food (4), Laporte on dairy and dairy products (5), Windhab on dairy and other products (6), Givens on the nutritive value of food and food processing (7), and Williams on cereal and flour (8). Some recent applications of NIR to on-line food analysis are reviewed below. For food, the definition of "on-line" has to be flexible enough to include extruders, fields, and even cows. Many of the applications show great ingenuity, and some show complex application of analysis techniques. Although organized here by food group, many of the reported techniques can be applied to other foods entirely, and success in science often comes from using ideas developed in one field in an entirely different field. The objectives of this chapter are to present and discuss some of the more interesting on-line applications of NIR for food analysis.

## FLOUR

A visible-NIR spectrometer with a fiber-optic probe was used to monitor both color and composition in a counterrotating fully intermeshing twin-screw extruder during the extrusion of yellow corn flour. Transflectance spectra were taken in the 800 to 2500 nm region. Screw speeds of 40, 50, and 60 rotations per minute (rpm) and temperatures of 130, 150 and 170 °C were used in a $3 \times 3$ factorial design to examine the effect of screw speed and temperature. During extrusion, starch breaks down to lower-molecular-weight macromolecules because of the thermal and mechanical forces. The largest branched molecules of amylopectin (with NIR bands at 2100 and 2280 nm) were found to break down through shearing mechanical forces. Evidence for macromolecular degradation of starch includes the decrease in intrinsic viscosity (measured by dilute solution viscometry) and changes in the molecular size distribution (measured by GPC and HPLC). Some molecular changes that take place during extrusion can be explained in terms of hydrogen bonding, which can be detected by NIR because of changes in bonded and free OH stretching. The thermal effects at temperatures below 100°C were slight (9). Other experiments showed that a fiber-optic-equipped vis-NIR spectrometer could be used to determine the residence time distribution in the extruder. The effect of feed rate, water content, and temperature on the residence time distribution were precisely determined and fit by the tanks in series (TIS) model (10).

## CHEESE

A NIR Instalab 600-Dickey-John was used for production process control of traditional Greek feta cheese. Suitable filters were chosen to measure absorbances of moisture (at 1940 nm), fat (at 2310 nm), and protein (at 2180 nm). Samples were drawn from well-defined critical points of production and quantified with

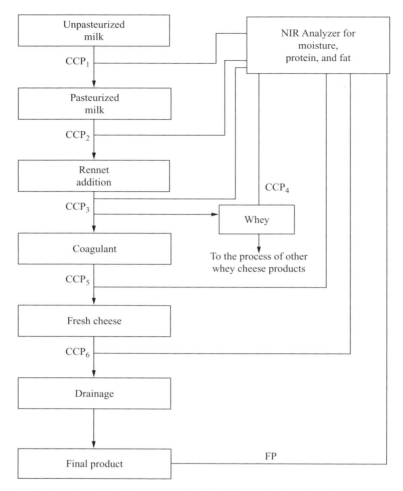

**Figure 9.2.1.** A schematic of the process for feta cheese showing the critical control points (CCP) and the points where NIR could be used for analysis (11).

conventional chemical methods and correlated to the NIR spectra. To make the NIR suitable for liquid products, a novel technique was developed using "quartz fine granular washed and calcined GR" (pro analysi). This technique proved to be reliable with a significant accuracy. The NIR could be used to continuously monitor the constituents during processing so that concentrations at critical control points could be quickly checked. Possible deviations could be directly calculated and immediately corrected to obtain constant quality as specified (11). Figure 9.2.1 shows a schematic of the process, the critical control points, and the points where NIR could be used for analysis.

Consistency in curd production is important in cheese manufacturing to ensure maximum yield and quality. Mertens et al. applied chemometrics and a continuously monitored NIR reflectance at a single wavelength to the rennet-induced coagulation of milk in the critical stage in cheese production. The methodology used to predict the proper cutting time must be able to account for the considerable variation in milk composition, particularly protein, that is characteristic of pasture-based dairying. Curd firmness can be continuously monitored by a controlled stress laboratory rheometer. Cutting time is then defined as the time by which a specific firmness or storage modulus $G'$ is reached. NIR has been used to monitor for cutting time by using the inflection point in the NIR signal vs. time plots, adjusted in a linear fashion where $t_c - a + bt_f$, where $t_c$ is cutting time, $t_f$ is the inflection time, and $a$ and $b$ are parameters determined from calibrations. An equation could be developed based on growth models that also considered process parameters. It was found that protein level affected cutting time through a quadratic transformation, whereas rennet had a linear effect. The final result was a much better prediction of cutting time, while still using the inflection point as a predictor, which the food engineers already were familiar with using (12).

## MILK

NIR was shown to have the potential for on-line measurement of rumen fluid composition by using specially designed fiber optics. A cow's rumen can be considered as a large fermentor in which the macromolecules of the feed are converted by microbes to low-molecular-weight acids. As noted above, the type of feed affects the quality of milk. The rumina of two cows were fistulated with a surgically cut hole that could be sealed by a cover with a small "window" for insertion of the probe. The NIR path length was 2 mm, 50 scans were coadded, and spectra were taken every 7.5 min. The usable wavelength region was 1100–1860 nm. GC, Kjeldahl, and pH were the reference methods, and partial least squares (PLS) calibrations were developed for acetic, propionic, $n$-butyric, $i$-butyric, $n$-valeric, and $i$-valeric acids, total acid, and ammonia nitrogen. Only the NIR analysis for acetic acid was sufficient for screening, whereas the others were sufficient for quality control analysis. The NIR-predicted acetic acid values could be used to monitor rumen fluid composition as a function of feeding (13).

As a substitute for repetitive tasks of a skilled operator, fuzzy logic was applied to real-time control of a spray-dried whole milk powder processing system making whole milk powder with greater than 90% free fat content and consistent color. The objective was to provide control of process startup and operation at a preset power consumption level while providing a quality product. For consistent quality, the system parameters of powder feed rate, lecithin injection rate, processor screw speed, and circulating water heating and cooling on-times were controlled with the feedback signals from sensors for product color, temperature, and power consumption. The reflected intensity ratios of red to NIR color values (red/NIR $*$ 100) were used as a measure of product color. The control algorithm performance was observed over time for 0.75 kW and 1.12 kW power consumption and 80, 85, and 90 color values. The algorithm controlled the process at $\pm 0.074$ kW of the desired power consumption and provided whole milk products with the desired color values within $\pm 3.0$-unit deviations. The free fat content was over 95%, and lactose was in crystalline form in the final dry milk product (14).

Brennan et al. used the NIR 800–1100 nm region for on-line monitoring of fat in milk processing. The system used microsystems optical components made with the LIGA technique, which is used in microelectronics to make high-definition structures in silicon. LIGA stands for *Lithographie, Galvanaplastie, Abformung*, which is deep-etch X-ray lithography, electroplating, and injection molding (15). Natsuga et al. carried out a preliminary experiment for the development of on-line monitoring of raw milk fat, protein, and lactose constituents with NIR. Wavelength range and light path length effects on the accuracy were investigated (16).

An approach to determine the content of fat, lactose, and protein in raw milk with NIR transmittance spectroscopy was developed, using the spectral range 700 to 1100 nm. This range allowed the use of a longer path length (more than 10 mm) than the 1-mm path length required for the 1100–2500 nm region and did not require homogenization. $Al_2O_3$ with a thickness of 2.5 mm was found to be optimum as a reference for acquiring the milk spectrum for this measurement. The calibration model was developed and predicted by using a PLS algorithm with multiplicative scatter correction (MSC) and/or second-derivative spectral pretreatment to reduce the scattering effect due to fat globules and casein micelles. This study resulted in SEPs of 0.06%, 0.10%, and 0.10% for fat, lactose, and protein, respectively, and showed the potential use of this method for real-time on-line monitoring in a milking process. Figure 9.2.2 shows the scatter plots of correlations between NIR prediction values and fat, lactose, and protein. These clearly show that not all of the components in a mixture can be determined to the same degree of accuracy and that applicability is a key to determining the success of a calibration. For example, the fat calibration can be used to predict samples in the range with relatively good accuracy, while the protein calibration may in practice only really be useful for prediction of high, medium, and low levels within the calibration range (17).

## YOGURT

Measurement data from an electronic nose (EN), a NIR spectrometer, and standard bioreactor probes were used to follow the course of lab-scale yogurt fermentation. The NIR spectra were taken from 400–2500 nm with a stainless steel immersion probe. The probe was top mounted and placed parallel to the reactor shaft at the same height as the stirrer blades. Spectra were taken every 5 min, referenced to the milk before inoculation. The pH was also measured in the reactor, but viscosity was measured off-line, and galactose, lactose, and lactate were analyzed off-line with high-performance liquid chromatography (HPLC). The sensor signals were fused with a cascade neural network; a primary network predicted quantitative process variables, including lactose, galactose, and lactate; and a secondary network predicted

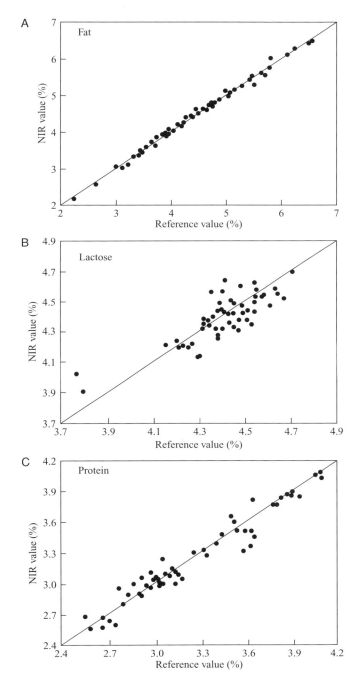

**Figure 9.2.2.** Plots showing the correlations between NIR predicted values and reference values for fat (A), lactose (B), and protein (C) (17).

a qualitative process state variable describing critical process phases, such as the onset of coagulation or the harvest time. PCA models were calculated with the software SIMCA-P$^{TM}$. A forward selection procedure was applied to find suitable subsets of sensor responses with high linear correlation to particular process variables. Although the accuracy of the neural network prediction was acceptable and comparable with the off-line reference assay, its stability and performance were significantly improved by correction of faulty data. The NIR signals correlated to the physical properties of the milk/yogurt system. The results demonstrated that on-line sensor fusion with the chosen analyzers improves monitoring and quality control of yogurt fermentation with implications to other fermentation processes (18).

## BEEF AND PORK

Anderson and Walker used on-line visible/NIR spectroscopy to measure fat content of ground beef streams. Thirty blocks of frozen beef from three truckloads each of nominally 50% and 15% fat were studied. The 27-kg blocks were ground and formed into a continuous stream on a conveyor with a custom molding head attached to the grinder. A diode-array type spectrometer was used to take reflectance measurements from the stream, which moved at 1.0 m/min. Measurements were made each 3.3 cm, and samples were collected from the stream such that each sample was represented by two successive measurements. The fat content of each sample was determined by ether extraction. SEP values for the calibration and validation data sets were 1.00% to 1.68% and 2.15% to 2.28% fat, respectively. Predictions for the fat content of an entire block gave SEPs of 0.70% to 1.05% fat. The results showed that the molding head was effective and that the NIR was suitable for accurate on-line measurements (19).

The amount and composition of the fat of Iberian swine are determinant factors in the quality of the meat and derived products. These qualities can depend on feed regime and genetics. A NIR spectrometer equipped with a standard 210/210 bundle remote reflectance fiber-optic probe, with a 5 cm × 5 cm quartz window, was used for the determination of water, protein, and infiltrated fat in pork loin muscle (longissimus dorsi), with good results. A comparative study was carried out on the determination of fat and protein in the same pork loin samples, using the fiber-optic probe without destruction of the sample and with conventional measuring cells using ground samples. The method was applied in the dissecting hall of a slaughterhouse. The content in infiltrated fat varied from 3% to 19% and that of protein from 21% to 31%. The regression method used was modified PLS. Absorption bands at 1510, 2060, and 2172–2186 nm were associated with protein, and bands at 1722, 1760, and 2308–2348 nm were associated with fat. The calibration results with the fiber-optic probe for 56 samples had corrected SEP of 0.74% for the fat (multiple correlation coefficient, RSQ = 0.94) and of 0.80% for the protein (RSQ = 0.881). The NIR and the Kjeldahl reference method were found to measure different types of protein, which affected the results (20).

Variability of meat raw materials can cause problems in processing and quality of meat products. Togersen developed and implemented on-line NIR applications for analysis of proximate composition of fresh and semifrozen ground meat at the outlet of meat grinders. The coarseness of the meat raw materials as well as manual sampling procedures effected prediction results. The prediction results were comparable with those of off-line applications, and NIR was an improvement over manual sampling for off-line analysis. Multivariable regression models based on NIR analysis of raw sausage batter could be used for estimation of instrumentally measured quality characteristics of cooked sausages (21, 22).

Swatland et al. discussed various methods for on-line analysis of meat quality, including NIR (23).

## POULTRY

Vis/NIR spectroscopy and PCA and discriminant analysis can be used to differentiate poultry samples. A Vis/NIR classification model, using nine principal components (PCs) and a linear discriminant function, correctly classified 100%, 90.0%, and 92.5% of the whole (skin and meat) samples for wholesome, septicemia, and cadaver categories, respectively. For skin-only samples, similar models using nine PCs resulted in lower accuracies. Examination of the PCA loadings for the whole samples suggested that the better discrimination of whole samples was dependent on spectral variation related to different forms of myoglobin (deoxymyoglobin, metmyoglobin, and oxymyoglobin) present in the chicken meat. In particular, key wavelengths were identified at 540 and 585 nm, which have been identified as oxymyoglobin bands, 485 nm as metmyoglobin bands, and 440 nm as deoxymyoglobin bands (24).

Park et al. used a hyperspectral imaging system for poultry safety inspection, particularly identification of fecal and ingesta contamination on poultry carcasses. The system included a camera with prism-grating-prism spectrograph, fiber-optic line lighting, motorized lens control, and hyperspectral image-processing software was developed, and although data were collected from 400 to 900 nm, only visible wavelengths were used in the analysis (25).

## FRUIT

A hyperspectral imaging system with a range of 450 to 851 nm was also used to examine reflectance images of fecal contaminated apples. Fecal contamination of apples is an important food safety issue. Fresh feces from dairy cows were applied simultaneously as a thick patch and as a thin, transparent (not readily visible to the human eye) smear to four cultivars of apples selected to represent the range of green to red colorations to address differences in coloration due to environmental growth conditions. The hyperspectral images were evaluated with principal component analysis with the goal of identifying two to four wavelengths that could potentially be used in an on-line multispectral imaging system. Results indicated that contamination could be identified using either three wavelengths in the green, red, and NIR regions, or using two wavelengths at the extremes of the NIR region under investigation. The three wavelengths in the visible and NIR regions offered commercial potential for color sorting of apples. Thick contamination could easily be detected with a simple threshold unique to each cultivar. In contrast, more computationally complex analyses, such as combining threshold detection with morphological filtering, would be necessary to detect thin contamination spots (26).

How to distinguish the stem end/calyx from a true defect is a persistent problem in apple defect sorting systems. In a single-camera NIR approach, the stem end/calyx of an apple is usually confused with true defects and is often mistakenly sorted. To solve this problem, a dual-camera NIR/mid-IR (MIR) imaging method was developed. The MIR camera can identify only the stem end/calyx parts of the fruit, while the NIR camera can identify both the stem end/calyx portions and the true defects on the apple. A fast algorithm was developed to process the NIR and MIR images. Online test results show that a 100% recognition rate for good apples and a 92% recognition rate for defective apples were achieved with this method. The dual-camera imaging system showed great potential for reliable online sorting of apples for defects (27).

Researchers have shown for years that NIR is a suitable technique for sugar content measurement in various fruits. However, most of the systems that have been developed have been hard to apply in automated industrial environments. Sanchez et al. studied external parameters that could affect the robustness needed for time-stable models. The sugar content of Golden Delicious apples was predicted using a portable glove-shaped apparatus named GLOVE, designed and set up in a European project (GLOVE, PL 97-3399). GLOVE is equipped with various miniaturized sensors to provide information on fruit quality, including sugar content, maturity, firmness, stiffness, and color. A Zeiss MMS1 NIR enhanced with a 256 silicon-element photodiode array detector was embedded on the GLOVE and was used to determine the sugar content. The parameters affecting the measurements, fruit, spectrometer temperature, and ambient light were studied. The fruit temperature had a nonlinear effect, and a spectrometer temperature between 4 and 30°C had a linear effect on the NIR predictions. Ambient light did not have any influence on the NIR model. These results should be taken into account in future measurements, in order to improve the robustness of the NIR-based model developed for apples (28).

Apple bruising normally happens to the tissue beneath the fruit skin, lowers the quality grade of the fruit, and can cause significant economic losses. Damaged cells are initially filled with water but then start to lose moisture, eventually becoming desiccated. Bruise detection is challenged by the fruit skin, time, bruise type, apple variety, and pre-and postharvest conditions. Lu reported a NIR hyperspectral imaging and a computer algorithm for detecting both new and old bruises on whole apples in the spectral region between 900 and 1700 nm. Apples were studied over a period of 47 days after bruising. The spectral region between 1000 and 1340 nm was most appropriate for bruise detection. Bruise features changed over time from lower reflectance to higher reflectance, and the rate of the change varied with fruit and variety.

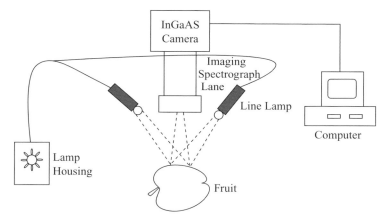

**Figure 9.2.3.** Schematic of the NIR hyperspectral imaging system for detecting bruises on apples (29).

With both principal component and minimum noise fraction transforms, correct bruise detection rate for both new and old bruises ranged from 62% to 88% for Red Delicious apples and from 59% to 94% for Golden Delicious apples, depending on the number of days after bruising and the spectral resolution. The optimal spectral resolution for bruise detection was between 8.6 and 17.3 nm. Although some of the classification results were relatively low, they were generally better than previous studies using different imaging techniques. With improvement in image acquisition speed and detector technology, the NIR hyperspectral imaging technique will have the potential for offline inspection and online sorting of fruit (29). Figure 9.2.3 shows the schematic of the NIR hyperspectral imaging system.

Multispectral imaging with a CCD camera was used to determine firmness ($r = 0.87$ and SEP = 5.8 N) and soluble solids content (SSC) ($r = 0.77$ and SEP = 0.78) of apples. This analysis was done using the 680–1060 nm region, 10-nm band pass, and neural network analysis (30).

He et al. used NIR transmission to determine on-line sugar content, acidity, and internal brown color changes in apples and oranges (31). NIR spectroscopy can be used to predict individual carbohydrates in orange juice with very high accuracy. The best prediction results were achieved with transmittance mode and 1-mm sample cells, but in-line transmittance and at-line transflectance measurements using 1-and 10-mm path lengths were also investigated (2).

## AVOCADO

NIR spectroscopy was applied to nondestructive determination of fruit dry matter (DM) of "Hass" avocados harvested at four different times during a growing season. Mean DM increased from 27.7% to 36.8% during the course of the study. A PLS model applied to data from fruit from all harvests combined ($n = 239$) gave a goodness-of-fit (RP) between predicted and actual DM measurements of 0.88 and an error of prediction (RMSEP) of 1.8% DM. MLR models incorporating four wavelengths returned equivalent validation statistics. Three of these wavelengths, in the vicinity of 900–920 nm, were consistent with the C-H absorbance band, leaving only a minor role for water-related absorbances. These observations suggest that on-line commercial NIR systems capable of operating in an interactance mode have the potential to usefully grade avocados on the basis of their DM content, yielding a product with improved taste and oil content (32).

## FIELD GRAIN

Timely information about the production of malting barley is of great significance for the malting industry. The supply situation depends on the area cultivated to spring barley, yield, protein content, and the graded proportion that is suitable for malting. Yield and quality are strongly influenced by environmental factors—weather being the most important. An investigation was carried out in Germany on the possible use of optical-NIR satellite images and nonremotely sensed data in an operational GIS-based system for the estimation of barley yields and quality. Phenological observations turned out to be a valuable source of information. The duration of the grain filling period derived from these data proved to be temperature dependent. Although the most meaningful parameter for yield estimations was the mean daily temperature between ear emergence and yellow ripeness, the relative humidity proved to be an informative parameter to assess the protein content. Knowledge about the length of this developmental period allowed assessments of the yield level, by means of which quality parameters could be reasonably estimated. By means of multiple linear regression approaches, using phenological and meteorological data, all three target factors, namely, regional average yields, protein contents, and screening percentage, were predicted with deviations well below 5% at the time of yellow ripeness. For an operational application, integration of the various information layers, remote sensing and weather data, into a GIS is recommended (33).

Lu et al. used NIR for barley analysis to maximize the exploitation of genetic variation as early as possible, allowing selection for malting quality to start in the $F_2$ generation. A NIR Infratec 1225 grain analyzer fitted with a modified sample transport module was used to collect transmittance spectra. Calibrations for protein content, potential malt extract, and diastatic power were used. Variation within and between $F_2$-derived families for grain yield and malting quality was investigated in $F_4$ breeding lines derived from $F_2$ families of four barley crosses with Franklin, Harrington, Skiff, and Tallon as parents. The variation between $F_2$-derived families was greater than within $F_2$-derived families for grain yield and all malting quality attributes. Superior segregates almost exclusively came from the best performing families. The greater similarity of lines eventually drawn from an $F_2$-derived family has significant implications for selection strategies in barley breeding programs, as it facilitates the early discard of $F_2$-derived families (34, 35).

## RICE LEAF CANOPY

Ground-based remotely sensed reflectance spectra in the range of 350–2400 nm were monitored during the growing period of rice under various nitrogen application rates. It was found that reflectance spectrum of rice canopy changed in both wavelength and reflectance as the plants developed. Fifteen characteristic bands were identified from the first-order differentiated spectra, measured over the growing season of rice. The bandwidths and frequencies changed with age and nitrogen application rates (36).

## CEREAL

NIR reflection spectroscopy can be used for noncontact determination of protein in wheat grain. Laboratory equipment was installed in a combine to record grain quality in the field in order to perform yield mapping during cereal combining. The data produced during the first year allowed indirect calculation of grain protein content and the production of a field protein map. On-line measurement was possible in the second year. The first on-line measurements had standard error of 0.55% protein (37).

## SOYBEANS

In a glycolipid fermentation, mannosyl erythritol lipid (MEL) is produced from soybean oil added to a medium as a source of carbon. NIR can be used to measure the concentrations of MEL and soybean oil extracted from the fermentation process with ethyl acetate. NIR bands at 1436, 1920, and 2052 nm were assigned to MEL, and a MLR calibration equation was developed using the second derivative spectral data at 2040 and 1312 nm. Thin-layer chromatography with a flame-ionization detector (TLC/FID) was used as the reference method. The regression coefficient ($R$) and the standard error of calibration (SEC) were 0.994 and 0.48 $g \cdot l^{-1}$, respectively. Absorption bands due to soybean oil were observed at 1208, 1716, 1766, 2182, and 2302 nm and the second derivative bands at 2178 and 2090 nm were used for the calibration, yielding values of $R$ and SEC of 0.974 and 0.77 $g \cdot l^{-1}$, respectively. The NIR method was applied to the measurement of the concentrations of MEL and soybean oil in an actual fermentation process with good results (38).

## CORN

Many published applications for food analysis are feasibility studies done toward the application of on-line analysis. For example, NIR was applied to the detection of fumonisin in single corn kernels infected with *Fusarium verticillioides*. However, the classification results for the corn kernels were generally better for oriented kernels than for kernels that were randomly placed in the spectrometer viewing area, making the transition from feasibility to on-line analysis more difficult (39).

## GRAIN MILLING

Current on-line measurement techniques for testing grain and milled products such as flour and semolina include NIR reflection, transmission, microwaves, and image analysis. Method selection is determined by product, objectives, suitability, and precision. The system must measure as many product parameters as accurately as possible, with reasonable setup costs. The bakery industry requires flours and semolinas of consistent quality of color, starch damage, protein, moisture, particle size, and ash content. NIR can be used to monitor wheat at the rate of 20 tons/h. The measured data can be summarized in a job report, using graphs, tables, and statistical analysis including maximum and minimum values and standard deviation. NIR can be used in a flour mill in the wheat-receiving silo and in the cleaning house. During wheat cleaning, the NIR analysis can be used to automatically change the mixing ratio for protein content, allowing optimal use of both poor- and high-quality wheats. Automatic control systems were being used to maintain constant ash content by adding low grade flour, stabilize protein content by adding gluten and maintain water content by adding water (40).

## MARGARINE

Isaksson et al. showed that NIR was equally good for in- and off-line measurements of moisture in margarine. Thirty-six batches with a moisture range of 14.28–21.91 weight-% were produced in a pilot plant study. For the in-line measurements, diffuse transmittance fiber-optic probes were used. For the off-line measurements, sample cups were used. PLS regression of multiplicative scatter corrected NIR spectra, in the 780–1100 nm region, gave prediction errors, expressed as root mean square error of cross-validation (RMSECV) of 0.13 weight-% moisture and correlation coefficient ($R$) 0.998 for the in-line measurements, and RMSECV of 0.14 weight-% and $R$ of 0.997 for the off-line measurements (41).

## SALMON

Isaksson et al. showed that VIS/NIR reflectance could possibly be applied in salmon production plants to classify fillets into broad texture classes before further processing or sale. Spectra of fillets of farmed Atlantic salmon were correlated to Kramer shear force measurement and texture profile analysis (TPA). Samples were analyzed prerigor (2 h after slaughter) and postrigor (6 days after slaughter). Classification using linear discriminant analysis gave up to 79% correct classification into three categories: low, medium, and high Kramer shear force. Using these class limits, no low-Kramer shear force sample was misclassified as a high-Kramer shear force sample, and vice versa (42).

## MELONS

The use of NIR spectroscopy for nondestructive determination of soluble solids (°Brix) in harvested fruits is well documented. The harvest time of some cultivars is difficult to judge from appearance, necessitating an instrumental method. In field determination of optimal harvest time would limit harvesting of immature fruits. Ito et al. reported using a portable NIR developed by Kubota Corporation for nondestructive estimation of °Brix in melons. The spectral data of growing melons cv. Earl's Knight SEIKA were measured in a greenhouse, and correlated with multiple linear regression analysis. The °Brix in the growing melons was estimated using a MLR based on harvested melons. Absorbances at 906 and 874 nm were key wavelengths for the analysis (43).

## WINE

Australian legislature demands the use of only grape-derived ethanol in the production of fortified wines. A major source of grape ethanol is from grape pomace (waste from a pressing step), predominantly skins, seeds and stems. The methanol content in this depends on storage conditions, bacterial, and fungal activity. Samples of distillates derived from the production of wine-fortifying spirit were analyzed for methanol by gas chromatography (GC) and NIR spectroscopy. Enzyme electrodes have been suggested as a possibility for on-line determination of ethanol and methanol, but are unlikely to remain stable at high ethanol concentrations. NIR calibration models were developed which could accurately predict methanol concentrations in samples of fortifying spirit that had been produced over a period of three years from four different commercial distillation facilities. The NIR method required validation with a wide range of samples to check for sample matrix effects. The best accuracy of the NIR predictive models, as measured by the SEP was 0.06 g/l methanol. The most useful calibration models used PLS regression on spectra from a scanning instrument, but it was demonstrated that calibrations could also be developed with a smaller number of fixed wavelengths, using MLR models. The ability of calibrations to predict other samples can be estimated by the ratio of the standard deviation of the reference analysis to the SECV. A ratio greater than three indicated a robust calibration. NIR spectroscopy offered the advantages of rapid analysis with simple routine operation. There may be some potential for in-line process control in the operation of a commercial distillation facility, though the feasibility study was done at constant temperature and temperature equilibration is important in a mixed alcohol system where hydrogen bonding significantly affects the NIR spectra (44).

## CONCLUSIONS

NIR has been shown to have great versatility for on-line analysis of food. Scientists have been very inventive in developing these applications. NIR has been put on extruders, molding heads, combines, stills, and fermentors and in gloves, mills, salmon plants, fields, greenhouses, and satellites. New techniques such as LIGA and hyperspetral imaging have been used. NIR spectra have been interpreted and used in combination with other process parameters, fuzzy logic, mid-IR, and electronic noses, and effects of path length wavelength region and environment have been studied to achieve improved process control and understanding. In all, there have been considerable achievements toward NIR on-line applications for food analysis.

## REFERENCES

1. P. J. Brimmer, F. A. DeThomas, J. W. Hall. Using online near-infrared spectroscopy for quantitative and qualitative analyses. *Cereal Foods World* **47(4)**: 138–141, 2002.

2. V. H. Segtnan. Thesis, Agricultural University of Norway Noragric—Centre for International Environment and Development Studies, 2002, p. iv + 30.

3. B. G. Osborne, T. Fearn, P. H. Hindle. *Practical NIR Spectroscopy with Applications in Food and Beverage Analysis*, 2nd. ed. Essex: Longman Scientific & Technical and John Wiley & Sons, Essex, 1993, p. 227.

4. A. O'Sullivan, B. O'Connor, A. Kelly, M. J. McGrath. The use of chemical and infrared methods for analysis of milk and dairy products. *Int J Dairy Technol* **52(4)**: 139–148, 1999.

5. M. F. Laporte, P. Paquin. Near-infrared technology and dairy food products analysis: a review. *Semin Food Anal* **3(2)**: 173–190, 1998.

6. E. Windhab, S. Bolliger, On-line use of NIR-spectroscopy in food processing. *European Food and Drink Review* 83–91, 1996(Autumn).

7. D. I. Givens, J. L. d. Boever, E. R. Deaville, J. L. De Boever. The principles, practices and some future applications of near infrared spectroscopy for predicting the nutritive value of foods for animals and humans. *Nutrition Research Reviews* **10**: 83–114, 1997.

8. P. C. Williams, J. L. Steele, O. K. Chung. *Recent advances in near-infrared applications for the agriculture and food industries.* Proceedings of the International Wheat Quality Conference, Manhattan, Kansas, USA, 109–128, 1997.

9. F. Apruzzese, S. T. Balke, L. L. Diosady. In-line Colour and Composition Monitoring in the Extrusion Cooking Process. *Food Res Int*, **33**: 621–628, 2000.

10. F. Apruzzese, J. Pato, S.T. Balke, L.L. Diosady. In-line measurement of residence time distribution in a co-rotating twin-screw extruder. *Food Res Int* **36(5)**: 461–467, 2003.

11. K. G. Adamopoulos, A. M. Goula, H. J. Petropakis. Quality control during processing of feta cheese – NIR application. *J Food Compos and Anal* **14(4)**: 431–440, 2001.

12. B. J. A. Mertens, C. P. O'Donnell, D. J. O'Caliaghan, Modelling near-infrared signals for on-line monitoring in cheese manufacture. *J Chemo* **16**: 89–98, 2002.

13. S. Turza, J. Chen, Y. Terazawa, N. Takusari, M. Amari, S. Kawano, J. Y. Chen. On-line monitoring of rumen fluid in milking cows by fibre optics in transmittance mode using the longer NIR region. *J Near Infrared Spectrosc* **10(2)**: 111–120, 2002.

14. A. B. Koc, P. H. Heinemann, G. R. Ziegler, W. B. Roush. Fuzzy logic control of whole milk powder processing. *Trans of the ASAE* **45(1)**: 153–163, 2002.

15. D. Brennan, J. Alderman, L. Sattler, B. O'Connor, C. O'Mathuna. Issues in development of NIR micro spectrometer system for on-line process monitoring of milk product. *Measurement*, **33(1)**: 67–74, 2003.

16. M. Natsuga, S. Kawamura, K. Itoh. Effects of wavelength range and light path-length on the accuracy of the constituent analysis of unhomogenized milk using near-infrared spectroscopy. *J Japanese Soc Agric Machi* **64(5)**: 83–88, 2002.

17. Y. A. Woo, Y. Terazawa, J. Y. Chen, C. Iyo, F. Terada, S. Kawano. Development of a new measurement unit (MilkSpec-1) for rapid determination of fat, lactose, and protein in raw milk using near-infrared transmittance spectroscopy. *Appl Spectrosc* **56(5)**: 599–604, 2002.

18. C. Cimander, M. Carlsson, C. F. Mandenius. Sensor fusion for on-line monitoring of yoghurt fermentation. *J Biotec* **99(3)**: 237–248, 2002.

19. N. M. Anderson, P. N. Walker. Measuring fat content of ground beef stream using on-line visible/NIR spectroscopy. *Trans ASAE* **46(1)**: 117–124, 2003.

20. I. Gonzalez Martin, C. Gonzalez Perez, J. Hernandez Mendez, N. Alvarez Garcia, J. L. Hernandez Andaluz. On-line non-destructive determination of proteins and infiltrated fat in Iberian pork loin by near infrared spectrometry with a remote reflectance fibre optic probe. *Anal Chimica Acta* **453(2)**: 281–288, 2002.

21. G. Togersen. Thesis, Agricultural University of Norway Noragric – Centre for International Environment and Development Studies, p. 32, 2002.

22. G. Togersen, J. F. Arnesen, B. N. Nilsen, K. I. Hildrum. On-line prediction of chemical composition of semi-frozen ground beef by non-invasive NIR spectroscopy. *Meat Sci* **63(4)**: 515–519, 2003.

23. H. J. Swatland, J. Kerry, D. Ledward. *On-line monitoring of meat quality*, in *Meat processing: improving quality*, Woodhead Publishing Ltd: Cambridge; UK, 2002, p. 193–216.

24. K. Chao, Y. R. Chen, D. E. Chan. Analysis of Vis/NIR spectral variations of wholesome, septicemia, and cadaver chicken samples. *App Engine Agri* **19(4)**: 453–458, 2003.

25. B. Park, K. C. Lawrence, W. R. Windham, R. J. Buhr. Hyperspectral imaging for detecting fecal and ingesta contaminants on poultry carcasses. *Trans ASAE* **45(6)**: 2017–2026, 2002.

26. M. S. Kim, A. M. Lefcourt, K. Chao, Y. R. Chen, I. Kim, D. E. Chan. Multispectral detection of fecal contamination on apples based on hyperspectral imagery: Part I. Application of visible and near-infrared reflectance imaging. *Trans ASAE* **45(6)**: 2027–2037, 2002.

27. X. Cheng, Y. Tao, Y. R. Chen, Y. Luo. NIR/MIR dual-sensor machine vision system for online apple stem-end/calyx recognition. *Trans ASAE* **46(2)**: 551–558, 2003.

28. N. Hernandez Sanchez, S. Lurol, J. M. Roger, V. Bellon Maurel. Robustness of models based on NIR spectra for sugar content prediction in apples. *J Near Infrared Spectrosc* **11(2)**: 97–107, 2003.

29. R. Lu, Detection of bruises on apples using near-infrared hyperspectral imaging. *Trans ASAE* **46(2)**: 523–530, 2003.

30. R. Lu. Multispectral iaging for predicting firmness and soluble solids content of apple fruit. *Postharvest Biology Techno* **31**: 147–157, 2004.

31. D. He, T. Maekawa, H. Morishima, D. J. He. Detecting device for on-line detection of internal quality of fruits using near-infrared spectroscopy and the related experiments. *Trans Chinese Soc of Agric Engin* **17(1)**: 146–148, 2001.

32. C. J. Clark, V. A. McGlone, C. Requejo, A. White, A. B. Woolf. Dry matter determination in 'Hass' avocado by NIR spectroscopy. *Postharvest Biology Techn* **29(3)**: 301–308, 2003.

33. K. Schelling, C. Weissteiner, K. Hunting, W. Kuhbauch, T. Benes. *Yield and quality estimation of malting barley based on remote sensing and GIS.* in *Geoinformation for European wide integration. Proceedings of the 22nd Symposium of the European Association of Remote Sensing Laboratories, Prague, Czech Republic, 4 6 June 2002. 2003, 549 555; 10 ref.* 2003: Millpress Science Publishers; Rotterdam; Netherlands.

34. M. Q. Lu, L. O'Brien, I. M. Stuart. Variation within and between F2-derived families for grain yield and barley malting quality. *Australian J Agric Res* **52(1)**: 85–92, 2001.

35. M. Q. Lu, L. O' Brien, I. M. Stuart. Barley malting quality and yield interrelationships and the effect on yield distribution of selection for malting quality in the early generations. *Australian J Agric Res* **51**: 247–58, 2000.

36. C. Yang, C. M. Yang, J. N. Galloway, E. B. Cowling, J. W. Erisman, J. Wisniewski, C. Jordan. *Estimation of leaf nitrogen content from spectral characteristics of rice canopy.* in *Optimizing nitrogen management in food and energy production and environmental protection. Proceedings of the 2nd International Nitrogen Conference on Science and Policy, Potomac, MD, USA, 14 18 October 2001. TheScientificWorld. 2001, 1: tsw.2001.387, 81 89; available at www.thescientificworld.com.* 2001.

37. J. Rademacher. Measuring protein content during combining. *Landtechnik* **57(6)**: 354–355, 2002.

38. K. Nakamichi, K. Suehara, Y. Nakano, K. Kakugawa, M. Tamai, T. Yano. Measurement of the concentrations of mannosyl erythritol lipid and soybean oil in the glycolipid fermentation process using near infrared spectroscopy. *J Near Infrared Spectrosc* **10(1)**: 53–61, 2002.

39. F. E. Dowell, T. C. Pearson, E. B. Maghirang, F. Xie, D. T. Wicklow. Reflectance and transmittance spectroscopy applied to detecting fumonisin in single corn kernels infected with Fusarium verticillioides. *Cereal Chem* **79(2)**: 222–226, 2002.

40. F. Gradenecker. NIR on-line testing in grain milling. *Cereal Foods World* **48(1)**: 18–19, 2003.

41. T. Isaksson, G. Naerbo, E. O. Rukke, N. S. Sahni. In-line determination of moisture in margarine, using near infrared diffuse transmittance. *J Near Infrared Spectrosc* **9(1)**: 11–18, 2001.

42. T. Isaksson, L. P. Swensen, R. G. Taylor, S. O. Fjaera, P. O. Skjervold. Non-destructive texture analysis of farmed Atlantic salmon using visual/near-infrared reflectance spectroscopy. *J Sci Food Agric* **82(1)**: 53–60, 2002.

43. H. Ito, S. Morimoto, R. Yamauchi, M. Hertog, B. R. MacKay. *Potential of near infrared spectroscopy for non-destructive estimation of soluble solids in growing melons.* in *Proceedings of the Second International Symposium on Applications of Modelling as an Innovative Technology in the Agri Food Chain, Model IT 2001, Palmerston North, New Zealand, 9 13 December, 2001. Acta Horti 2001, No.566, 483–486, 9 ref.* 2001.

44. R. G. Dambergs, A. Kambouris, L. Francis, M. Gishen. Rapid Analysis of Methanol in Grape-Derived Distillation Products Using Near-Infrared Transmission Spectroscopy. *J Agric Food Chem* **50**: 3079–3084, 2002.

# Disease Diagnosis Related to Food Safety in Dairy

ROUMIANA TSENKOVA

## INTRODUCTION

In modern times, food quality, safety, and bacterial contamination have become very important issues. Milk composition and safety are germane to successful dairy farming as well as being important to the dairy industry as a whole. The quality of raw milk has a direct influence on the performance of processed milk and milk products. Therefore, establishing a *Hazard Analysis Critical Control Points* (HACCP) in the very beginning of the milk food chain is needed to monitor both cow health and milk abnormality (1, 2). Mastitis is an intramammary bacterial infection, which is a well-spread disease among dairy cattle (3, 4). It causes decrease in milk yield and reduces the synthesis and secretion of milk components. Chemical changes in milk composition due to mastitis reduce milk production of individual cows and alter milk quality. Milk from mastitic cows is often characterized by low proportions of casein, high concentrations of serum proteins and an altered mineral balance (4–6). Somatic cell count (SCC) is a recognized indicator of cow health and milk quality. Milk SCC reflects the level of infection and resultant inflammation in the mammary gland of dairy cows concurrent with mastitis diagnosis. Milk somatic cells are primarily leukocytes that depend on intensity of the cellular immune defense. Some of these cells come from mammary ducts.

Mastitis is one of the most common dairy cow diseases. It can cause considerable economic losses to dairy farmers. The losses are incurred from several sources including a decreased milk yield, marked compositional changes in milk that reduce milk quality, treatment and labor costs, and increasing risk of early culling of cows. In addition, milk with a high somatic cell count can have a deleterious effect on cheese

*Near-Infrared Spectroscopy in Food Science and Technology*, Edited by Yukihiro Ozaki, W. Fred McClure, and Alfred A. Christy.
Copyright © 2007 John Wiley & Sons, Inc.

production. Healthy mammary glands free from infection almost always have SCC less than 100,000 cells/ml, whereas unhealthy cows infected with mastitis will have SSC greater than 300,000 cells/ml. Hence, mastitis infection of mammary gland is indicated by elevated SCC, electroconductivity and presence of a pathogen in milk from the individual quarters.

Usually, a composite sample of raw milk is analyzed once or twice per month for each dairy farmer with the Milko Scan (Foss-Electric A/S, Hillerod, Denmark), a mid-infrared spectrometer designed specifically for this job (1). Farmers are paid according to the results from the Milko Scan. Milko Scan measurements are made on samples brought from the dairy farm into the lab, quite often several hours after the sample is collected. Unless handled properly, the samples can increase in SCC during the interval between the time the sample is collected and the analysis is performed. It should be obvious that a method for making more timely measurements would be in order.

Near-infrared spectroscopy (NIRS) has been applied to fat, protein, milk urea nitrogen (MUN), and lactose measurement of homogenized milk (7–12) and nonhomogenized milk (13, 14) as indicators of milk quality and feed management. NIR spectroscopy has been successfully employed (15–19) for milk composition measurement and somatic cell determination using spectra of composite (from all four quarters of the udder) samples of nonhomogenized from individual cows. Further studies (involving determinations of SSC, electrical conductivity, and pathogens) of milk from individual quarters had to be made to more precisely locate the udder infection. The udder of a dairy cow is made up of four udder quarters. A composite sample of cow's milk contains equal portion from all four udder quarters, assuming that all of them are healthy and could be milked. Currently, common dairy practice involves milk quality control performed daily at the herd level, namely, measurements of samples drawn daily from a bulk tank. Farmers usually retain monthly records of SSC for each cow in his herd. These records do not contain information related to the individual cows. More recently with the introduction of robotic milking parlors, it would be advantageous to monitor the health of each individual cow. This advanced approach would require the development of fast, even noninvasive, sensors. NIR technology appears to be the only technology to fulfill the needs of modern-day milking parlors. Having NIR sensors for on-line analysis of both composite and/or udder quarter milk during the milking process should provide the data necessary for maintaining health records for individual cows.

NIR spectra of cow's composite and udder quarter milk subjected to multivariate data analysis contain information about milk abnormality and cow health disorders. Models for NIRS measurement of SCC in composite and udder quarter cow's milk have been developed. Models for simultaneous measurement of electrical conductivity and somatic cell count, as well as for identification of the main bacterial pathogens causing mastitis are described further. NIR spectroscopy has proved to be a valuable tool for mastitis diagnosis and for milk quality evaluation.

# MILK QUALITY EVALUATION AND DIAGNOSIS OF COW MAMMARY GLAND INFLAMMATION USING NIR SPECTRA TO MEASURE SCC IN COMPOSITE MILK

## Instrumentation

In NIR spectroscopic studies, when models are developed, each milk sample is divided into two subsamples. One is subjected to spectral analysis, and the other is analyzed by reference methods.

## Samples

A total of 196 composite milk samples from seven Holstein cows have been analyzed (20). The samples have been collected for 28 days, consecutively, beginning on the seventh day after calving. Cows were fed a ration containing corn silage, Timothy hay, commercial concentrate mixture (corn, barley, alfalfa meal, beet pulp, and $CaCO_3$), trace minerals, and vitamins premix, and soybean meal provided 48% of total CP of the diet. The average BW of the cows was 552 kg. Animals were fed twice daily ad libitum and always had access to drinking water. Three of the examined cows had been healthy, with SCC lower than 137,000 cells/ml. One cow had been mastitic during the entire experimental period—the measured SCC varied from 204,000 to 11,876,000 cells/ml. Three cows had mastitic and healthy periods (SCC between 80,000 and 4,737,000 cells/ml).

## NIR Instrument

NIR transflectance $(T)$ spectra were obtained by the InfraAlyzer 500 spectropho-tometer, (Bran+Luebbe, Nordestedt, Germany), in terms of optical density log $(1/T)$ in a wavelength range from 1100 to 2500 nm. A flow cell with pathlength of 0.2 mm, connected with automated liquid sampling system and taking alternatively milk samples and cleaning solution was used. Before the spectral analysis each sample was warmed up to 40°C in a water bath with temperature control of ±0.1 °C. During the analysis, the same temperature has been controlled through the use of an integrated water-jacketed holder of the flow cell connected with the water bath.

Before the spectral analysis each sample was warmed up to 40°C in a water bath, and spectra were acquired at 37°C.

## Reference Analysis of SCC

Each milk sample was divided into two subsamples. One was subjected to spectral analysis, and the other was analyzed for SCC by fluoro-opto-electronic method using FOSSOMATIC 400 (Foss-Electric A/C, Hillerod, Denmark). SCC standards were used to calibrate the Foss instrument throughout the study. The repeatability coefficients of variation of this method are 4–5% for the region between 400,000 and 500,000 cells/ml and 5–10% for the region between 100,000 and 200,000 cells/ml (7). $Log_{10}$ SCC was calculated to normalize the SCC distribution, and further quantitative analysis was made on the transformed data. Samples were also analyzed for fat, total protein, and lactose content (AOAC, 1990) by Milko Scan (Foss-Electric A/S, HillerØd, Denmark).

## Data Analysis

A commercial software program (Pirouette Version 2.6, Infometrics, Inc., Woodinville, WA) was used to process the data and to develop regression equation for log SCC determination.

The data were randomly divided into a set of 128 calibration samples and a set of 64 validation samples. The data sets covered similar ranges of each investigated parameter.

Methods used for preliminary examination of the data included smoothing the spectral data, multiplicative scatter correction, standard normal variance correction, baseline correction, and first- or second-derivative transformation of log $(1/T)$ data. The smoothing and derivative transformations were based on the Savitzki-Golay second-order polynomial filter (22).

Calibration for SCC was performed by partial least-square (PLS) regression using the calibration set of samples. The PLSR utilizes both the spectra and the respective reference data for the examined samples to determine latent variables (PLS factors). A criterion for calculating PLS factors is to describe the possible maximum amount of variance of NIR data, as well as to be maximally correlated to the dependent variables. PLS performs both the calculation of PLS factors and regression of the reference data simultaneously. After all PLS factors are calculated, the loadings of these factors are combined into a calibration (regression) equation. In this equation there is one coefficient for each wavelength. When the calibration equations have been developed, 15 PLS factors have been set up as maximum.

The optimum number of PLS factors used in the models was determined by a cross-validation method. In cross-validation, five samples were temporarily removed from the calibration set to be used for validation. With the remaining samples, a PLS model was developed and applied to predict the respective milk component for each sample in the group of five. Results were compared with the respective reference values. This procedure was repeated several times until prediction for all samples had been obtained. Performance statistics were accumulated for each group of removed samples. The validation errors were combined into a standard error of cross-validation. The optimum number of PLS factors in each equation was defined to be that which corresponded to the lowest standard error of cross-validation. The performance of each regression was tested with independent validation set of samples.

Calibration and validation statistics for each regression include standard error of calibration, coefficient of multiple correlation, standard error of prediction,

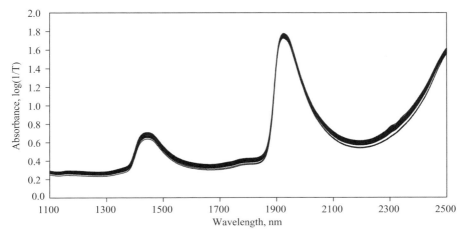

**Figure 9.3.1.** Near-infrared spectra of nonhomogenized milk samples.

and correlation coefficient between measured, and NIRS predicted value. Bias and variation coefficient for calibration and validation set of samples were also calculated. The statistical parameters were used for evaluation of the NIRS determination accuracy.

Correlation coefficients between PLS factors and regression coefficients, log SCC, fat, protein, and lactose content of milk samples, respectively, were also calculated. F-ratio of variance was used to determine significance of correlation coefficients (23).

## Measurement of SCC in Composite Cow's Milk

Raw log $(1/T)$ spectra of nonhomogenized cow's milk samples are shown in Figure 9.3.1. Because of strong absorbance by O-H groups in water, two bands around 1445 and 1930 nm dominate the spectra. The characteristic absorption bands of fat and other milk components such as protein and lactose are very weak in comparison with the water bands and are difficult to visualize. This fact has required the application of multivariate analysis to extract the spectral information connected with milk composition.

Results for the range, mean values, and standard deviation of log SCC, fat, protein, and lactose content of milk samples used in the calibration and the validation sets measured by reference methods are presented in Table 9.3.1. Calibration and validation statistics for regression equations are summarized in Table 9.3.2.

The best accuracy (the smallest standard error of prediction and the highest validation correlation coefficient) for determination of log SCC for the validation set of samples has been found for smoothed log $(1/T)$ data and 10 PLS factors.

**TABLE 9.3.1. Range, Mean and Standard Deviation (SD) of Fat, Protein, Lactose and Log SCC Content of the Examined Milk Samples Measured by Reference Methods**

| Parameter | Set of Samples | $n$ | Mean | Min | Max | SD |
|---|---|---|---|---|---|---|
| Fat, % | Calibration set | 128 | 3.724 | 1.36 | 6.02 | 0.850 |
| | Validation set | 64 | 3.762 | 1.28 | 6.25 | 0.864 |
| Protein, % | Calibration set | 128 | 3.106 | 2.50 | 3.92 | 0.281 |
| | Validation set | 64 | 3.106 | 2.49 | 3.86 | 0.284 |
| Lactose, % | Calibration set | 128 | 4.458 | 4.04 | 4.86 | 0.162 |
| | Validation set | 64 | 4.443 | 3.51 | 4.79 | 0.190 |
| Log SCC | Calibration set | 128 | 4.964 | 3.845 | 7.075 | 0.693 |
| | Validation set | 64 | 5.004 | 3.778 | 7.275 | 0.775 |

Figure 9.3.2 illustrates the relation between log SCC data obtained by the reference method and the respective values predicted by NIR regression based on smoothed log $(1/T)$ data. The equations obtained with multiple scatter correction or first-derivative spectral data transformation had similar accuracy. The obtained standard error of prediction of 0.382 and variation coefficient of validation of 7.63% allow screening of milk samples and differentiation of healthy and mastitic milk samples. These results present better standard errors of prediction than the value of 0.60 reported by Whyte et al. (24) for the spectral region from 400 to 1100 nm.

The equations based on spectral data transformed as second derivative or by using baseline correction showed lower accuracy, suggesting that the baseline variations, corrected by the second-derivative transformation or baseline correction, contained information that is significant for SCC determination.

PLS modeling has been used as a tool for discerning the location of spectral information related to SCC. Regression equation for log SCC, based on smoothed

**TABLE 9.3.2. NIRS Calibration and Validation Statistics for Log SCC Determination**

| | Calibration Set | | | | Validation Set | | | |
|---|---|---|---|---|---|---|---|---|
| Spectra transf. | PLS factors | SEC | $R$ | VCC, % | SEP | $r$ | Bias | VCV, % |
| Smooth | 10 | 0.361 | 0.868 | 7.27 | 0.382 | 0.854 | −0.007 | 7.63 |
| MSC | 12 | 0.325 | 0.896 | 6.55 | 0.400 | 0.841 | −0.005 | 7.99 |
| SNV | 11 | 0.342 | 0.883 | 6.89 | 0.423 | 0.820 | −0.018 | 8.45 |
| BC | 9 | 0.362 | 0.865 | 7.29 | 0.445 | 0.800 | 0.016 | 8.89 |
| 1D | 12 | 0.334 | 0.890 | 6.73 | 0.407 | 0.834 | −0.044 | 8.13 |
| 2D | 11 | 0.356 | 0.872 | 7.17 | 0.472 | 0.771 | 0.002 | 9.43 |

MSC, multiple scatter correction; SNV, standard normal variate; BC, baseline correction; 1D, first-derivative transformation; 2D, second-derivative transformation; SEC, standard error of calibration; $R$, coefficient of multiple correlation; SEP, standard error of prediction; $r$, validation correlation coefficient; VCC, %, variation coefficient for calibration set-(SEC/Mean value) × 100; VCV, %, variation coefficient for validation set-(SEP/Mean value) × 100.

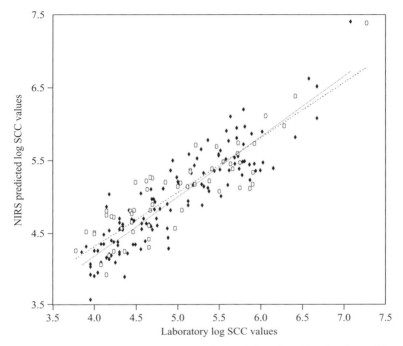

**Figure 9.3.2.** Realationship between actual and near-infrared predicted values of log SCC (dark squares-samples from calibration set, □-samples from validation set, —regression line for calibration set,---regression line for validation set).

log $(1/T)$ spectral data and loading of factors included in the model have been studied (Figs. 9.3.3–9.3.5). Correlation coefficients between loading weight of each PLS factor and regression coefficients, and between scores of each PLS factor and log SCC, fat, protein, and lactose content of milk samples, respectively, were calculated. Results are presented in Table 9.3.3.

PLS factor 1 was found to have some relationship with log SCC, but the contribution of that factor to the regression was very small. Usually, factor 1 loading follows the water spectra, showing that most variability in the sample spectra is caused by the water spectral changes (see the very high negative correlation coefficient for fat, which is negatively related to water). The result could be explained with both the water changes and some increase of the scattering caused by presence of SCC. No correlation was found between the rest of the factors and log SCC. Therefore, NIR determination of log SCC was based on relative changes in milk composition affecting milk spectral changes.

The highest significant correlation between PLS factors and fat content was found for factors 1 and 2, but contribution of these factors to the log SCC regression was very

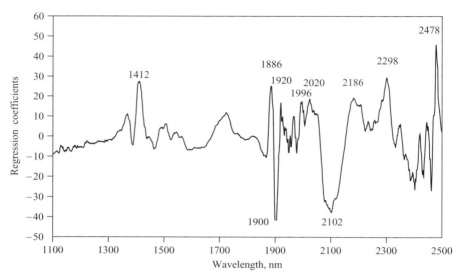

**Figure 9.3.3.** Regression coefficients for each wavelength in PLS model determining log SCC and based on smoothed log $(1/T)$ milk spectra.

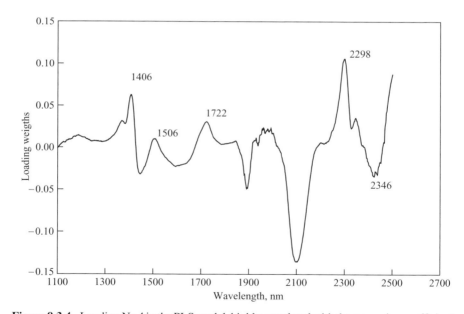

**Figure 9.3.4.** Loading No 4 in the PLS model, highly correlated with the regression coefficients and the lactose content of milk.

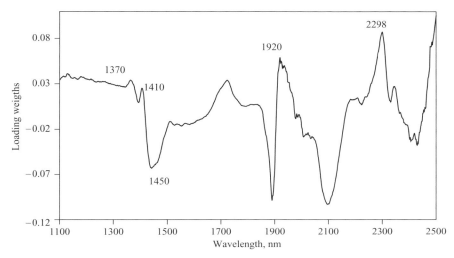

**Figure 9.3.5.** Loading No 5 in the PLS model, highly correlated with the regression coefficients and the protein content of milk.

small. Therefore it could be concluded that fat content had little influence on log SCC determination. The percentage of variations explained by each PLS factor decreased with its number. The third factor had the highest correlation with the protein content when compared to the rest of milk components.

Factor 4 had a higher contribution than factor 3 to the log SCC regression vector and the highest correlation with lactose when compared to the rest of the factors.

**TABLE 9.3.3. Correlation Between Loadings of PLS Factors and Regression Vector (1) and Between Scores of PLS Factors and Each Analyzed Milk Component, Respectively (2)**

| PLS Factor No | Correlation Coefficient | | | | |
|---|---|---|---|---|---|
| Loadings | Regression vector(1) | Fat, % (2) | Protein, % (2) | Lactose, % (2) | Log SCC (2) |
| 1 | 0.045 | $-0.850^{**}$ | 0.152 | $-0.082$ | 0.166 |
| 2 | 0.091 | $0.715^{**}$ | $-0.097$ | $-0.077$ | $-3.7 \times 10^{-9}$ |
| 3 | $0.234^{*}$ | $0.248^{*}$ | $0.323^{*}$ | $0.292^{*}$ | $-2.9 \times 10^{-7}$ |
| 4 | $0.319^{**}$ | $-0.340^{*}$ | $-0.314^{*}$ | $-0.345^{*}$ | $-9.6 \times 10^{-7}$ |
| 5 | $0.502^{**}$ | 0.179 | $0.324^{*}$ | 0.117 | $-2.9 \times 10^{-6}$ |
| 6 | $0.386^{**}$ | 0.147 | $-0.304^{*}$ | 0.008 | $3.4 \times 10^{-6}$ |
| 7 | $0.271^{*}$ | $-0.146$ | $-0.031$ | $-0.070$ | $-3.4 \times 10^{-6}$ |
| 8 | $0.402^{**}$ | $-0.088$ | $0.235^{*}$ | 0.015 | $2.1 \times 10^{-6}$ |
| 9 | $0.307^{*}$ | 0.090 | $-0.403^{*}$ | $0.250^{*}$ | $2.3 \times 10^{-7}$ |
| 10 | $0.352^{**}$ | $-0.230$ | $0.334^{*}$ | $-0.085$ | $-1.8 \times 10^{-5}$ |

$^{*}P < 0.05$, $^{**}P < 0.01$.

Mastitis is well known to decrease lactose content. This fact explains the relation between lactose content and determination of log SCC (4). This emphasizes the possibilities of detecting changes with lactose when analyzing milk spectra and proves its strong relation with SCC. Factors 5, 6, 8 and 10, which showed high correlation with regression coefficients, had the highest correlation with protein content. This is consistent with the fact that mastitis causes alteration of protein fractions in milk. Mastitic milk has more proteolytic activity than normal milk, due to increase of proteinase plasmin, which hydrolyzes the casein (25, 26). Harmon (6) and Urech et al. (25), have reported decreased αs-casein and β-casein content and elevated whey proteins and γ-casein in the total protein of mastitic milk.

On the other hand, mastitis increases capillary permeability, which facilitates passage of proteins from blood to milk. These proteins are mainly serum albumin and immunoglobulins, which are implicated in udder defense mechanisms (27).

High positive coefficients in the regression plot (Fig. 9.3.3) were found at 1412, 1886, 1920, 1996, 2020, 2186, 2298, and 2478 nm. The wavelength 1412 nm could be interpreted as the O-H water absorption band (28). The importance of this wavelength is related to an influence of altered ionic concentration in milk on the water band caused by mastitis. Mastitis increases sodium and chloride content of milk (4). The changes in ionic concentration could provoke a shift in the position of water absorbance bands (29, 30). The other significant peaks at 1886 and 1920 nm are in the region of O-H stretching mode, first overtone and C=O stretch vibration, second overtone (31).

Two reasons might explain the importance of these wavelengths. One is the influence of changes in quantity of water-soluble proteins and ionic concentration on water absorption bands. The C=O absorption is likely due to absorption of urea, the main nonprotein nitrogen compound in milk. Ng-Kwai-Hang et al. (32) have reported significant contribution of SCC to variation of nonprotein nitrogen in individual milk samples from 3600 cows. The absorption at 1996, 2020, 2186, 2298, and 2478 nm might be explained with protein and urea absorption. Diaz-Carrillo et al. (11) have reported peaks in NIR spectra of albumin at 2170 and 2300 nm, and in spectra of casein at 1968, 2166, and 2282 nm, respectively. The obtained peak in the regression vector at 2298 nm is nearer to the reported albumin peak at 2300 nm than to the casein peak at 2282 nm, which might show greatest influence of water-soluble proteins like albumin for SCC determination. The absorption at 2186 nm in milk has been assigned with Amide A+Amide III absorption by Charnik-Matusewicz et al. (9), using two-dimensional correlation spectroscopy. Murayama et al. (32) assigned absorption at 2290 nm to albumin in water solution with the contribution of C-H and amide III vibrations. Peaks at 2020 nm could be associated with absorption due to the C=O stretch, second overtone, which corresponds to urea.

High negative coefficients in the regression vector plot have been presented at 1900 and 2102 nm (Fig. 9.3.3). Absorption at these wavelengths is explained with absorption of O-H stretch/C-O stretch combination vibration at 1900 nm and O-H bend/C-O stretch combination vibration at 2100 nm, associated with carbohydrates (33). In the case of milk, the main carbohydrate component is lactose. Lactose content has been negatively correlated with SCC (4) and therefore has explained the high negative coefficients obtained in the regression vector at 1900 and 2102 nm.

Convincing evidence that determination of log SCC was based mainly on changes in lactose content, ionic concentration, and alteration of protein fractions of milk was obtained by further analysis of the loading plot of significant PLS factors. Strong contribution of 1406, 1506, 2102, 2298, and 2346 nm in the regression vector was also found to be very prominent in the loading of factor 4 (Fig. 9.3.4), related to lactose. The loading of factor 5 (Fig. 9.3.5) has shown 1900 and 1920 nm to be related to protein.

The accuracy of SCC determination in composite cow's milk by NIR spectroscopy has allowed health screening of cows and differentiation between healthy and mastitic milk samples. It has been found that SCC determination by NIR milk spectra is based on the related changes in milk composition. The most significant factors that influence NIR spectra of milk are the alteration of milk proteins and changes in ionic concentration of mastitic milk.

## MILK QUALITY EVALUATION AND DIAGNOSIS OF COW MAMMARY GLAND INFLAMMATION

Finding a cow with mammary inflammation requires taking fast measures to cure the cow and avoid collecting milk from the sick udder quarters. Therefore, analyzing udder quarter foremilk has been presented (20) as the next step in the dairy management to guarantee the health of each cow and the production of high-quality milk.

### Milk Samples and SCC Measurements

A further experiment was conducted after detailed study of all factors that might have caused milk spectral variations. A total of 200 quarter foremilk samples, from morning and afternoon milking, were collected from 5 Holstein cows for 10 consecutive days. One of the cows was in her third lactation, one in the fourth, two in the fifth, and one in the sixth lactation.

All samples were analyzed for SCC by FOSOMATIC (Foss-Electric A/C, Hillerod, Denmark) and for absolute electrical conductivity (AEC) by Milk Checker (Oriental Instruments Ltd., Tokyo, Japan). Electrical conductivity measurements were carried out simultaneously with NIR spectral measurements for each milk sample. $Log_{10}SCC$ was calculated to normalize the SCC distribution.

### Absolute Electrical Conductivity, AEC, Measurement of Quarter Milk Samples

AEC of each quarter foremilk was measured by Milk Checker. Equal quantities of milk samples were acquired before milking and after stripping. Milk samples were divided in two and subjected to spectral and electroconductivity measurements, respectively.

## NIR Spectra Acquisition

NIR milk spectra were obtained from all samples by InfraAlyzer 500 spectrophotometer (Bran+Luebbe, Norderstedt, Germany), using a transflectance measured mode with 0.2-mm path length in the spectral region from 1100 to 2500 nm with 2-nm intervals. Spectra were recorded in the linked computer as absorbance, that is, $\log(1/T)$. Before spectral analysis each sample was warmed up to $40°C$ in a water bath.

## Data Analysis

A commercial program Pirouette Version 2.6 (Infometrics, Inc., Woodinville, WA, USA) was used to process the data and to develop models for AEC and log SCC determination. Every third sample in the data set was used to organize a separate test set. The rest of the samples formed the calibration subset.

The methods for data treatment included first-derivative transformation of $\log(1/T)$ data with window size of 25 points, based on the Savitzki–Golay (22) polynomial filter. Calibration for quantitative determination of log SCC and absolute electrical conductivity was performed with PLS regression as described above.

For qualitative analysis, that is, classification of samples, they were divided in two classes: class 1, healthy and class 2, mastitic, based on their SCC content or/and AEC data. First, the samples were divided by their SCC: with SCC lower than 300,000 SCC cell/ml (class 1) and with SCC higher than 300,000 SCC cell/ml (class 2). Second, samples were divided by their AEC: with AEC lower than 5.7 mS (class 1) and group of samples with AEC higher than 5.7 mS (class 2). Third, samples were divided concerning both criteria: SCC and AEC. The healthy group contained samples with SCC lower than 300,000 cell/ml or AEC lower than 5.7 mS (class 1), and the mastitic group contained samples with both SCC higher than 300,000 cell/ml and AEC higher than 5.7 mS (class 2). SIMCA (33) was implemented to create models for milk sample classification regarding NIR spectra.

To classify samples with various pathogens, SIMCA (21) was employed to develop models for milk sample classification regarding NIR spectra. SIMCA develops models for each class based on factor analysis, that is, principal components that describe the variations of the spectral data. Once each class has its own model new samples are classified to one or another class according to their spectra. Samples from the calibration set were used to develop SIMCA models for class 1 and class 2, respectively. The obtained models were evaluated with samples from the test set, and different models were compared. SIMCA identified variations that were quite different from the inherent variance of the training set. First, the training set data matrix was decomposed by principle component analysis (PCA) and the optimum number of factors was determined. Mahalanobis distance calculations were applied to the score matrix, for the primary set of factors, to compare unknowns to the training set. If an unknown sample was not a member of the groups, it was rejected. A spectrum was classified as a member of a respective class if its Mahalanobis distance was less than 3 standard deviations from the cluster's centroid.

**TABLE 9.3.4. Range, Mean and Standard Deviation (SD) of SCC and AEC of the Examined Milk Samples Measured by Reference Methods**

| Parameter | Data Set | $n$ | Mean | Min | Max | SD |
|---|---|---|---|---|---|---|
| Log SCC | Calibration | 134 | 4.641 | 3.00 | 6.796 | 0.759 |
| Log SCC | Test | 66 | 4.663 | 3.477 | 7.038 | 0.792 |
| Absolute Electrical Conductivity, mS | Calibration | 134 | 5.186 | 3.50 | 9.30 | 0.941 |
| Absolute Electrical Conductivity, mS | Test | 66 | 5.176 | 3.20 | 8.20 | 1.011 |

## Quantitative Determination of SCC and AEC of Udder Quarter Milk

The results for the range, mean values, and standard deviation of log SCC and AEC of milk samples in the calibration and the test sets measured by reference methods are presented in Table 9.3.4. Twenty-two milk samples had SCC higher than 300,000 SCC cell/ml, and thirty-six samples had AEC higher than 5.7 mS. Statistical information for the NIRS calibration models that were developed for quantitative determination of log SCC and AEC is shown in Table 9.3.5.

The relation between laboratory measurements of log SCC and the predicted values by the NIRS model for independent test set of samples is shown in Fig. 9.3.6. Results obtained for AEC calibration and prediction by NIRS were similar to those for log SCC. Calibration correlation coefficient $R$ between NIR spectral data and laboratory data was 0.850 for log SCC and 0.856 for AEC, and correlation coefficient $r$ for prediction with the test set was 0.884 and 0.874, respectively (Table 9.3.5). The relation between laboratory measurements of AEC and the predicted values by the respective NIRS model for independent test set of samples is shown in Figure 9.3.7.

## Milk Sample Classification by Qualitative Multivariate Analysis

Results of SIMCA classification of milk samples based on NIR spectra after they were divided into healthy or mastitic according to log SCC are presented in Table 9.3.6. For the calibration set of samples, SIMCA models (model for healthy quarters

**TABLE 9.3.5. NIRS Calibration and Validation Statistics for Log SCC and Absolute Electrical Conductivity Determination in Nonhomogenized Cow Quarter Foremilk**

| Parameter | Calibration Set | | | Test Set | | |
|---|---|---|---|---|---|---|
| | PLS Factors | SEC | $R$ | SEP | $r$ | CV, % |
| Log SCC | 8 | 0.412 | 0.850 | 0.368 | 0.884 | 7.89 |
| AEC | 8 | 0.502 | 0.856 | 0.491 | 0.874 | 9.48 |

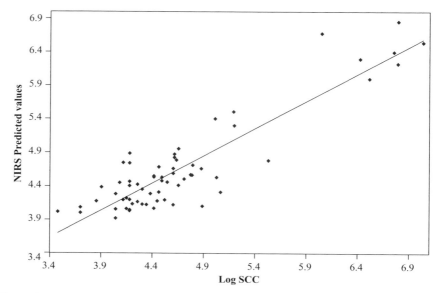

**Figure 9.3.6.** Relationship between measured and near infrared predicted values of log SCC for the independent test set of samples.

and model for mastitic quarters) incorrectly classified only 1 of 134 samples. The prediction performance of the models for unknown samples from test set was also very successful. From 66 samples in the test set, 63 samples (95.45%) were correctly classified and only 3 samples (4.54%) were misclassified. From these three samples, one was false negative and two were false positive.

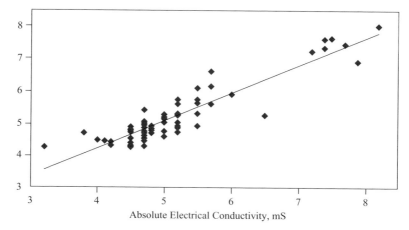

**Figure 9.3.7.** Relationship between measured and near infrared predicted values of absolute electrical conductivity for the independent test set of samples.

**TABLE 9.3.6. SIMCA Classification Results Based on Milk Spectra According to Their SCC (class 1, samples with SCC<300,000 cell/ml; class 2, samples with SCC>300,000 cell/ml)**

| Calibration Set | | | | Test Set | | | |
|---|---|---|---|---|---|---|---|
| Correct Classification | | Incorrect Classification | | Correct Classification | | Incorrect Classification | |
| n | % | n | % | n | % | n | % |
| 133 | 99.25 | 1 | 0.25 | 63 | 95.44 | 3 | 4.54 |

Results of qualitative SIMCA classification of milk samples based on their NIR spectra after they were divided into healthy or mastitic according to their AEC are presented in Table 9.3.7. Incorrect classification was obtained for five samples from the calibration set (1 false negative and 4 false positive) and for four samples from the test set (2 false negative and 2 false positive). The false negative samples from the calibration set and the test set had SCC<160,000 cell/ml, which showed that according to their SCC content they were healthy, but their AEC reference value was higher than the threshold set up for AEC.

When the class of mastitic samples was organized according to both criteria, SCC and AEC, only one incorrect (false negative) classification has been observed for the test set (Table 9.3.8).

**TABLE 9.3.7. Results for SIMCA Classification of Milk Samples According to Their AEC (class 1, samples with AEC<5.7mS; class 2, samples with AEC>5.7mS)**

| Calibration Set | | | | Test Set | | | |
|---|---|---|---|---|---|---|---|
| Correct Classification | | Incorrect Classification | | Correct Classification | | Incorrect Classification | |
| n | % | n | % | n | % | n | % |
| 127 | 96.27 | 5 | 3.73 | 62 | 93.94 | 4 | 6.06 |

**TABLE 9.3.8. SIMCA Classification Results Based on Milk Spectra According to Their SCC and AEC (class 1, samples with SCC<300,000 cell/ml or AEC<5.7mS; class 2, samples with both SCC>300,000 cell/ml and AEC>5.7mS)**

| Calibration Set | | | | Test Set | | | |
|---|---|---|---|---|---|---|---|
| Correct Classification | | Incorrect Classification | | Correct Classification | | Incorrect Classification | |
| n | % | n | % | n | % | n | % |
| 134 | 10 | 0 | 0 | 65 | 98.48 | 1 | 1.52 |

## MILK QUALITY EVALUATION AND DIAGNOSIS OF COW MAMMARY GLAND INFLAMMATION WITH NIR SPECTRA OF UDDER QUARTER FOREMILK TO IDENTIFY PATHOGENS IN MILK

It has been proposed that identification of cows with abnormal composite milk by NIRS should be followed by localizing the mammary inflammation and finding the udder quarter with mastitis. For that purpose, NIR spectra of composite and quarter foremilk, respectively, have been used to measure somatic cells and absolute electroconductivity. For further medical treatment, the main pathogen in quarter foremilk has been identified, again, by its NIR spectra.

### Milk Samples

For pathogen identification by NIRS, 33 spectra of quarter milk samples from 11 mastitic cows were analyzed. The pathogen for each mastitic quarter was previously known and acquired as a reference data. The cows were from different farms, and the samples were collected at different times of the year. The examined data set included 6 spectra of milk from 2 cows with *Staphylococcus aureus* (SA), 15 spectra of milk from 3 cows with *coagolase-negative Staphylococcus* (CNS), and 12 spectra of milk from 6 cows in the experiment with other than *Streptococcus agalactie* (OS). Milk spectra from each cow's udder quarters, healthy and mastitic, were acquired with triplicate repetition.

### Pathogen Identification by Referent Method

Pathogen identification was carried out in a microbiological laboratory, by a standard procedure and the results were used as a reference data for NIR spectral analysis to develop diagnostics models.

### NIR Spectra Acquisition

Transmittance ($T$) spectra of 1-mm-thick quarter foremilk samples were obtained by NIRSystem 6500 spectrophotometer (FOSS NIRSystems, Silver Spring, MD, USA) in the spectral range of 400–2500 nm, using quartz cuvettes with 1-mm-thick walls. Before spectral analysis each sample was warmed up to 40°C in a water bath.

### Data Analysis

To classify samples with various pathogens, SIMCA (21), described above, was employed to develop models for milk sample classification regarding NIR spectra.

The difference between raw spectrum of each mastitic quarter and the spectrum of the respective front or rear healthy quarter of the same animal (Fig. 9.3.8) was calculated to compensate for the variations observed, because of differences in front and rear quarters, farms, seasonal variations, etc.

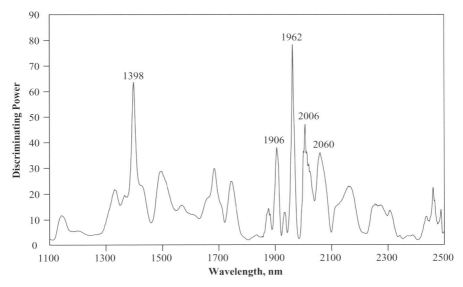

**Figure 9.3.8.** Discriminating power plot in near infrared region for differentiation between the class with low SCC and the class with high SCC content.

When SIMCA was applied to raw milk spectra without any data pretreatment, Machalanobis distance for CNS/OS was less than 3 and only a few misclassified samples were found. This result was due to the strong influence of individuality of each cow and the difference between front and rear quarters, farms, and the time when samples were collected.

Spectrum subtraction proved to be a very useful data pretreatment approach that brought amplification of the differences in spectra with similar background. After applying it, there were no misclassified samples found by the SIMCA model (Table 9.3.9) and the interclass distances were more than 3 for all classes corresponding to the investigated bacteria species.

**TABLE 9.3.9. SIMCA Misclassification Results for Bacteria Identification** (*Staphylococcus aureus* (SA), *coagolase-negative Staphylococcus* (CNS), **other than** *Streptococcus agalactie* (OS))

| | Misclassification | | | |
|---|---|---|---|---|
| | SA-Negative Prediction of 3 PCs | CNS-Negative Prediction of 5 PCs | OS-Negative Prediction of 5 PCs | NO Match |
| SA-Negative | 6 samples | 0 | 0 | 0 |
| CNS-Negative | 0 | 15 samples | 0 | 0 |
| OS-Negative | 0 | 0 | 12 samples | 0 |

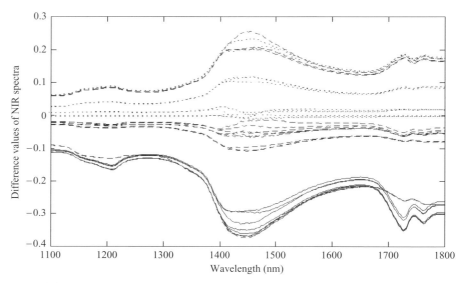

**Figure 9.3.9.**  Raw spectrum differences in the spectral region around 1450 nm after subtraction of the respective "healthy" spectrum from the "mastitic" one.

The largest interclass distance was obtained between SA and OS, followed by SA/CNS. Fewer principal components (PC = 3) were needed to describe the variations in SA class. The shortest distance was found again for CNS/OS. In Figure 9.3.9, distinctive differences in baselines and slopes of subtracted raw spectra of milk with different types of bacteria are observed. Negative baseline was registered for *Staphylococcus aureus* and *coagolase-negative Staphylococcus*. After subtraction, it was possible to see clear differences in the CH region of milk fat absorbance around 1200, 1700, and 2300 nm and changes at water absorbance bands around 1100–1350 and 1450 nm. The second derivative of subtracted spectra showed changes in the region around 1390, 1408, and 1450–1550 nm, suggesting changes with water in milk and milk protein hydration (8, 9).

Abnormal milk is defined by increased SCC. If AEC is increasing and there is a presence of bacteria in milk, this is a good indication for mastitis disease at a very early stage. NIRS calibration models developed for somatic cell count, log SCC, and AEC, measured in quarter milk samples, together with the models for pathogen identification has made unified mastitis diagnosis possible. The results have shown good correspondence between NIR determination of SCC to somatic cell count measurement that is now accepted as a world-standard method to define abnormal milk (6) and the AEC that has been used as a parameter for fast screening and early mastitis diagnosis (2, 6). Obtained accuracy allowed distinguishing samples with low and high levels of SCC or/and AEC, making possible abnormal milk identification, as well as early diagnosis of mastitis based on one or both of the investigated milk parameters.

The NIR spectrum, being rich in physical and chemical information about the analyzed sample, contains enormous information about analytes and their interactions. Diagnosis, as a decision making process, is always based on a multitude of symptoms. An analogy between various symptoms and the absorbance at no correlated wavelengths explained by the PC of the examined data set has been presented. Milk samples have been correctly classified as healthy and mastitic by their NIR spectra according to their SCC and/or AEC, and the results have been more accurate when compared to the calibration results for a single parameter, SCC or AEC. Correct classification of the samples in the calibration set has been higher then 96%, and for samples in the independent tests set, higher than 93.9%. A discriminating power plot (Fig. 9.3.8) indicates the "wavelength symptoms" with the highest discriminating power that have contributed most in distinguishing between the two classes of samples (healthy and mastitis). These wavelengths corresponded to absorbance bands of milk components changing with mastitis. The most significant wavelengths have been found to be 1398, 1906, 1962, 2006, and 2060 nm. These wavelengths have been associated with O-H stretch/O-H bend combination vibration, connected with lactose absorption (21). Other important wavelengths around 1500, 2060, and 1700 nm have been assigned with protein and water absorption and their interaction, CH bonding absorbance, respectively. Good consistency of these results with the well-known fact that mastitis decreases milk lactose content and changes types of proteins presented in milk and ionic concentration (3, 4, 5) has been found.

The changes in baseline and slope caused by different bacteria types seen in Figure 9.3.9 suggest changes with bulk water in mastitic milk samples, as well as changes in their scatter properties, that is, in molecular structure and particle size. This result is consistent with the well-known fact that mastitic milk becomes "watery" and milk turbidity changes according to the bacteria that have caused the mastitis. When different proteins change their structure with the disease, it is expected that water-protein interaction bands change their absorbance, which has been seen in the experiments described. With microbiological tests it is difficult to see clear differences between milk from healthy quarters and, especially, those with mastitis caused by CNS. On the PCA score plots, the healthy class and CNS class formed independent classes, but the distance between them was small. After spectrum subtractation, Mahalanobis distance between CNS and other classes became larger than 3. The highest weights observed on the PC loading plots belong to OH absorbance bands, most of them in connection with water absorbance and water-protein interactions. The result is consistent with the raw spectra baseline changes observed, as well as with the discriminating power plot. These results bring new insight into disease understanding and visualizing the differences in mastitic milk caused by different pathogens.

NIR spectroscopy has proved to be a powerful means for nondestructive milk quality evaluation and a screening tool for improving animal health. NIRS allows milk composition measurement including somatic cell count, electrical conductivity, and mastitis diagnosis to be done simultaneously, with the same technology. It opens new frontiers for on-line milk analysis in dairy farming, at the very beginning of the milk production food chain, to guarantee good milk quality for the dairy industry. Mastitis diagnosis, as an example, demonstrates that NIRS combined with multivariate analysis and chemometrics provides completely new molecular diagnosis approach including noninvasive bacteria identification, very much needed in contemporary medicine and food safety evaluation.

## REFERENCES

1. K. Svennersten-Sjaunja, L. O. Sjaunja, J. Bertilsson, H. Wiktorsson. *Livestock Prod Sci* **48**: 167, 1997.

2. J. Hamann, V. Kromker. *Livestock Prod Sci* **48**: 201, 1997.

3. O. W. Shalm, E. J. Carrol, N. C. Jain. *Bovine Mastitis.* Lea and Febiger, Philadelphia, 1971.

4. R. J. Harmon. *J Dairy Sci* **77**: 2103, 1994.

5. M. J. Auldist, S. Coast, B. J. Sutherland, G. H. McDowel, G. L. Rogers. *J Dairy Res* **63**: 269, 1996.

6. J. E. Hillerton. *Bulletin of the International Dairy Federation* No. **4**: 345, 1999.

7. International Dairy Federation Standard. International Dairy Federation, Brucells, Belgium, 148A: 1995. 1995.

8. K. Svennersten-Sjaunja, L. -O. Slaunja, J. Bertilsson, H. Wiktorsson. *Livest Prod Sci* **48**: 167, 1997.

9. B. Czarnik-Matusewicz, K. Murayama, R. Tsenkova, Y. Ozaki. *App Spect* **12**: 1582, 1999.

10. S. Šašic, Y. Ozaki. *Appl Spectrosc* **9**: 1327, 2000.

11. E. Díaz-Carrillo, A. Munoz-Serrano, A.Alonso-Moraga, J. M. Serradilla-Manrique. *J Near Infrared Spectrosc* **1**: 141, 1993.

12. M. Laporte, P. Pacuin, *J Agric Food Chem* **47**: 2600, 1999.

13. T. Sato, M.Yoshino, S. Farukawa, Y. Someya, N. Yano, J. Uozumi, M. Iwamoto. *Jpn J Zootech Sci* **58**: 698, 1987.

14. Z. Schmilovitch, E. Maltz, M. Austerweil. *EAAP Publ No 65* PUDOC Sci Publ Wageningen, The Netherlands, p. 193, 1992.

15. A. Purnomoadi, K. K. Batajo, K. Ueda, F. Terada. *Int Dairy J* **9**: 447, 1999.

16. R. N. Tsenkova, K. I. Yordanov, Y. Shindle. *EAAP Publ No 65* PUDOC Sci Publ Wageningen, The Netherlands, p.185, 1992.

17. R. Tsenkova, S. Atanassova, K. Itoh, Y. Ozaki, K. Toyoda. *J Anim Sci* **78**: 515, 2000.

18. R. Tsenkova, S. Atanassova, K. Toyoda, Y. Ozaki, K. Itoh, T. Fearn. *J Dairy Sci* **82**: 2344, 1999.

19. R. Tsenkova, S. Atanassova, S. Kawano, K. Toyoda. *J Animal Sci* **79**: 2550, 2001.

20. R. Tsenkova, Near Infrared Spectroscopy for cow's biomonitoring. Ph.D. Thesis, Hokkaido University, Japan 2004.

21. S. Wold, *Pattern Recognition* **8**: 127, 1976.

22. A. Savitzky, M. J. E. Golay. Smoothing and differentiation of data by simplified least squares procedures. *Anal Chem* **36**: 1627, 1964.

23. R. G. D. Steel, J. H. Torrie. "The Principles and Procedures of Statistics. 2nd ed." McGlaw-Hill Co., Inc., New York, USA, 1980.

24. D. Whyte, R. Claycomb, R. Kunnemeyer. Measurement of somatic cell count, fat and protein in milk using visible to near infra-red spectroscopy. Presented at 2000 ASAE

Annual International Meeting, July 9–12, Milwaukee, Wisconsin, Paper No. 00-3010. ASAE, St.Joseph, USA, 2000.

25. E. Urech, Z. Puhan, M. Schallibaum. Changes in protein fraction as affected by subclinical mastitis. J Changes in protein fraction as affected by subclinical mastitis. *J Dairy Sci* **82**: 2402, 1999.

26. R. J. Verdi, D. M. Barbano, M. E. Dellavalle, G. F. Senyk. Variability in true protein, casein, nonprotein nitrogen, and proteolysis in high and low somatic cell milks. *J Dairy Sci* **70**: 230, 1987.

27. B. Poutrel, J. P. Caffin, P. Rainard. Physiological and pathological factors influencing bovine serum albumin content of milk. *J Dairy Sci* **66**: 535, 1983.

28. H. Maeda, Y. Ozaki, M. Tanaka, N. Hayashi, T. Kojima. Near infrared spectroscopy and chemometrics studies of temperature-dependent spectral variations of water: relationship between spectral changes and hydrogen bonds. *J Near Infrared Spectrosc* **3**: 191, 1995.

29. J. F. Binette, H. Buijs. Fourier transform near infrared process monitoring of multiple inorganic ions in aqueous solution. In: Davies, A. M. C. and Williams, P. (eds.), "Near Infrared Spectroscopy: The Future Waves", NIR Publication, Charlton, UK, 286, 1996.

30. K. Molt, S. Berentsen, V. J. Frost, A. Niemoller. NIR spectrometry-an alternative for the analysis of aqueous systems. *J Near Infrared Spectrosc* **10**: 16, 1998.

31. K. F. Ng-Kwai-Hang, J. F. Hayes, J. E. Moxley, H. G. Monardes. Persentage of protein and nonprotein nitrogen with varying fat and somatic cells in bovine milk. *J Dairy Sci* **68**: 1257, 1985.

32. K. Murayama, B. Charnik-Matusewicz, Y. Wu, R. Tsenkova, Y. Ozaki. Comparison between conventional spectral analysis methods, chemometrics, and two-dimentional correlation spectroscopy in the analysis of near-infrared spectra of protein. *Appl Spectrosc* **54**: 978, 2000.

33. S. Wold. Pattern recognition by means of disjoint principal component models. Pattern Recognition **8**: 12, 1976.

34. M. W. Woolford, J. H. Williamson, H. V. Henderson. Changes in electrical conductivity and somatic cell count between milk fractions from quarters subclinically infected with particular mastitis pathogens. *J Dairy Res* **65**: 187, 1998.

35. JR. Workman. Handbook of organic compounds. Academic Press, London, NW1 7BY, UK, 2001.

*Near-Infrared Spectroscopy in Food Science and Technology*, Edited by Yukihiro Ozaki, W. Fred McClure, and Alfred A. Christy.